CONSTRUCTION EQUIPMENT GUIDE

CONSTRUCTION EQUIPMENT GUIDE

David A. Day, P.E.
Professor of Civil Engineering
University of Denver

in the Wiley Series of
Practical Construction Guides
M. D. Morris, P.E., Editor

A Wiley-Interscience Publication
John Wiley & Sons
New York   London   Sydney   Toronto

Library of Congress Cataloging in Publication Data
Day, David A   1924–
  Construction equipment guide.

  (Wiley series of practical construction guides)
  "A Wiley-Interscience publication."
  Includes bibliographical references.
  1. Construction equipment. I. Title.

TH900.D38   624'.028   72-10163
ISBN 0-471-19985-0

Printed in the United States of America

10 9 8 7 6 5 4 3 2 1

SERIES PREFACE
■■■■■■■■■■■■■■■■■■■■■■■■■■■■■

The construction industry accounts for about 14% of the United States' Gross National Product. It will soon exceed ninety billion dollars a year, and the worldwide need for new building to keep pace with increased population and its technology seems endless. Without one until now, this activity needs a literature of its own.

In the past the construction industry gained its information from vendor's literature, trade magazines, theoretical journal papers by engineers or architects, warmed-over engineering texts, and the spoken word. Much of this lot has been vague or contradictory. The maturity of construction from a hand-crafted trade to a major industry governed by laws, economics, socio-environmental elements, and technology (from equipment to computer systems) demands an accurate, factual, standard literature in a language understandable to the people who must use it as tools to get the daily job done efficiently and economically.

This Series intends to provide professional builders with up-to-date practical information written by working contractors whose special knowledge falls into those information areas most needed by the industry. Without frills they tell "how-it-is-done", and "how-to-do-it." They do no philosophizing, and little theorizing unless it actually helps to explain an item directly, or to accomplish the practice. With these books the constructor can approach his job and his problems with more knowledge and confidence.

While the Series is directed to the working contractor, these volumes also can be of valuable advantage to engineers, architects, specification writers, materials and equipment manufacturers, project planners, insurance and legal people; and, the source of them all—educators and students.

New York, N.Y.

M. D. Morris, P.E.
Series Editor

PREFACE
■■■■■■■■■■■■■

Construction Equipment Guide is one of the Wiley Series of Practical Construction Guides and follows the concept of being factual and useful. A reader and user is likely to be interested in or concerned for the successful planning or supervision of construction performed with equipment.

This book stresses the equipment fundamentals as they relate to the construction operation. Many sections deal with the physical concepts of the work, the surrounding conditions, and, therefore, the equipment requirements. With such an approach it is also necessary to stress the controls governing the equipment's performance. The viewpoint is that of an engineer. However, it is not necessary for the reader to have an engineering background, though an ability to visualize physical concepts or fundamentals will be helpful.

The kinds of equipment covered in this book are commonly and widely used for construction work. They are regularly produced by more than one equipment manufacturer today. All have had earlier generation models and give every indication of being perpetuated with competitive improvements or innovations. Special pieces of equipment not in any manufacturer's stock line are avoided. The variety of equipment included will be found useful on building, highway, or "heavy" construction. One familiar with the construction industry will recognize that only home construction is not included. Generally speaking, the variety included in this book is any common construction equipment which is big enough to have its own power unit for operation or prime mover for mobility. Yet the large equipment that is more often special or custom-built for a specific application is avoided as being too specialized. The specifications for all models of the equipment included would fill many books. What, then, will be included in this book about such equipment? The primary use, the governing physical conditions, and the design features that insure performance on the job will be discussed. Design details that distinguish one model or make from another are avoided, as well as those details that do not add particularly to an understanding of the primary use. However, design features that lead to the determination of production rates for the equipment are included.

The book will stress the factors that govern or control the equipment productivity on construction work. It will deal not only with the "what" of the equipment, but more importantly, with the "whys" for its performance. The costs of owning and operating the equipment will be discussed sufficiently for estimating purposes. Then the production rate and direct costs can be combined to find a unit cost for the equipment to do an operation. No attempt will be made to show additions for overhead and profit to result in a bid price. The arrangement of chapters is for understanding of the factors governing productivities and, thus, unit costs for the equipment. Types of equipment are grouped in the separate chapters by their common function (Compressors and Pumps) or by the commonality of their operation (Equipment for Earthwork).

Many specifications for construction equipment were reviewed for this guide book. They were very helpful but are too numerous to credit individually. Certain booklets published by leading equipment manufacturers were particularly enlightening to explain the principles of their types of equipment. These have been listed along with other informative books and articles at the end of appropriate chapters.

The first three chapters are intended to develop an overview, an approach, and a bag of tools for equipment planning. The overview is on the equipment and its treatment in the book. Particular attention on the approach to construction equipment planning is needed because the subject is very broad and has too many important parts for quick focusing. A few guiding principles for the suggested approach should prove helpful for concentrating on the subject. In addition to understanding the direct cost items, it is wise to know about cost accounting, maintenance systems, computers, and the critical path method (CPM) as a few key tools to control equipment for construction operations.

Construction Equipment Guide is written for people who need and/or want to understand the engineering know-how for the performance and satisfactory completion of construction using commonly recognized equipment. This can be meaningful for:

1. the engineer entering construction-related work—field testing, inspection, supervision, or operations;
2. the person studying engineering or construction technology to enter the construction industry;

3. the engineering aid or technician who is working on engineered construction and is interested in understanding better the equipment being used;

4. the construction man who wants to know more about the application of engineering in the equipment for his work; and

5. construction equipment distributers, dealers, and salesmen.

The book should help these people and others with a genuine interest if they study it carefully and refer to it as specific applications are encountered.

Acknowledgments

A book with broad coverage would not be possible without the help and cooperation of many interested people. Certainly, the encouragement from Editor M. D. Morris of the Wiley Series of Practical Construction Guides was needed for this book. Also, I want to recognize the generous help and encouragement from my talented friend, Joseph C. Kellogg, a construction management consultant.

Many of the equipment manufacturers have been generous with their time and contributions for input to the book. I wish it were possible to list all the people from these companies who have given me information or inspiration. However, certain companies should be recognized for the special help their personnel have given: Barber-Greene Company, Caterpillar Tractor Company, Construction Machinery Company, Galion Iron Works & Mfg. Co., the Koehring Company, the Robbins Company, TEREX division of General Motors, and several divisions of WABCO. Credit has been given to these and many other equipment manufacturers where their contributions have been used in the book.

The writing, typing, and checking of the copy has taken many hours, not only for the author, but for a number of other generous and tolerant people. My appreciation and thanks are extended to Mrs. Clarice Lubchenco, who typed the initial drafts of early chapters, and to Mrs. Emily Durham for later typing. The final draft would not have been possible without the talented typing and efficiency of my daughters Marilyn and Suzanne. I also thank the other members of my family, and particularly my wife Mary, who have been unusually patient and tolerant of all the time and effort put into the writing of this book.

Denver, Colorado   *David A. Day*

CONTENTS

■■■■■■■■■■■■■■■■

xi

Chapter 3 BASIC COMPONENTS AND FUNDAMENTALS — 67

Chapter 4 COMPRESSORS AND PUMPS 119

Chapter 5 EQUIPMENT FOR EARTHWORK 161

xiv ■■■ Contents

CONSTRUCTION EQUIPMENT GUIDE

Chapter 1

∎∎∎∎∎∎∎∎∎∎∎∎∎∎

∎∎∎∎∎∎∎∎∎∎∎∎∎∎∎∎∎∎∎∎∎∎∎∎∎∎∎

CHOOSING CONSTRUCTION EQUIPMENT

∎∎∎∎∎∎∎∎∎∎∎∎∎∎∎∎∎∎∎∎∎∎∎∎∎∎∎∎∎

Introduction to Equipment for Construction

The equipment to carry on construction work is an extremely vital force in modern competitive operations, particularly for the so-called heavy and highway construction. The production planning for a project often focuses on the productivity of equipment which governs the operational output. Furthermore, the financial planning for an entire construction business often stems from the investment in equipment, since the total of this element constitutes the largest long-term capital investment in the business.

It is not intended that this book should contain all the facts and data on performance to be considered and used for selecting the equipment for construction work. Other up-to-date references contain that sort of information. Later chapters of this book will cover significant points about the selection and use of commonly available pieces of equipment.

In this chapter the accent will be on the general equipment selection problem.

The main endeavor for successful construction management is to have the operations it plans achieve a satisfactory final product according to the plans and specifications at the least possible cost. One vital determinant in planning for construction equipment is its total cost to management. The total cost includes not only the original investment or the rental charge but also the cost of operating it and repairing or maintaining the equipment in good running order. These costs of equipment for construction are covered in detail in the next chapter.

Every piece of construction equipment is designed to deal with material in one form or another. Raw material of the earth may be broken up, dug out, or sucked in from its natural location. In that move the material properties will be changed from a natural state to a new one caused by the equipment. Other construction equipment is designed to handle loose or flowing material—to weigh, proportion, or mix it for use in a more finished product of construction. This is an important part of processing some construction materials. They will be finished by molding, vibrating, and compacting them with the use of specially designed equipment. Still other materials for construction will be ready-made or manufactured in designed pieces. These will be handled by another type of construction equipment.

Important properties and points about the behavior of materials for construction will be discussed where these need to be understood for satisfactory equipment selection. A basic discussion of the natural earth materials will be found in Sec. 3.3. The properties to know about air and water are introduced in Chapter 4. Other important material properties are mentioned in subsequent chapters of the book. Each reference to the materials of construction is for the sake of the equipment that will be dealing with them.

1.1 Identification of Construction Equipment

The various types or pieces of construction equipment to be discussed in this book will be identified in a convenient way. They could be grouped for the sake of discussion in several ways. One might be a classification by the action the piece of equipment takes or function it performs for construction. In this way, a scraper would be classified as a loose-material mover, i.e., equipment that scoops up, loads, carries, and dumps loose material. Another way would be to identify a piece of equipment by the

FIGURE 1-1 Equipment working on highway construction (courtesy of Northwest Engineering Company).

construction project's operation in which it performs. For instance, a scraper generally works in an earthwork operation. This would also be true of dozers, belt loaders, etc.

To help the reader focus on the sorts of equipment to be discussed in this book, the equipment included will be mentioned in the following sections. These will use the two mentioned schemes for identifying equipment—by function it performs and by project operation in which it works.

1.1.1 Functional Identification of Equipment

Identifying construction equipment by its function may be like a cafeteria selection. It starts with power units and prime movers, then offers loaders, haulers, and other material handlers, and includes many pieces of material-processing equipment. In this arrangement each piece of equipment has a separate identification; however, they can be grouped.

An initial group important to practically all construction equipment is the power units. They may be internal combustion engines operating on gasoline or diesel fuel. The form of power may be an electric generator which supplies energy to electric motor power units, or the electric motor may operate directly from an available source of electricity without the separate generator. The old steam boiler and engine form a power combination. An air compressor and hydraulic pump are also power units which need an engine or motor to generate the power.

Prime movers make up another functional group of construction equipment. These are the parts or pieces of equipment that make it possible for the machinery to do its construction operation. They translate the output of power units to working forces. One form of prime mover is the crawler tracks or rubber-tired mounting that moves a self-propelled mobile piece of equipment. For stationary equipment the prime moving parts may be cables, chains, or belting from a power unit. Or the prime mover may be the forcefulness of compressed air or hydraulic fluids moving in pipes or hoses. A more complete prime mover is needed for movable parts or pieces of equipment that are not self-propelled. Then a total crawler or rubber-tired tractor is a prime mover. It may be used for towing a scraper, a hauling wagon, or a trailer for moving other equipment.

Another group of equipment is the variety that functions to do excavating. This includes the several forms of tractor-mounted dozers—the bulldozer, angledozer, etc. A similar piece of equipment is the motor grader, which also functions to move excavated material for the sake of shaping or grading a surface. Either of these can be equipped to do ripping or scarifying of the material to be excavated. To excavate and load material for moving it, there are the versatile pieces of equipment known as front-end loaders, power shovels, backhoes, draglines, and clamshell excavators. Several more specialized pieces of equipment that provide this function are the elevating graders, belt excavators, trenchers, dredgers, and tunnelers.

Another variety of equipment works to install construction elements into the ground or breaks material out of it. In this group are the pile drivers and extractors, caisson drillers, well drillers, and wagon drills for breaking out rock.

A wide variety of equipment is used on construction operations to pick up solid pieces of material and move it. There might be several groups, depending on what the equipment does with the material it handles. Rather than identify different groups, which would be somewhat arbitrary, the equipment with this function included in the book is the following: the various lifting cranes—mobile cranes, tower cranes, etc.— and, related to them, the derricks; fork lifts and the similar high-lifts serve for simpler, shorter moves. More specialized equipment of this sort includes the lumber and pipe carriers.

Another group of construction equipment serves the function of moving loose or processed material such as earth, sand, and wet concrete. This sort of equipment generally cannot load itself without special arrangements, but it can dump the loose material. Identified in this group are the belt, bucket, and screw conveyors and the haulers—dump-trucks, rear- and bottom-dump wagons.

Material processing equipment constitutes another group to be discussed in this book. The materials to be processed for construction are natural rock and gravel to make aggregates for graded fill, concrete, and asphalt. Then the aggregates and the cementatious and other ingredients are mixed in processing soil-cement, concrete, or bituminous paving materials. With these processes in mind one can identify the equipment involved. To process aggregates there are grated feeders, screens, and a variety of crushers—jaw, gyratory, roll, impact, and hammermill. For the mixing processes there are aggregate bins, cement silos, batchers, concrete mixers, pavers, asphalt mixing plants and holding hoppers.

To finally place the processed materials requires another group of construction equipment. These assure that the material is placed uniformly to be homogeneous and compactly to be strong. The equipment to do this includes concrete spreaders and screeding machines, asphalt spreaders and finishing machines, and the variety of compaction equipment.

1.1.2   Operational Identification of Equipment

In the previous section all the equipment to be discussed in this book was identified by its functional use for construction. As mentioned earlier, the variety of equipment might be identified by the construction operation on which it most frequently works. This is appropriate because equipment selections are based primarily on the construction operations where they will be used. Specifics of an operation must be known before a good equipment selection for it can be made. The planner must have a mental picture of the operation and the key data for it to plan his methods and equipment.

There is one disadvantage to identifying equipment by the operations where it will be used rather than its functional purpose. That is the tendency to think of a piece of equipment in relation to only one type of operation. For example, if a front-end loader is discussed with earthwork operations only, the possibilities for that type of equipment to be used in an aggregate production plant might be overlooked. However, a knowledgeable and innovative planner will be well aware of the functions of each piece of equipment and be ready to apply it to various operations.

If the construction operations are thought of in terms of the material involved and what is to be done with it, there should be no fixations about the operational identification of equipment.

One set of construction operations has equipment working on natural material, putting it in a loose or fluid condition and moving it. This allows the grouping of equipment as follows:

1. compressors and pumps to work on air and fluids;
2. excavators and earthworking equipment to work on and move earth material;
3. trenchers, dredgers and tunnelers working in special ways on natural materials of the earth; and
4. conveying and hauling equipment to move the natural or processed material in a loose condition.

The other set of operations calls for equipment to process and install material for a final construction product. Therefore, the equipment in this set will be grouped as follows:

1. aggregate production equipment to process and grade natural materials;
2. concreting equipment to process, move, and place concrete material;
3. asphalt production and paving equipment to process and place paving materials;
4. material handling equipment to receive and move finished material for construction products; and
5. foundation and erection equipment to take processed material and install it in the ground or in space for a finished construction product.

Identifying construction equipment in the way suggested in the two sets above is desirable. That method will be followed in this book. In fact, the groupings shown are the subjects of following chapters. The order is varied somewhat for the convenience of efficient coverage.

1.2 Productivities and Efficiencies of Equipment

The productivity of a piece of construction equipment is a term used in this book to mean its rate of production during an hour's time. In other words, a piece of equipment's productivity tells one how many units of output the equipment produces in an hour. This is not a fixed quantity for a given piece of equipment. It will depend on the job conditions and management as well as the operator's skill, persistence and coordination with other construction forces.

The very best productivity that can be expected, generally governed by the designed limitations in the equipment, will be called the peak productivity, q_p. That is based on the equipment working the entire 60 minutes in an hour. To allow for the human factor in operating equipment that is not automated, there will be a somewhat lower rate of production which will be called the normal productivity, q_n. This takes into account the fact that most operators will not work a piece of equipment to its limits at all times and will take a break from their operation about once every hour. Normal productivity might be assumed to equal peak productivity for 48 or 50 minutes of each hour. This means that $q_n \approx (48/60)q_p = 0.8q_p$. The factor of 0.8 can be thought of as an average normal, working efficiency factor, f_w. In addition to the f_w factor, which is quite predictable, it is necessary to use a job-management factor, f_j, to take care of the interruptions of equipment operation due to factors governed by the job and management. When these two factors are combined, they result in an overall operational efficiency. This factor for the actual productivity, f_a, is the product $f_w \times f_j$, and

$$q_a = f_a q_p = f_w f_j q_p = f_j q_n, \text{ where } q_a = \text{actual productivity.} \tag{1-1}$$

A listing of these and other symbols used in the book is given in Sec. 1.5.

The Highway Cost Section of the U.S. Bureau of Public Roads found in an extensive study of construction operations conducted in the 1950's that the average productive time of equipment during regular "net available" working hours was 44 minutes, or 73% of that time. This is working efficiency, f_w. It varied on highway projects from a high productivity of 53 minutes for crawler-tractor scrapers to a low of 38 minutes for power shovels. The reasons for the losses in productivity will be evident by the discussion in subsequent chapters.

Also, it must be recognized in productivity determinations that if the construction operation is not ready for the equipment to perform its work, the average q for the overall project will be reduced still further. Some studies have determined that the actual average productive time on construction is less than 50% of the total available time. This reduction to an average of less than 30 minutes out of an available 60-minute clock hour takes into account major delays of 15 minutes or more due to equipment repairs, weather, poor planning, etc.

Some representative values for working efficiencies and job-management factors will be suggested. They are to be used in estimating equipment productivity for a given condition. These efficiencies take into account the human element, job layout, weather, machine failures, and parts and service availability.

TABLE 1-1 Equipment Efficiency Factors[4]

General Condition	Working Eff., f_w	Job-Mgt. Eff., f_j	Work Condition	Combined Efficiency, f_a		
				Job-Management Condition		
				Good	Average	Poor
Good	0.90	1.00	Good	0.90	0.77	0.59
Average	0.80	0.85	Average	0.80	0.68	0.52
Poor	0.70	0.65	Poor	0.70	0.60	0.45

The use of productivities in planning for equipment must be logical and realistic. The planner needs to use sound understanding and good judgment in their application. When comparing one piece of equipment to another, or figuring several pieces working together, the productivities used must be comparable. Generally, this means that the peak or normal productivity for each will be used in such determinations. In that way, the job-management condition that is not dependent on the equipment will not affect the comparison. On the other hand, if the equipment used will affect the job-management factor, that f value should be included in the comparison.

Productivity of construction equipment is an important basis for its selection to do an operation. Therefore, the determination of q values will be a major emphasis in following chapters on specific equipment. To select the right equipment for doing a total operation involving many hours of work, the efficiencies must be taken into account. These concepts will be understood better with subsequent applications.

1.3 Equipment Selection Factors

The foremost factors in the selection of equipment to do a construction operation are cost and maintainability. That is, equipment will be chosen that can do the job at the least total cost, other things being equal. Cost elements for equipment use will be covered in the next chapter.

There are a number of other significant factors in the selection of equipment. These must be considered in every equipment selection. They include accounting for the:

1. specific job or operation to be done,
2. specification requirements,
3. mobility required of the equipment,
4. weather's influence on equipment performance,
5. time scheduled for doing the job,
6. balancing of interdependent equipment,
7. versatility and adaptability of the equipment, and
8. operator's effectiveness with the equipment.

A feasible solution to the equipment selection problem for actual field conditions will undoubtedly involve a number of these factors. In fact, it would be a strange construction operation if the choice were dependent on only one factor. However, for the sake of clarity in understanding and being able to apply the various factors in a selection, we will study them one at a time. That is, while we are studying one factor, it will be assumed that the other factors remain subordinate in their effect.

1.3.1 Specific Job or Operation to be Done

Obviously, the specific construction operation is the primary factor in the selection of equipment to get the job done. The effect of the specific job or operation to be done in the selection of equipment has several general aspects. The problem involves knowing:

1. the physical work to perform in doing the operation;
2. the availability of working space; and
3. the power requirements and availability.

The design features of a piece of equipment necessarily dictate certain things about where it can be used. For instance, if the operation is to dig a 15-foot-deep trench, a wheel-type trencher could not be used because that type of equipment has a limit of about 8 feet of depth. Many other examples could be cited. Any further mention of specific operations will be saved for discussions on the specific types of equipment in following chapters.

When there are working-space limitations, the construction planner must call on a knowledge of the variety of possible equipment and the specifications for any unit he might select. For instance, the working space may have a limited headroom. This would suggest, for a loading operation, selecting a piece of equipment with a telescoping or hydraulic boom as opposed to a fixed cable-controlled boom requiring more total swinging room. Perhaps the material should be moved by a conveyor system with beeline delivery instead of any swinging crane unit. Another example to call attention to this equipment selection factor is that of a concreting operation in a congested work space. It might be advisable under such circumstances to choose concrete pump equipment, which requires a minimum of space in its pipe run compared to a 5- to 7-foot-wide runway for concrete buggies.

If the job is in a remote, undeveloped wilderness, there will be different concerns for the needs in powering equipment than there would be for a job in an urban setting. Therefore, equipment selection is based to a certain extent on the availability of power sources to run the equipment. The planning for the power requirements of an individual unit of equipment is frequently governed by its specifications. The general variety of power units to consider were introduced in Sec. 1.1.1. Power units for different types of construction equipment will be more fully discussed in Chap. 3. In this general explanation of selection factors the discussion of power requirements and availability will concentrate on those points that depend on the specific job and construction-site conditions.

The most commonly selected power units for construction equipment are internal combustion engines. For equipment that will be operated for long periods of time, a diesel engine is generally more economical. Whether the engine is diesel or gasoline powered, there is need for a ready supply of the required fuel. The availability of either fuel in the United States varies inversely with the distance from the construction site to the nearest commercial-industrial area. Of course, several days' supply of either diesel fuel or gasoline can be transported to the construction site by a supplier's tank truck or in drums hauled on the contractor's truck. To minimize inconvenience, advanced planning is needed to take care of the refueling times.

If the specific job or construction site is in close proximity to an electrical power source and the equipment need does not call for much power variation and mobility, there are advantages in considering an electrical power unit. Where electrical energy is already available and can be easily tapped, this form of power is the most economical possible. Electrical power is the cleanest form and may be required for environmental reasons. It does present safety hazards to be considered. In some construction situations electricity can be generated economically to be used for a variety of equipment and facility uses. Some real ingenuity has been shown in providing such economical generation of power using the natural force of gravity similar to hydroelectric power generation.

The other forms of power that might be considered to satisfy the construction job conditions are chosen more for the specific jobs or operations to be done. As such they are selected more for the individual equipment or tools to operate. Thus, compressed-air power is generally chosen for powering drilling tools operated by hand where there is a concern for the operator's safety. This is particularly significant in tunneling operations. In contrast, steam power could be considered with certain pile-driving hammers where close proximity of the workers is not such a problem and the impact benefits of this form of power are advantageous.

1.3.2 Specification Requirements

Ideally, the specifications of a construction contract should dictate or express desired end results only. In that case no equipment would have to be specified so long as there is a variety of equipment that could be chosen to produce desirable finished products. However, what might appear to be a desirable product and what will stand the tests of load, stress, time, and the elements of nature may be two entirely different products (Fig. 1-2).

FIGURE 1-2 Equipment proven successful for tunneling (courtesy of the Robbins Company).

In order to avoid the likely chance of an undesirable finished construction product under certain circumstances, it is necessary to specify intermediate steps or equipment. For instance, the density and load capacity of an earthfill are not properties to be checked only when the fill is finished. If no attention is given to these properties as the fill is being built up, there is bound to be a lack of homogeneity in it—no matter how conscientious the contractor may be. Consequently, specifications generally call for the earthfill to be placed and compacted in layers, each of about six inches' compacted thickness, which can be tested by the resident engineer or his soils engineering technician. To further insure adequate control of the results, certain types and sizes of compacting equipment have been specified as well as the procedure in their use, i.e., speed of traveling and number of passes over the fill. This was intended to insure the desired end results which would otherwise be difficult to control or obtain. It is not hard to understand, then, why engineers find it necessary to specify methods and equipment for building an earthfill.

A very similar line of reasoning could be used regarding the necessity of specifying intermediate steps and equipment requirements for producing quality concrete on a construction project. Thus, there is justification for specifying certain types or features of equipment as a means of control in construction work to insure the desired end results. The sort of equipment that would likely be specified under these conditions is equipment to process or shape materials.

Unfortunately, somewhat poorer specifications will often indicate the specific type or size of equipment to do an operation where such restriction is not necessary. For instance, the contract's specifications might call for the use of a ladder-type trenching machine for excavating all pipe trenches. Actually, another type of trencher or a backhoe could possibly do the operation as well as, and more economically than, a ladder-type trencher for an experienced contractor under the given conditions. The point is that no particular equipment type, size, or make should be specified if others could also produce a satisfactory end result.

The fact remains that a variety of equipment is specified under differing circumstances for various particular operations. Whether the specification of equipment or method for accomplishing an operation is justified is not the responsibility of the construction planning engineer. The construction planner must abide by the specification for particular equipment, unless he could have alternative equipment approved under the contract and then plan accordingly. A specification pertaining to equipment would forcibly influence his equipment selection and the solution to his planning problem.

1.3.3   Mobility Required of the Equipment

The question of mobility of the equipment to be used on construction work focuses on two general concerns. These are:

1. the necessary movement of working equipment and materials for any one operation; and

2. the planned movement from one operation to another on any one project or from one construction project to another.

The importance of equipment movements in either situation of concern depends on the time it takes to make each move and the frequency of such moves. If the piece of equipment is taking part in a full day's concrete pour, and then has to move only to the other side of the building for its next day's operation, it would not have to be highly mobile. That is, the equipment used for such conditions would not need to be truck mounted, but could move on crawler tracks. To move from one project to another, it would be loaded onto a trailer. Such movements as frequent as one a week or more often, should be made by equipment that does not require time-consuming and expensive use of some other equipment to move it.

Generally speaking, the following statements can be applied. Heavy equipment that has to move at least once a week should be self-propelled. Lighter equipment which is mounted on its own traveling axles can be moved economically by being towed, if it can be easily handled by a vehicle that is readily available to the project. If the moves are lengthy, say, greater than 1000 feet, and can be made on a prepared or hard surface, truck-mounted or wheel-type equipment should be considered. For shorter hauls or rough, overland routes, crawler-mounted equipment should be selected, since the real and costly advantage of the rubber tired unit—high speed operation—is greatly reduced. If the equipment or plant item is generally used in one position with very infrequent moves—consuming, say, less than 2% of its working time—it will probably be more economical to not have it mounted on wheels or tracks. It can rest and be tied down to a temporary and inexpensive but firm base, thus saving on extra stability and mounting framework and accessories. For the infrequent moves it would be more economical to call on the services of regular hoisting and hauling equipment, than having the power for transportation standing idle, unused for the long periods between moves.

A generalized expression may be derived which can be helpful in solving the mobility problem. Moves during an operation or between operations cost money. The total expense during each move will definitely include the sum of all the ownership expense components of all the equipment involved, the operator expenses of each piece involved, and the other operating expenses for the pieces of equipment contributing power for the move. We will call this total expense of dollars per hour, C_m. Then, the effect of moves on the cost per unit of production, c_m, is

$$c_m = \frac{dC_m}{60vQ_m},\tag{1-2}$$

where d = average distance moved, in feet;

v = average speed during move, in feet per minute; and

Q_m = number of production units completed between moves.

This additional expense for moves as a part of the total production cost for constructed units should be held to a minimum consistent with the plans for all related operations. It will be of particular concern in selecting equipment for such operations as the following:

1. excavating for foundations scattered over a large industrial or plant site;
2. pile-driving operations; or the
3. placing of concrete for separated pours.

1.3.4 Weather's Influence on the Equipment

The influence or effect of weather on the selection of equipment is important. Conditions of weather that must be considered are temperature, moisture, wind, and air pressure. They all affect the desired performance of equipment in their various ways. In different ways some conditions of weather affect the operator's ability to run his equipment efficiently.

Temperature has a significant effect on the efficiency of an internal combustion (I.C.) engine's operation. At low temperature, perhaps below freezing, the air taken in has to be heated more to help its atomization for mixing with the fuel and ignition in the power stroke. Also oil and other viscous fluids will be thicker. When the temperature is very high, say, 100°F or more, the air taken in for combustion is thinner, i.e., has less oxygen per cubic foot, than normal and has to be compressed to obtain the needed oxygen-and-fuel mix. This is even a greater factor at altitudes much above sea level. The effect of altitude is discussed in Sec. 4.1.3 with compressed-air equipment. In that power form, as well as in I.C. engine performance, the standard atmospheric conditions are 60°F at sea level. For every 10-degree variation from the standard temperature there is about a 2% change in the output of the power source.

The temperature surrounding process mixing equipment must be considered in another way. The temperature of materials for asphalt or concrete mixes must be within a reasonable working range. If the aggregates are near or below 32°F, due to the surrounding atmosphere, their moisture will be frozen. Those aggregates would act adversely on all materials with which they might be mixed. Equipment for mixing cementatious materials often must have means for preheating the separate materials to insure a good, uniform mix.

Rain, snow, or excessive moisture in the atmosphere and on the ground where equipment travels can cause problems. Generally, a wet ground surface will cause poorer traction for the equipment driving contacts. Equipment propelled by rubber tires will have 5 to 10% less traction on wet ground. The reduction is much greater on snow or on an icy surface. An intermediate situation occurs when there is a heavy dew on the ground early in the morning.

Precipitation on the ground material also frequently increases the rolling resistance. This effect and that of traction can be noted by checking Tables 3-4 and 3-5. The general conclusion to derive from these observations is that for self-propelled equipment, rain and wet conditions will hinder their travel.

Other construction operations involving equipment may be affected by too much precipitation. An open cut or trench in natural ground may be more likely to cave in, if the earth gets too wet. The subgrade for a foundation or pavement should not have water standing there when processed material is placed on it. Unmixed cementing material must be protected from any moisture. Furthermore, the mixed material itself cannot be poured when there is more than a sprinkle of rain falling. All unsheltered construction operations are generally stopped by inclement weather.

Windy, dry weather also has an adverse effect on construction operations. Hot, windy weather will have a drying effect which necessitates control of the water content during material processing. This is a problem in road-mix soil aggregate, bituminous road construction, and concrete mixing and curing. The rate of evaporation from the surfaces of the material will be important in each of these operations. It will be more rapid for higher temperature, drier and windier surrounding air. The exact effect on any given material will vary with its composition and other variables. Consequently, the practical method of accounting for the effect of evaporation in any operation is to study it under the given field conditions. The pre-planning, however, should allow for an ample supply of necessary water, if the conditions suggest a high rate of evaporation.

One more significant point about weather's influence on equipment operation has to do with its effect on the operators. Adverse weather conditions such as hot, dry, and dusty air will tend to decrease the operator's ability to put out a maximum production effort. Also, freezing temperature will have its effect unless the operator has a heated place to work. The relative effect of this influence will be greater for high-speed, highly maneuverable equipment, or those units requiring greater skill and, therefore, more fatiguing for the operator to handle.

1.3.5   Time Scheduled for Doing the Job

The selection of equipment is directly dependent upon, and may be decided by, any of the following time considerations:

1. the time allowed by the construction contract;
2. necessary and economical timing of sequential operations;
3. relative effect of overhead on economy of operation; and
4. variation of equipment rental rates with the time it takes pieces of equipment to do the operation.

The relative effect of these important time concerns will be analyzed in the following discussion.

There are generally three ways that the allotted time for completion of construction work may be specified. The construction contract may set a completion or due date, a certain number of calendar days to complete the work, or a certain number of working days. If the number of working days is specified, only the regular working days, D_w, during which construction could reasonably and satisfactorily progress, should be charged to the contractor. Knowing the planned quantity or number of units, Q, to construct, the necessary average production, q_a, to meet the contracted time limit can be found as

$$q_a = \frac{Q}{H_{aw}D_w} \text{ constructed units/hour,} \qquad (1\text{-}3)$$

where H_{aw} = the average work time in hours per day, and
D_w = available working days.

The average productivity thus determined allows for possible repairs or breakdown of the equipment, lack of material or interdependent equipment on hand in time for efficient operation, and, generally, any anticipated reduction from peak or normal productivity.

If the number of calendar days or a due date is specified, the determination of necessary productivity is not quite so easy. In this case the average production must also allow for the loss of whole days due to weather, strikes, and other uncontrollable interruptions to the work of the construction forces. An estimate could be made of the probable days lost due to these contingencies during the construction period. Then the operation might be planned to end close to but before the due date. Because of the probable heavy financial penalty for delay beyond the due date and the uncertainty of weather delays, it would be well not to plan for completion too close to the final contract due date. The same reasoning would apply if a certain number of calendar days were specified.

In these latter specified cases the number of calendar workdays, D_c, for the duration of the contract is of extreme importance. To allow for some unpredictable delays, a contingency factor, f_c, might be used. This factor should be lowered farther below the ideal value of 1 as the likelihood of inclement weather, strikes, etc., increases. Then it is necessary to plan for an actual calendar production rate, q_a', obtained from

$$q_a' = \frac{Q}{H_{aw}D_c f_c}. \tag{1-4}$$

With the required productivity determined in this way based on the allotted contract time, the equipment selection is limited to those possibilities which can meet or exceed a q_a' production rate.

The concern for necessary and economical timing of sequential operations is of special importance for certain construction operations. These are, generally, the repetitious operations which cannot be economically carried on as a single continuous operation without the interruptive steps, which consume from an hour to a day or so, in each cycle. The most obvious example is a repetitious concreting operation involving considerable formwork. Other examples include the drilling, blasting, and removal of rock excavation, and the casting, curing, and post-tensioning of concrete girders in the field.

A specific example of the repetitious concrete operation involving a great deal of formwork could consist of poured-in-place concrete walls that can be built in half a dozen separate sections. Since the formwork is an expensive part of such an operation, it would be desirable to use only enough forming material to form one or two sections at a time. Thereby the form costs are reduced by several reuses of the form material and the carpentry crew is more economical by working in a smaller group for more days and becoming more efficient by repeating their work. This will be represented and discussed further with the graphical schedule shown in Fig. 1-3.

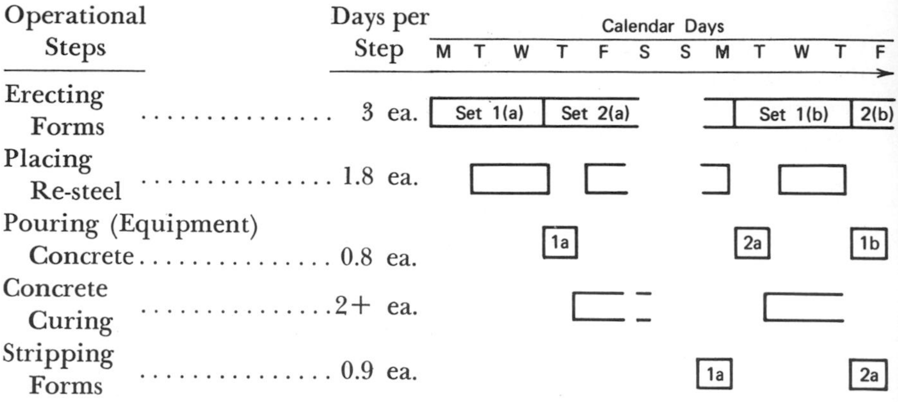

Operational Steps	Days per Step	Calendar Days											
		M	T	W	T	F	S	S	M	T	W	T	F
Erecting Forms	3 ea.	Set 1(a)		Set 2(a)						Set 1(b)			2(b)
Placing Re-steel	1.8 ea.												
Pouring (Equipment) Concrete..............	0.8 ea.			1a						2a			1b
Concrete Curing	2+ ea.												
Stripping Forms	0.9 ea.								1a				2a

FIGURE 1-3 A concretework schedule for
sequential steps.

While the sort of construction plan scheduled above would undoubtedly help reduce the form costs, it may lead to poor equipment economy. For instance, the equipment for batching, mixing, transporting, and placing the concrete would be needed only every third or fourth day. This would suggest using as inexpensive and, consequently, probably as small equipment as readily available. Yet the selected equipment must be big enough to handle the planned yardage of pour within one day's shift. Therefore, it appears that the equipment productivity will not always govern an operation, such as this concreting which is governed by the form erection forces. In other words, the equipment planned for this sort of noncontinuous, sequential operation would not be selected on the basis of economy alone but on special timing considerations. Of course, using ready-mixed concrete from a central plant serving many projects or customers would eliminate this concern.

The noncontinuous, sequential operation of concreting simple repetitious wall sections should be planned for optimum economy. By one method the planner would use simple mathematics or a computer to find the combination of equipment, labor, and material expenses that would reduce to a minimum. The planning and determination involved here suggest the use of the new, dynamic management tool of critical path method (CPM) and its application to resource allocations. These methods will be discussed later in this chapter.

Several other concerns for time scheduling influence the selection of construction equipment. These include overhead expenses in the overall costs of the construction operations. As far as equipment expense is involved, the overhead concern is more important when there is a carrying charge for equipment owned by the contractor. Another expense concern related to the time scheduled for construction has to do with the variation of equipment rental rates with the duration of its use and operation. The details of ownership and operating costs will be taken up in Chap. 2. Therefore, these points of this equipment selection factor based on time scheduling will be deferred for now.

1.3.6 Balancing of Interdependent Equipment

Many construction operations have two or more types of equipment working together to get their parts of the job done. These types are said to be interdependent equipment. In order for them to work together effectively and economically, their rates of production must be as compatible as possible. The measures taken in the planning and selection of interdependent equipment to insure their compatibility are for "balancing" them. That is, equipment working together is balanced in size and productivity to be sure of an economical operation.

One part of balancing interdependent equipment is to have the sizes of their working elements compatible. For example, a loading bucket should be able to just fill a hauling container with a certain number of bucketfuls. Or a concrete truck should be able to be filled with an even number of full drums from the mixer.

If the loader is a 3/4-yard loader, then the hauling units should have capacities of 3-, 4 1/2-, or 6-yard dump containers. Then it would take an even number of bucketfuls—4, 6, or 8, respectively—to load the haul containers. An extension of this part of balancing such a combination of interdependent equipment, frequently suggested for the sake of economical operation, is that the hauling container should be filled with a minimum of three and a maximum of eight loader bucketfuls. The reasons for this measure to balance the equipment will be seen in Chap. 5 on earthwork equipment and Chap. 7 on power excavators.

Another part of balancing interdependent equipment is by matching productivities of the several types of equipment. A loader or mixer or other key piece of equipment will have a certain productivity for a given operation. Of course, as explained in Sec. 1.2, that piece of equipment has peak, normal, and actual productivities. Now if there is just one unit of that type of equipment, say, a loader, on the job, its peak productivity will be the maximum for that part of the work. The hauling equipment working with it cannot cause a higher production to be reached. This is so no matter how many haulers are used, a lower maximum production rate would result if the best that the haulers could manage is less than the peak productivity of the loader.

An ideal balance of productivities would result if the maximum q value of the loader were exactly matched by the sum of the corresponding q values for the haulers planned. This notes that there will be a number of haulers working with the loader. An expression can be written to show the possible relationship of reasonable balance.

Let q_{p1} = peak productivity for the loader, mixer, etc.;

$\quad q_{p2}$ = peak productivity for one of the hauling units; and

$\quad N$ = number of hauling units to work with the loader, etc.

Then a balance of this interdependent equipment based on using peak productivities is, ideally,

$$q_{p1} = Nq_{p2}. \tag{1-5}$$

Since the actual operation is done by a whole number, N, of units, the above equality is seldom, if ever, realized. The actual number of whole units used will determine which type of equipment controls the production rate for the operation. Therefore, the planner selects the number of units to use based on

$$N_1 > \frac{q_{p1}}{q_{p2}}$$

for the loader, mixer, etc., to govern the operation. Or, in the less frequent case,

$$N_2 < \frac{q_{p1}}{q_{p2}}$$

for the hauling units to govern the rate of production. In these expressions $>$ means *greater than* and $<$ means *less than*. Reasons for using N_1 or N_2 will be given in the later chapters with discussion of specific equipment.

The selection of the right equipment and number of units of interdependent equipment could be based on normal productivities, q_n, as well as q_p values. It must be based on logical and compatible values. To have a well-balanced operation, the number selected should be within one of the theoretical number, N, found from Eq. (1-5). For example, if the loader's peak productivity q_{p1} = 160 cu yds/hr and the haulers' q_{p2} = 45 cu yds/hr, then $N = 160/45 = 3.55$ units, theoretically. A reasonable balance based on these peak productivities is gained by using three or four haul units. In this case $N_1 = 4$ and $N_2 = 3$. In fact, it is always true that the right values of N_1 and N_2 are within one unit of each other.

1.3.7   Versatility and Adaptability of the Equipment

Another equipment selection factor that is not dependent on cost is the versatility and adaptability of the equipment. The planner should consider this factor when he has a number of operations requiring similar equipment. If these operations are all for one project, some equipment may be able to work on various operations. The same would be true if there were several nearby projects being done during the same weeks or months. To make use of pieces of equipment on a variety of operations, an "equipment-use chart" can prove most helpful. When all the equipment is listed in order of their need on the job and then scheduled by operations, moves, etc., a very graphic picture is made for the use of the equipment throughout the life of the job.

A versatile piece of equipment is one designed to be multipurpose. It functions in a variety of operations. The most obvious example among types of construction equipment is the tractor. It can be used as a prime mover, a pusher, a dozer, a ripper, etc. Another good example is represented in the basic equipment for the power shovel, dragline, clamshell, and mobile crane. Changing from one function to another should be a fairly simple move. Other types of construction equipment are designed for versatility and adaptability.

Frequently, the purchaser of a piece of equipment will be interested in these selection factors. He will own the piece for a number of years. Over that time of his work he cannot be sure just what operations he will have to do. Therefore, the owner of equipment is interested in having a piece that can do some variety of operations. In that way he can be more certain of keeping it busy and getting his money's worth from the equipment.

1.3.8   Operator's Effectiveness with the Equipment

All pieces of construction equipment are made to be run by an operator. If the equipment is highly automated, this person merely has to push the right buttons to run it. However, most construction equipment is not so simple to operate, especially the mobile pieces. Much skill is required of the operator to work the piece of equipment effectively. The design features to make the equipment's operation easier are important in selection of the right piece to use. This involves the location of operating controls, the operator's visibility of his work, and the comfort and safety devices in operating the equipment.

Another aspect of the operator's effectiveness having an important bearing on the economy of selected equipment has to do with the size of equipment that he can handle safely and easily. There used to be a limitation with the hand controls and mechanims operated by levers, pedals, and steering columns with a small amount of mechanical advantage. A single person could not operate a very large piece of equipment without excessive fatigue or help. Now with electrical and hydraulic powered controls and boosters, the size of equipment an operator can handle is almost unlimited. Since the operator's wage is usually fixed for any locale and time, the planner can realize some definite economy in having the operator direct as large a piece of equipment as can be used.

The effectiveness of a good operator with a piece of equipment can be even better, if its maintenance and repairs are easy to make. Of course, the first hope on this point is that the equipment will require a minimum of maintenance; next, that maintenance to be performed on a regular basis is easily done; and finally, that minor adjustments and repairs can be made with easily replaced standardized parts and a minimum of equipment "down" time.

1.4 CPM and PERT Management Tools

Several decades ago the managers of construction planned their operations and then kept track of them as best they could using bar charts. A bar chart has a linear bar for each separate operation or activity. Each bar's length is proportional to the time required to do the operation. With a calendar scale parallel to the bar lengths, the chart shows the dates planned for the start and finish of each operation. A chief drawback to the bar chart is that it does not by its basic design show the dependence of one operation directly on another. Therefore, if one operation is delayed and its bar has to be slid back to later dates, the user must understand the project well enough to know the effect of that change of plans on other operations.

An example of a bar chart for a simple job is shown graphically on the right side in Fig. 1-4. Some of the key terms and types of data common to the newer CPM and PERT tools are included on this presentation.

Activity Duration, days	Scheduled		Earliest		Slack		Working Days
	Start*	Finish	Start*	Finish	Total	Free	*An activity (operation) can get underway the day following the indicated "start" date
2			0	2	5	0	
4			2	6	5	0	
2	0	2	0	2	0	0	
5			2	7	1	0	
1			6	7	5	4	
4			7	11	1	0	
1			0	1	6	0	
8	2	10	2	10	0	0	
3			1	4	6	6	
3			11	14	1	1	
5	10	15	10	15	0	0	

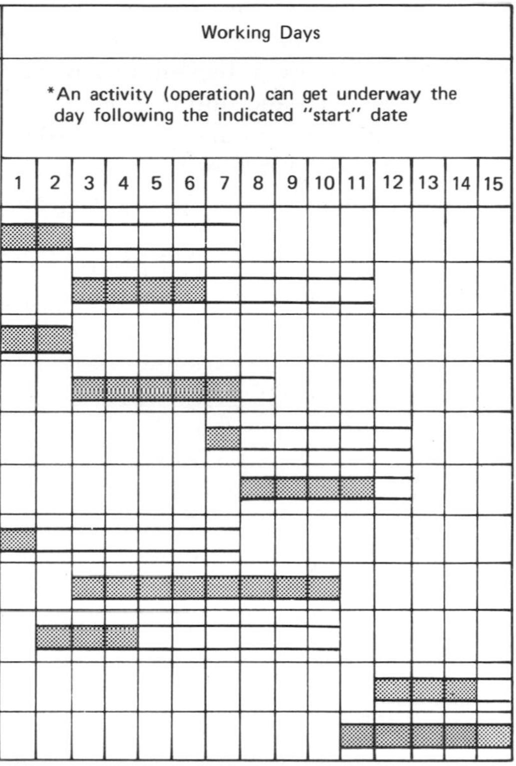

(Working Days columns numbered 1 2 3 4 5 6 7 8 9 10 11 12 13 14 15)

FIGURE 1-4 Example of Data and Bar Chart

In the late 1950's several new management tools were developed to help with the planning and control of operations. The Critical Path Method, generally abbreviated CPM, was devised by J. E. Kelley, Jr., of the Sperry Rand Corporation doing a job for the E. I. duPont de Nemours Company. CPM was developed to improve the planning, scheduling, and coordinating of duPont's plant construction. Independently and shortly after disclosure of the CPM tool, the Project Evaluation and Review Technique (PERT) tool was made known by the U.S. Navy Special Projects Office. The CPM and PERT management tools are basically very similar, but they have certain important differences. These will become evident in the following discussion.

1.4.1 Fundamentals of CPM and PERT

The CPM and PERT techniques both identify the separate operations to be done and the dependency of each on other operations. Each technique uses arrows or lines connecting the dependent operations or events, showing the beginning and ending of these. The lines and connecting points form a network diagram that is the CPM or PERT skeleton. When the necessary facts and information are added to give the network values, it can be used for planning and scheduling. In both techniques there is a critical path of operations and events that will govern the others. This shows what must be done as originally planned to keep the total project on schedule.

The differences between CPM and PERT exist in the relative importance of and approach to the required operations and events as activities of the project. CPM emphasizes the performance of individual operations. It is used where each operation can be planned based on past experience and estimating the time to do one in the future. This suggests that the CPM technique is a deterministic one.

On the other hand, the PERT technique emphasizes the occurrence of events in the course of completing a project. An event is sometimes called a milestone that occurs when the operations or multitude of activities leading to the event have all been accomplished. There is a sense of high variability of these activities and their various parts in the PERT technique. This would be true in a research-and-development program. In that case the performance and time to accomplish an operation can only be speculated. PERT is based on making three estimates for each operation's duration. They are the:

1.   optimistic time—the shortest possible time if no problems occur and efficiency is high;

2.   most likely time—the average time if the operation were repeated often; and

3.   pessimistic time—the longest time that could be anticipated if many problems interfere with getting the operation accomplished.

This suggests that the PERT technique is a probabilistic one and it might use statistical methods, which are introduced briefly in Chap. 2.

1.4.2. Procedure for Using CPM

Construction operations are activities that have strong similarity to ones which have been done before. As such the methods for doing each operation can be carefully planned and the time needed can be closely estimated. This means that construction operations are deterministic and the CPM technique can be used. Before discussing the procedure to use this management tool, it will be helpful to describe the elements of CPM more fully.

The parts of a CPM diagram are called arrows and nodes. Examples of these symbols are shown here.

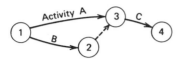

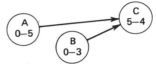

Arrow Diagram—
    nodes (1, etc.) are events

Node Diagram—
    activity at each node

A network diagram starts and ends at nodes and has others in between these terminal points. Nodes are generally shown as circles within which some notation can be written. From a node one or more lines are drawn to tie in with other nodes. A line will begin, or have its tail, at one node and end with its head point at another node. The idea of tail and head ends gives the concept of an arrow, and often an arrow point is shown on the head end to show which direction the activity progresses. These do not need to be straight lines or arrows, but each one has to connect two nodes. Connecting nodal points shows the dependency of the later one (at the head of the line) to the earlier node or event.

The CPM technique may use either an arrow diagram or a node diagram. With the latter the nodes represent the activities to be done. All information about the activity's estimated time to begin and end the operation are shown on its node. The lines between nodes merely show the dependencies of the activities at adjacent nodes. The node-diagram approach can be confusing with so much information given at each node.

An arrow diagram for the CPM technique is preferred by many users. In this form each solid line represents a construction operation or activity. The nodes on an arrow diagram are the events that show the start or finish of one or more activities. The starting or finishing times are shown at the nodes. Basically, there are earliest start (ES) and earliest finish (EF) as well as latest start (LS) and latest finish (LF) times for each activity. The planned time to do the activity, or its duration, can be shown on the arrow in this form of diagram.

Now an example of the CPM technique with an arrow diagram can be used to help extend the explanation. It will be assumed that the methods for doing each operation have been planned to arrive at their initially required time (duration). Also their dependencies on preceding and following operations are set by the plans. The data for the original plan of the operations are given as:

Activity (Oper.)	Dependencies	Duration, days	Day for Scheduling				Float Days
			ES	EF	LS	LF	
AB	BC,BD,BE	1	0	1	0	1	0
BC	AB,CE,CF	2	1	3	1	3	0
BE	AB,EG	8	1	9	5	13	4
BD	AB,DE	3	1	4	7	10	6
CE	BC,EG	10	3	13	3	13	0
DE	BD,EG	3	4	7	10	13	6
CF	BC,FI	7	3	10	17	24	14
EG	BE,CE,DE	5	13	18	13	18	0
GH	EG,HK,IJ	9	18	27	18	27	0
HI	a dummy	0	27	27	27	27	0
FI	CF,IJ	3	10	13	24	27	14
IJ	GH,JK	5	27	32	27	32	0
HK	GH	7	27	34	28	35	1
JK	IJ	3	32	35	32	35	0

Completing the columns for each activity, the ones it depends on or are dependent on it, and its planned duration are the first steps of this tabulation. A dummy activity, such as HI, has no duration but is used to show the dependencies of activities not connected by common nodes or events.

In the next step the earliest start (ES) and earliest finish (EF) times can be determined. The ES time for any activity is equal to the latest EF of any activity that must be done before it. The EF is found by adding the activity's duration to its ES time. This procedure is followed for all activities to find the earliest finish for the total project. The time thus found is the minimum for the construction project as originally planned. In this example the total project takes 35 days.

Now the planner is ready to find the latest start and finish times for each activity. This is done by working backwards from the activity with the latest EF, which is its LF also if the project is not to be prolonged. And its LS will be the LF less its duration. Taking the activities one by one up the list, one checks to see what event (last letter) or following activity sets its LF time. Thus, event K set the LF = 35 for operation HK, whereas the LF for operation IJ is set by the LS = 32 of activity JK. If the activity has two or more activities that must follow it directly, then its LF is the earliest (lowest number) LS of those following operations. Thus, activity BC's LF = 3 = LS for activity CE and not CF's (LS = 17).

The last step for this tabulation of CPM data is to find the planned slack in the timing for any activity that does not need to be done exactly on schedule. This slack is called the *float time* for an activity. It is equal to LS minus ES or LF minus EF for the activity. Float time is the time that an activity can be delayed from its ES time without disturbing the critical activities that follow it.

The activities or operations that have no float time (FT) are critical to the successful scheduling of the total project. The critical path is drawn along the connecting activities with FT = 0. This is best shown on the CPM arrow diagram for this example (Fig. 1-5).

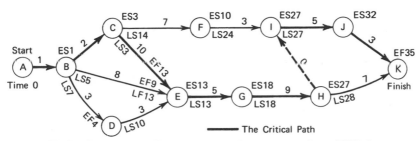

FIGURE 1-5 An arrow diagram for CPM.

The nodes or events are labelled by capital letters A, B, etc. Each line representing an activity is described by the nodal letters at either end of its arrow. The duration of an activity is shown along its arrow. The controlling times, ES, LS, etc., are shown at the nodes. It will be a good exercise to check the times shown by following the ES and EF (= ES + duration) through from the beginning at A to the end at event K. Then back up from K to check all LS or LF times shown.

A complete CPM analysis for a building project may have hundreds of activities. Furthermore, there may be several ways to do each of the variety of operations. So the number of data bits for a CPM tabulation and diagram can be sizable. The only efficient means for handling a really complicated analysis is with a high-speed computer. Computer programs for the CPM management tool are available.

1.4.3. Equipment Planning with CPM

The CPM tool is a very effective technique developed to plan and control operations. The operations to be controlled may involve a variety of forces and resources. In using CPM for construction operations, the resources may be labor, materials, equipment, or money. More than likely, there will be a combination of these resources for each operation.

The equipment to do an operation is a reasonably predictable resource.

Its rate of production for a given operation can be fairly well estimated. How such a productivity can be determined is the subject of many following sections in this book on specific types of equipment. When the rate of production is estimated for a given quantity to be done, the time or duration can be figured for the operation. This will be based on the original plan and method with a specific set of equipment. A question arises that should be answered with the help of CPM. Would another method or set of equipment be better or lead to economies if used for a given operation? That question should be asked for each operation when the whole project is the basis of concern.

Some idea of the use of CPM for planning equipment will be suggested with the previous example. If that is a small earthwork and paving project, many of the operations or activities will be planned with equipment as the main resource. Let us assume that activity BE is cut-and-fill earthwork along the route planned for a set of six scrapers loaded with the help of two pusher tractors. The activity CE may be borrow excavation loaded by a shovel into four haul units. CE is on the critical path. If there is any reason to want the project finished in less than 35 days, production on the CE operation could be increased. This can be simply done by using two of the same size shovels and doubling the number of haulers. If that would cause too much congestion or require more operators than are available, larger equipment could be used. Doubling the set of equipment on CE should cut its duration in half. If the original plan for activity BE is not changed, that one will become critical. To keep that from happening, the productivity for BE can be increased by using a set of nine scrapers with three pushers to get the operation done in five to six days.

This discussion is intended to give some idea of how the CPM tool can be used for equipment planning. CPM is most useful in the planner's concern for his equipment resource allocations. It is highly desirable and sometimes required for controlling the plans and schedules for construction operations. Following chapters of this book will deal with specific types and components of equipment. With the included data and discussion a construction planner can gather the information he needs to use CPM. A thorough discussion of the CPM technique and a variety of applications can be found in books specializing on this management tool.

1.5 Abbreviations, Symbols, and Units

Throughout the book certain abbreviations, symbols, and quantitative units will be used. These are for quick identification and for construction quantities, production rates, efficiencies, times, and costs.

The abbreviations and symbols used frequently in this book and not necessarily defined in the context will be listed here.

$\approx$	approximately equal to
$>$	greater than
$<$	less than
$\geq$	greater than or equal to
$\leq$	less than or equal to
α	angle of repose for a granular material
AASHO	American Association of State Highway Officials
ASTM	American Society for Testing Materials
bbl	barrel or 4 sacks = 376 lbs of cement
bhp	brake horsepower output
cfm	cubic feet per minute
CIMA	Construction Industry Manufacturers Association
Btu	British thermal units of energy
C_t	coefficient of traction
cu yds/hr or cy/hr	cubic yards per hour
DBPP	drawbar pull power for crawler tractor
°F	degrees Fahrenheit temperature
δ	density of material in lbs/cu ft
EVW	empty vehicle weight, generally for trucks
F	force required or to overcome
fpm	linear speed, feet per minute
G	grade of incline in percent (%)
gpm	flow rate, gallons per minute
GR	grade resistance, lbs/ton
GVW	gross vehicle weight, total loaded
I.C.	internal combustion
lbs/cu ft	pounds per cubic foot (pcf)
N	prime mover's drive shaft speed, rpm
π	pi = 3.14
psi	pounds per square inch
psig	pounds per square inches gauge
r	radius of a driving wheel
rpm or RPM	revolutions per minute
RR	rolling resistance, lbs/ton

T	torque of drive shaft in lb-ft
TE or F_T	tractive effort at drivers of mobile equipment
θ	angle
v	velocity of linear motion
V	volume of material in cu ft or cu yds
vpm	rate of vibrations, impacts per minute
w	weight per unit length, such as lbs/lin ft
W	total weight of an object
SAE	Society of Automatic Engineers standard rating
PCSA	Power Crane and Shovel Association
E	pile-driving energy, ft-lbs

Letters are used as abbreviations for estimating quantities, efficiencies, times, or costs of interest in dealing with construction equipment. A common basis will be followed for the identification of the construction units. Lower-cased letters are values: for a part of a total, a rate per hour, or a cost per unit of production. Capital letters identify total quantity, total time, or total cost. "Total" is intended to mean any amount that can be identified separately and will generally take more than an hour to charge or complete. The common units used in a general way in this book are:

A	=	average annual investment
ADB	=	acceleration, deceleration, and braking
c	=	cost per unit of production
C_e	=	total cost for equipment
C_o	=	original investment in the equipment
CT	=	total cycle time, generally in minutes
D	=	working days in a year
DT	=	time to dump a load
e	=	an hourly expense for the equipment
E	=	an annual expense for the equipment, or efficiency of equipment operation
eff.	=	effective portion of total potential
e_{op}	=	operating expense per hour
e_{ow}	=	ownership expense per hour
e_r	=	rental rate per hour
f	=	factor showing efficiency of production
f_a	=	overall or actual job efficiency

f_c = factor for unpredictable contingencies

f_j = job-management factor

FT = fixed time

f_w = normal working efficiency

H = working hours for the equipment

LT = load time, i.e., time to get a load

N = number of similar pieces of equipment (also rpm of prime mover's driving shaft)

q = production per hour or productivity

Q = total quantity to handle or produce

q_a = actual productivity = $f_a q_p$

q_n = normal productivity = $f_w q_p$

q_p = peak productivity working 60 minutes

S_h = shrinkage factor

S_w = swell factor

TPH = tons per hour

TT = turning time

U = useful life of equipment or component

VTE = variable travel time when empty

VTL = variable travel time when loaded

References

1. Farrell, Fred B., "Some Equipment-Management Problems on Highway-Construction Jobs," reprinted from *Highway Research Abstracts,* Vol. 23, No. 10 (Washington, D.C., Nov. 1953).

2. Drevdahl, E. *Profitable Use of Excavating Equipment,* Desert Laboratories, Inc. (Tucson, Arizona, 1961).

3. Shaffer, L. R., Ritter, J. B. and Meyer, W. L., *The Critical-Path Method,* McGraw-Hill Book Company (New York, N.Y., 1965).

Chapter 2

∎∎∎∎∎∎∎∎∎∎∎∎∎

∎∎∎∎∎∎∎∎∎∎∎∎∎∎∎∎∎∎∎∎∎∎∎∎∎∎∎∎∎

COSTS AND
CONTROL OF
EQUIPMENT

∎∎∎∎∎∎∎∎∎∎∎∎∎∎∎∎∎∎∎∎∎∎∎∎∎∎∎∎∎

Introduction to Equipment Costs

The main object for successful construction management is to have the operations they plan finish with a satisfactory end product, according to the plans and specifications, at the least possible cost. Therefore, a most important factor in planning for construction equipment is its total cost to management. The total cost includes the original investment or the rental charge and the cost of operating and maintaining the equipment in good running order.

In the case of owning equipment and operating it for doing construction work there are two well-defined categories of time which should be recognized for cost reasons. There is the time when the piece of equipment is operating and actually producing or contributing to the finished product. On the other hand, there is the time when the piece of equipment is owned, and so an expense to the business, but it is standing idle and not

33

working. There are parts of the total equipment cost that are chargeable only to the operating time. These charges will be called the *equipment operating costs* in this book. Other charges that should be made during the idle time as well as the operating time are mainly attributable to the ownership of the piece of equipment. Therefore, these will be called the *equipment ownership costs.*

The basic difference between operating and ownership costs should be recognized so that any item not specifically enumerated in the following could be properly assigned. It should be obvious that the equipment operating expenses, such as for the fuel and operator, are those costs that mount up during any working hour when the piece of equipment is operating. For the hours that the equipment is not operating or being used these expenses do not logically happen. On the other hand, the equipment ownership expenses are those costs that mount up and are chargeable over a period of months regardless of the operation of the piece of equipment. Thus, one type depends on the actual productive time and the other on the passage of calendar time. With this distinction in mind we can easily see why the ownership expenses are called the fixed costs by some analysts.

2.1 Equipment Operating Costs

The equipment operating costs or expenses are those that are necessary to power and make possible the smooth, effective use of the piece of equipment. They will include those necessities required essentially only during the time when the equipment is in operation. These expenses are listed and explained in the following paragraphs.

(1) *Charges for the operator* are his wages and the extras for taxes on the wages, workmen's compensation insurance, and fringe benefits. On some equipment there will be more than one worker involved in operating it. The outstanding example of this occurs on power cranes and excavator equipment where an oiler is frequently required. Operator union contracts and conditions for efficient use of the equipment should be checked to ascertain the possible need for more than one man to operate any given piece of equipment. The current local union contracts should be used to determine the wages involved and the provisions for fringe benefits that will add to this operating charge. The extras based on these operating wages will vary from year to year and from one labor contract to another, but generally they will amount to from 10 to 20% of the wages and even more in some cases.

(2) *Cost of fuel* is the expense of input for powering the equipment. The input may be gasoline or diesel fuel, electrical power, or other forms of power source. The cost of the various forms of fuel as well as their effectiveness per unit of power delivered will vary considerably. The rates of common fuel consumption can be based on the following:

$$\text{Gasoline, gph} = \frac{0.7 \text{ (bhp)} \times \text{load factor}}{6.2} \qquad (2\text{-}1a)$$

$$\text{Diesel fuel, gph} = \frac{0.5 \text{ (bhp)} \times \text{load factor}}{7.2} \qquad (2\text{-}1b)$$

where the consumption is in gallons per hour (gph);

bhp = brake horsepower (to be explained in Sec. 3.1.2); and the load factor is found in Table 2-1.

TABLE 2-1 Load Factors for Fuel Consumption[8]

Type-of Equipment-Use	Operating Conditions		
	Excellent	Average	Severe
Wheel-type, on paved road	0.25	0.30	0.40
Wheel-type, off highway	0.50	0.55	0.60
Crawler-track type	0.50	0.63	0.75
Power excavators	0.50	0.55	0.60

The form of power source to be used will not necessarily be based on the lowest fuel cost for operating a given piece of equipment, but depends on many other power considerations which will be discussed in the next chapter.

(3) *Cost of lubricating oils* is an expense which includes daily or otherwise regular lubrications and periodic complete changes of oil. This part represents a minor item in the total equipment operating cost, but is separated from the fuel cost since it can be easily determined. The lubricating cost is usually estimated as being directly proportional to the actual hours of operation and dependent on the frequency of the complete oil changes. The consumption of lubricating oil can be figured using the following equation.

$$\text{Lubricating oil, gph} = \frac{0.6(\text{bhp})(.007)}{7.4} + \frac{\text{crankcase capacity}}{H_o}, \qquad (2\text{-}2)$$

where bhp = brake horsepower; the crankcase capacity is in gallons; and H_o = the hours between oil changes.

(4) *Cost of minor repairs and adjustments* is an expense which mounts up on the jobsite where the piece of equipment is to be operated. The repairs involved here would be of such a nature that necessary spare parts are readily available to the field operations. The operators and other dependent workers would be kept standing by for further work. These minor repairs should take no longer than fifteen minutes to make. They are to be distinguished from the major repairs that are among the equipment ownership expenses. The adjustments included here as a part of the operating expenses are those alterations readily made in the field as anticipated in the original design of the piece of equipment. An example of such adjustment would be the manual change of the dozer's blade angle or the change from one type of crane pickup device to another.

(5) *Cost of tire repairs and replacement* is an operating expense for a wheel-type unit. This is not the original cost of the tires which might be considered a part of the initial investment. The money invested in the piece of equipment pays for an original set of tires for the wheel-type unit. It is an established fact that none of the original tires on mobile units will have a useful life as long as that of the equipment on which they came. The tire is not made to stand its heating, friction, flexing, and abuse to the same time limit set for the resistance to wear and tear built into the equipment on which it serves. The main difference can be related to (a) the hours of operation or (b) the miles traveled.

One basis for estimating the tire life on a piece of equipment is based on constant use in mining operations as presented in the Drevdahl reference.[8] The object is to establish the maximum possible tire life with the least abusive conditions of use and the best maintenance. Then to find the shorter estimated life, apply factors for each condition of use and maintenance expected. A possible optimum or maximum life of a tire might be 5,000 hours or 50,000 miles. Then the reduction factors are found from Table 2-2.

An example will help us to understand this method of estimating tire life and, consequently, the operating expense of tires.

Given: a rubber-tired front-end loader operating in a gravel pit moving a maximum of 300 feet and averaging 10 mph. Tires are watched carefully to keep the specified inflation, but frequently they are overloaded by 10%.

Find: the estimated life for the front-end loader's driving tires (use is between dump-drivers and self-loading scraper's). From Table

TABLE 2-2 Tire Life Reduction Factors

Condition of Use	Factor to Apply				
	1.0	0.9	0.8	0.7	0.6
A. Tire inflation, psig, compared to specified	100%	90%	80%	75%	70%
B. Loading on tire, compared to specified	100%	110%	130%	150%	—
C. Average speed, mph	10	15	20	25	30
D. Wheel position	Trailing	Front	Dump-drivers		Scraper's
E. Traveling surface conditions	Soft earth	Maintained gravel road	Rough gravel		Sharp rock

2.2, one finds the factors to be A = 1.0, B = 0.9, C = 1.0, D = 0.7, E = 0.7, so estimated tire life, U_t, is

$$U_t = 5{,}000 \ (1.0)(0.9)(1.0)(0.7)(0.7) = 2200 \text{ hours.}$$

The equipment operating expense for these tires can be found if the cost of one is known. Say that such a tire costs $2,500 new. Therefore, the two drive tires of this front-end loader give an operating expense, $e_t = 5000/2200 = \$2.27$ per hour.

The expense of tire replacement or retreading is an equipment operating expense since it is mainly dependent on the actual operation of the equipment. If the wear and tear of tires is too variable, the above method is difficult to apply. A rougher estimate must be made. The tire expense may amount to the equivalent of one, or more, new tires every twelve to eighteen months. The shorter time is due to hard use, i.e., rough loads, poor maintenance, or excessive speed and loads. A normal year for most equipment amounts to 1500 to 2000 working hours. More will be discussed about tire use and costs in Secs. 3.2.2 and 9.5.4.

All of these equipment operating costs are either given, estimated, or easily converted to cost per working hour. For the sake of convenience in preparing estimates and charges, and since these costs are chargeable only to the operating time, it is recommended that each of them be originally reduced and kept in terms of cost per working hour. Then the total equipment operating expense per working or operating hour can be readily obtained. This part of the equipment expense should be essentially the same regardless of whether the piece of equipment involved is owned or rented. It can be shown mathematically by the equation,

$$e_{op} = (1 + k_w)\, e_w + e_f + e_l + e_m + e_t, \tag{2-3}$$

where e_{op} is the equipment operating expense per hour;

$\quad e_w$ is the hourly wage for operating the equipment;

$\quad k_w$ is the factor to account for wage extras, 0.1 to 0.2;

$\quad e_f$ is hourly cost of fuel consumed;

$\quad e_l$ is hourly cost for lubrications and oil changes;

$\quad e_m$ is hourly cost for minor repairs and field adjustments; and

$\quad e_t$ is hourly cost for tire retreading or replacement.

2.2 Equipment Ownership Costs

The equipment ownership costs are those expenses that the owner of a piece of equipment must take into account to evaluate and protect his investment. They are the concern of the construction business management as well as the equipment rental dealer. First, there is the original cost of the equipment which is everything spent on, for, and on behalf of getting the equipment into the contractor's yard or to the construction site ready to operate. That is accepted by the U.S. Internal Revenue Service (IRS) as the basis for initial investment cost.

Several of the equipment ownership cost components can be figured based on the average yearly value of the equipment investment over its useful life. The IRS guidelines show the depreciable life for general contract construction as five years. In reality, the average value will decrease during the life of the equipment, so probably the ownership cost should vary with such change. However, to facilitate the determination of a constant regular charge for any hour during the equipment's life, it is convenient to use a constant average value for the entire useful life. This value is called the *average annual investment, A*. It is found by taking the sum of the beginning-of-the-year investment book value for each year of the piece of equipment's useful life and dividing by the estimated useful life in years. The assumption here is that the value of the equipment carried in the accounting books throughout any one year, and consequently the basis for the charges for that equipment during the year, is the value at the beginning of the year. If the decreasing value of the piece of equipment follows a straight line, then its average annual investment can be found easily. In this case the average annual investment is found as

$$A = \frac{1}{2} \frac{(U + 1)}{U} C_o, \tag{2-4}$$

where U = the estimated useful life of the equipment in years and
$\quad C_o$ = the original initial cost of the piece of equipment.

The components of the equipment ownership expense are listed and explained in the following paragraphs.

(1) *Interest* is the charge for borrowed money or the return for money invested. Certainly, if money is borrowed from a bank or another lending house to pay the original cost of a piece of equipment, the owner will be charged interest on the loan. This charge accounts for the cost of money invested by the lender. Therefore, even if the purchaser uses his own money to buy a piece of equipment he should include the interest charge in calculating the equipment ownership expense. If he had not used the money for purchasing the equipment, he could have invested it in stocks or bonds to earn interest. The annual interest charge, E_i, generally should be taken as a percentage, $P_i = 5$ to 15%, of the original investment, C_o, in the equipment.

(2) *Taxes* are the costs charged by the federal, state, or local governments based on ownership of the equipment property. This charge, E_x, generally is figured as a percentage, $P_x = 1$ to 5%, of the A value of the equipment for any year.

(3) *Insurance* is the cost of the premium for insurance to protect the owner against financial loss in case of loss or damage to his equipment. This annual charge, E_p, generally is figured as a percentage, $P_p = 1$ to 3%, of the book value of the equipment. To find a charge for this over the equipment's lifetime, the average annual investment (A value) can be used.

(4) *Storage* is the cost of keeping the piece of equipment in a safe, protected place during the extended time when it is not working or on a jobsite under the contractor's control. This is the charge for extra land rented or a covered shed for storing the equipment. The storage charge, E_s, is generally taken as a nominal percent, one or less, of the A value of the equipment during an average year.

(5) *Major repairs and overhauling* result in the cost of making major repairs or replacement of parts and overhauling the piece of equipment in a shop by mechanics or other equally skilled service people. Such a charge is not expected to be a regular amount each year, but tends to increase with the equipment's increase of age. The only certain aspect of such a charge is that it will amount to a substantial proportion of the original investment. Records kept of the cost of these charges during the

life of the particular or similar equipment will indicate the proportionate expense to anticipate. The equipment owner's charge for this item is based on a total cost for major repairs and overhauling generally amounting to 50 to 100% of the original cost of the equipment. A common practice for finding this percentage is to take a five-year average of the repair and overhaul costs and divide that by the average equipment depreciation for those same five years. Depreciation for a piece of equipment is discussed in the next section.

The major repair and overhaul cost can be minimized by contracting for repairs through a maintenance agreement with the dealer. That saves the owner from having a shop with spare parts and trained mechanics. Higher percentages should be used for pieces that have more moving parts or are subjected to greater vibration, shock, or wear and tear in general. This charge for major repairs and overhauling each year would be found as:

$$E_j = \frac{k_j C_o}{U},\qquad(2\text{-}5)$$

where E_j = year's charge for major repairs and overhauling;
$\quad k_j$ = percentage of original equipment cost (C_o) which will be reinvested in major repairs and overhauling; and
$\quad U$ = useful life of equipment, in years.

2.2.1 Depreciation of Equipment

The regular charge to account for the decrease in value from the original investment or previous value recorded in the business accounting books is called depreciation. In the United States the IRS watches this accounting closely. Eventually, when the piece of equipment serves for its estimated useful life, the sum of these regular charges will have reduced the equipment's value to its worth when retired—salvage value, or zero in rare cases. Thus, it should be recognized that the depreciation charge accounts for the decrease in the original money value invested in the equipment over the time of its use. The difference between original, initial cost and final, salvage value will be called the depreciable cost (C_d).

The reduction in value of equipment accounts for various factors such as used-up life, obsolescence or outmoding of the piece in comparison to newer models or equipment, and the resale value of the used machinery. Contractors complain that design changes and improvements made every year cause recently purchased equipment to be outmoded too quickly. The owner is forced to bear the cost of experimenting that should be a manufacturers' expense. This is similar to the automobile industry rather than aircraft practice. The accurate appraisal of these factors is quite arbitrary and so involves good judgment or an educated guess. The variety of possible appraisals can be nearly satisfied by one of the following methods for determining depreciation:

1. straight-line depreciation;
2. constant-percent or declining-balance depreciation;
3. depreciation by the sum-of-the-digits method; or
4. depreciation covered by a sinking fund.

All of these methods are initiated by determining the estimated useful life, U, for the piece of equipment. The useful life of construction equipment is commonly taken to be from 3 to 12 years, with the shorter life for highly mobile or portable equipment subjected to frequent shock, vibration, and abuse, i.e., greater wear and tear.

The straight-line depreciation per year, E_{dsl}, can be determined by dividing the original cost (C_o) less final value or salvage (S), which is the depreciable cost (C_d), of the equipment by its useful life; thus, for (1),

$$E_{dsl} = \frac{C_o - S}{U} = \frac{C_d}{U}. \tag{2-6}$$

This is the easiest method to apply and the one accepted by the federal tax officials for many years.

The constant-percent or declining-balance method calls for a percent to be established as the constant factor to be applied to the previous balance or book value for determining the depreciation charge, F_{dc}, and so the new book value. Thus, for (2), $E_{dc} = K_c$ times previous book value, which equals $C_d(1 - K_c)^{n-1}$ for any year, n; therefore,

$$E_{dc} = K_c C_d(1 - K_c)^{n-1}, \tag{2-7}$$

where K_c is the constant-percent depreciation rate, which cannot be more than twice the straight-line rate by IRS rules. For example, assume $K_c = 0.4$, or 40 percent, and $U = 5$ for a $10,000 piece of new equipment, assuming no salvage value, i.e., $C_d = C_o$:

the beginning book value = $10,000;
the next year's value   = $10,000 − 0.4(10,000) = $6,000;
the third year's value   = $ 6,000 − 0.4(6,000) =   3,600;
the fourth year's value   = $ 3,600 − 0.4(3,600) =   2,160;
the fifth year's value   = $ 2,160 − 0.4(2160) =   1,296;
at end of useful life   = $ 1,296 − 0.4(1296) =   778.

Obviously, the value of the equipment will never depreciate to zero by this method. It is necessary to make some change of method at the end of the useful life to depreciate to zero, or the salvage value, a piece started by the constant-percent or declining-balance method. The IRS tax laws allow changing to straight-line depreciation to do this.

If a reasonable salvage value, S, can be estimated for the piece of equipment, the constant percent method can be used to arrive at the salvage value at the end of its useful life. This is shown by equating the book value at the end of the useful life to this salvage value, thus:

$$C_o (1 - K_c)^u = S. \tag{2-8}$$

Then this equation can be used to calculate the constant percent desired:

$$K_c = 1 - \left[\frac{S}{C_o} \right]^{1/u}. \tag{2-9}$$

The sum-of-the-digits method (3) necessitates determining "the digits" after deciding on the estimated useful life. For instance, if $U = 5$, the sum of the digits is $5 + 4 + 3 + 2 + 1 = 15$. Then the amount of the original value that is apportioned each year as depreciation, E_{dsd}, would be 5/15 the first year, 4/15 the second year, etc. Therefore, the value of the originally $10,000 piece of equipment would depreciate according to method (3) as follows:

the beginning book value = $10,000;
the next year's value   = $10,000 − 5/15(10,000) = $6,667;
the third year's value   = $ 6,667 − 4/15(10,000) =   4,000;
the fourth year's value   = $ 4,000 − 3/15(10,000) =   2,000;
the fifth year's value   = $ 2,000 − 2/15(10,000) =   667;
at end of useful life   = $ 667 − 1/15(10,000) =   0.

The sinking-fund method (4) is used when money is put away in equal periodic installments, usually each year, to earn interest and amount to a total of installments and interest in the fund at the end of the equipment's useful life equal to its original value. Thus, there will be money available from the sinking fund at the end of the useful life. The scheme is that this money will be used to replace the piece of equipment which has served its useful life. Therefore, we realize that several assumptions must be made when the sinking fund is established. One assumption is that an amount of money equal to the original cost of the piece of equipment will be sufficient to replace it at the end of its useful life. The second assumption is that the earning power of the money in the fund, represented by the interest rate, will remain the same during the life of the fund. If these are not safe assumptions, some others based on good judgment and predictions must be made for the sinking fund.

Using the two assumptions stated above, the equal periodic payments, a, to the sinking fund would be

$$a = S_f \left[\frac{i}{(1 + i)^u - 1} \right], \tag{2-10}$$

where S_f = sinking fund to be accumulated = C_o less salvage, if any, and
$\quad i$ = annual interest rate to be earned by the fund.

The depreciated value of the equipment at any time according to method (4) will be the original value minus the installments and the interest that they have earned up to that time. Thus, the depreciation charge for any one year, E_{dsf}, is equal to the periodic payment, a, plus the interest earned in the sinking fund during the past year.

The four methods of accounting for the depreciation of a piece of equipment can be represented graphically (Fig. 2-1). In each method the charge is often made just once a year, probably at the end of each year of useful life; or, at the most, the depreciation charges would be made each month. Consequently, the graph of each method should be a stepped line. However, for clarity in analyzing the advantages and disadvantages of these methods we will connect each successive depreciated value by a

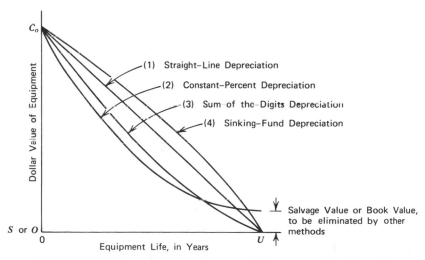

FIGURE 2-1 Comparison of four methods of computing depreciation.

smooth curve. These curves are based on no salvage value, except to take advantage of the problem mentioned with the constant-percent method.

Obviously, the straight-line method is the easiest to apply, but it probably does not lead to a realistic depreciated value for the piece of equipment during the first years of its life. This method has the distinct advantage over the others that the annual or periodic depreciation charge, E_{dsl}, is a constant value throughout the useful life of the piece of equipment. The constant-percent or declining-balance method leads to a more realistic depreciated value of the equipment and should be considered if an early resale of the used piece is anticipated. This method has the obvious disadvantage of not arriving at a book value of zero at the end of the useful life. However, this is perhaps desirable if any salvage value, which must be estimated at the beginning of the life, is too indeterminate. The sum-of-the-digits method also leads to a more realistic depreciated value for the equipment and eliminates the disadvantage of the constant-percent method. Furthermore, the depreciation charge, E_{dsd}, can be calculated easily for any one year. The sinking-fund method probably gives the most unrealistic depreciated value for the equipment during the first years of its life. However, the intent of the sinking fund would imply keeping the equipment for its full useful life. Certainly, if a fund is to be built up to cover the depreciation, this method for determining the equipment's depreciated value should be used.

Now that we have enumerated and discussed the important components of the equipment ownership expense, let us summarize by putting them together in mathematical form. The total annual equipment ownership expense based on straight-line depreciation and $S = 0$,

$$E_{ow} = E_p + E_x + E_i + E_s + E_j + E_{dsl} \qquad (2\text{-}11a)$$

$$= (P_p + P_x + P_s) A + \left[P_i + \frac{(K_j + 1)}{U} \right] C_o, \qquad (2\text{-}11b)$$

where P_p is % for insurance to find annual protection cost, E_p;
    P_x is % for taxes to find annual tax charge, E_x;
    P_s is % of average investment (A) to find annual storage charge, E_s;
    P_i is % of original cost (C_o) to find annual interest charge, E_i;
    K_j is % of original cost (C_o) to find annual repair charge, E_j; and
    U is the useful life of the equipment.

When a method other than straight-line depreciation is used, the average annual investment, A, will have to be calculated as mentioned prior to Eq.

(2–4), and the annual depreciation charge will be changed. For instance, if we are interested in the annual equipment ownership expense of a piece of equipment good for five years using sum-of-the-digits depreciation, we find the average annual investment,

$$A_{sd} = \frac{\left(1 + \dfrac{10}{15} + \dfrac{6}{15} + \dfrac{3}{15} + \dfrac{1}{15}\right) C_o}{U} = \frac{7\,C_o}{3\,U}; \text{ and}$$

$$E_{ow} = (P_p + P_x + P_s)\,A_{sd} + \left(P_i + \frac{K_j}{U} + K_s\right) C_o; \text{ or} \qquad (2\text{-}12)$$

$$E_{ow} = \left[\frac{7\,(P_p + P_x + P_s)}{3\,U} + \left(P_i + \frac{K_j}{U} + K_s\right)\right] C_o;$$

where $K_s = \%$ or proportion of depreciation.

The next important consideration is the estimation of the number of possible working hours per year for a given piece of equipment. This is necessary to be able to charge a job or user on an hourly basis for the time it has control over the use of the equipment. The number of hours generally estimated and shown in tables is 1200, 1400, 1600, or 2000 per year. The larger numbers are for equipment of greater versatility or that which would generally be used on longer-duration or more frequent jobs, whereas the lower number is for equipment of more special use or used on small odd jobs. Of course, after a piece of equipment has been used by a contractor or other owner for a year, it is possible to make an estimate of the working hours based on experience. Certainly, after several years of experience and records with a certain type of equipment used in the business, the owner should use these rather than take the arbitrary number of working hours per year suggested by a published table. Whichever method is used, we shall show the possible or charged working hours by H_{cy}. Therefore, the hourly charge for equipment ownership expense,

$$e_{ow} = \frac{E_{ow}}{H_{cy}}. \qquad (2\text{-}13)$$

It should be noted that the charged working hours are all the normal, conceivable hours of work that the piece of equipment might realize while it is charged out to any particular job. Thus, if the contractor-owner of a certain tractor normally should have it working on single-shift jobs for five days a week and nine months a year, $H_{cy} = 1560$ hours. The actual worked time, H_{wy}, may be only 1440 hours. Then during the actual working hours the total equipment expense per hour would be

$$e_o = \frac{E_{ow}}{H_{cy}} + e_{op} = e_{ow} + e_{op}. \tag{2-14}$$

And for the hours that the jobs should have used the equipment but did not because of delay due to breakdown, lack of material, etc., the charge would be e_{ow}. This part of the expense of owning the equipment might be charged to job overhead as an indication of something less than perfect management. Of course, if the delay was no fault of the job management, such as a general area or nationwide transportation stoppage due to strike, then the charge should go to general overhead for the company.

2.3 Economic Life of Equipment

In the previous section the idea of a piece of equipment's estimated useful life was discussed. The IRS has a guide of life periods for different classes of depreciable property. The estimation of useful life is necessary in order to establish from the very beginning of its life a reasonable cost or charge for the use of the equipment. Now we are interested in a more realistic way to determine the practical life of a piece of equipment based on economic reasons. Two methods with quite different objectives will be suggested here. The first arrives at an economic life based on the minimum hourly cost to use the piece of equipment. The second method claims the economic life is found when the maximum financial return occurs. Both of these methods require accounting for the costs of equipment as it is used, as well as other costs to be explained. So neither can tell the economic life when the equipment is new.

2.3.1 Minimum-Cost-per-Hour Basis

The economic life of a piece of equipment can be based on the minimum actual cost per hour for owning and operating it. The actual cost will include the original capital cost and other ownership expenses, as well as the operating expenses such as the fuel and the operator, the cost of insurance, taxes, interest, storage, and repairs and overhauling. The key to finding the most economical replacement time is to watch the cumulative cost per hour over the equipment life. It will be high at first when there is the initial investment and the equipment has not worked many hours. As it works more hours over the months and early years, the cumulative cost per hour will continue to decrease. But when the piece of equipment gets older and has increasingly frequent repairs and overhauls, its cost mounts up. At some time in the equipment's life these higher maintenance and repair costs will make the cumulative cost per hour higher than for its previous hours. When the cumulative cost per hour was at its lowest level, the equipment reached the end of its economical life according to this basis.

An interesting way to look at the various ownership and operating expenses that make up the piece of equipment's cumulative cost per hour is to consider their differing effects on the question of the equipment's economic life or replacement time. Some of the expenses favor keeping the machine indefinitely. Depreciation and investment costs, such as insurance and taxes, are examples that favor keeping it. This is because they all diminish as the book value of the equipment decreases with time.

Other expenses in the cumulative cost per hour favor early machine replacement. Maintenance and repair costs, along with the expense of equipment breakdown time, are the main examples in this category. They favor an early trade because these expenses will increase with the equipment's age. A third category of expenses does not affect the economic life or replacement time for a piece of equipment. The operator's wages and fringe benefits are the prime examples.

The first two categories of expenses in the cumulative cost per hour effect the economic life of the equipment. Their variation is shown graphically in Fig. 2-2.

Of course, the cumulative cost per hour is the total of expenses that favor retaining the equipment and those favoring early replacement. So the top

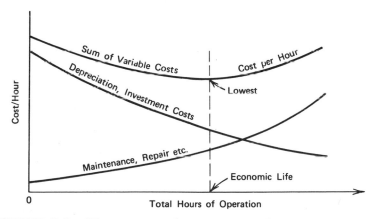

FIGURE 2-2 Variation of equipment's variable costs.

curve in the set of curves shows the low point of cumulative cost to be found. The total hours of operation to that time gives the economic life for that piece of equipment.

A more practical way to find the lowest cumulative cost per hour keeps a running graph of the expenses and hours of operation. Figure 2-3 is an

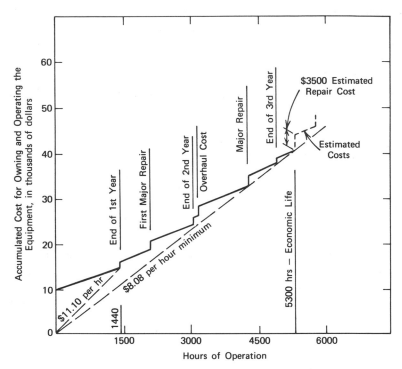

FIGURE 2-3 A graphical method of finding the lowest cumulative cost per hour.

example of such a representation. It is based on a piece of equipment with an original investment of $10,000 for the machine at the beginning of its first year and having an estimated useful life, $U = 4$ years and $H_{cy} = 1600$ hours. We will assume that insurance for it is bought at the beginning of each year and the taxes, interest, and storage are calculated and added at the end of each year of service. The sloping lines of the accumulated cost curve between steps indicate the operating expenses for the equipment.

We note on the graph as should be expected that the sloping operating expense lines generally become steeper after the first year or two of use. Also, the occasions and costs for repair and overhaul are more frequent or greater. Consequently, as the owner approaches major repairs when the machine is getting older, in the third or fourth years for this example piece, he should analyze the economic situation before going ahead with the repair work. The equipment may have reached its economic life for the owner. With an estimate of the necessary repair or overhaul cost and the future operating expenses to the end of the current year the graph can be projected forward. If we have reached the economic life based on such a projection, we should consider trading the piece for a newer model. Of course, if the trade-in would be financially better by going ahead with the major repair, it should be done; otherwise, the money for the repair would be wasted.

The example above assumes that the rate of production by the piece of equipment is relatively constant for each hour of its useful life. An owner might anticipate a somewhat declining productivity as the equipment gets older. In that case a more accurate determination would be based on cost per unit production. This requires knowing the productivity each hour, month, or year and correlating that data with the costs at the same time. Allowing for declining productivity will mean the economic life is somewhat less than found by the minimum cost per hour.

The example just given is for a single piece of equipment. It is always best to decide on the economic life of an individual piece rather than use an average for a particular type of equipment. Of course, an estimated economic life must be based on typical or average experience. However, the owner must make his final decisions to keep or replace individual pieces of equipment, just as he has to decide for his private automobile.

2.3.2 Maximizing-Return Basis

Some equipment owners prefer to decide the economic life based on the current value of future revenues and costs for the possessed piece of equipment or its replacements. This is chosen instead of the least unit cost method, which does not account for improved productivity and better earning capacity of a new piece of equipment to replace the old one. It is not good business to keep an inefficient, obsolete machine. One that is inadequate to do an operation should not be kept just because a newer, bigger, and better designed one will cost much more. The newer piece of equipment, if properly selected, should produce enough more units and be enough more profitable to cover the difference in cost.

The current value approach to economic life is based on the future need and present value of profits from the demand for such equipment. It does not account for past earnings or costs of the possessed piece of equipment, or any like it, except to use them for predicting the future. The question of how far into the future to look comes up immediately. This can be answered reasonably by noting the effect of determining the current value or present worth of a dollar at different times in the future.

On the basis of compound interest and established economic theory, the present worth factor (PWF) will give the current value of the amount of money that will grow to a dollar in n years, compounding at interest rate, i. For example, we can assume a reasonably attractive interest rate of 10% per year. The amount that grows to $1.00 with:

Time to Grow, n, years	Present Worth Factor, PWF, $i = 10$	Means a Current Value
0	1.00	$1.00
10	0.3679	$0.37
15	0.2231	$0.22
20	0.1353	$0.14
30	0.0498	$0.05

Values like these for PWF can be found in books on economics and finance or published tables for annuities, interest, and present values.

The above tabulation of values shows that more than three-quarters, or 75%, of the growth is in the first 15 years. If the interest rate is reduced to 8% ($i = 8$), this amount of growth is realized in 17 years. So it is reasonable to make current-value cost analyses on the basis of estimated revenues and costs of the possessed piece of equipment and its replacements during the next 15 to 20 years. This is the approach used by James Douglas in his results on the "Optimum Life of Equipment for Maximum Profit."[2]

The current values found by Douglas' determination take into account the time value of money; technological advances in equipment; and monetary matters such as taxes, depreciation, inflation, increasing cost of borrowing money, and the increased cost of future machines. Current value that is calculated for a piece of equipment and its replacements is the sum of revenues less costs, which amounts to profit. The object of this determination is to find the piece of equipment's economic life based on maximizing the profit.

The estimated revenues take into account the decline of earning power for a machine growing older. It suffers from loss of horsepower, wear and tear, and similar factors that increase cost per unit of production.

Chargeable costs should include the operating expenses for fuel, lubrication, and wages with fringe benefits. These are applied directly to the hours the piece of equipment is used. Consequently, to get the current value of these future expenses is a complex task since each addition has a different time element. It is reasonable to neglect such hourly expenses from a comparison between an existing machine and its replacement because they will be close to the same for both. Of course, such hourly expense cannot be forgotten when finding the total cost for using a piece of equipment.

The costs to include in a comparative study of current value will emphasize the choice between the possessed piece of equipment and its replacement. These costs include loss in market value, allowance for inflation, interest on investment, and insurance, license, and taxes to be paid for owning the equipment. Since most of these are recognized by payments or determinations once a year, the time element for the current value of these costs can be figured from a single, appropriate time for each future year. Other costs such as major repairs, tire replacements, or overhauls must be estimated with a certain number of hours (say, in 1000-hour increments) of use for the piece of equipment. Depending on the owner's experience with his equipment's use, these variable costs will come at different times of the future years. Some careful estimating is needed to include these costs in the current-value determination.

Now the economic life of a piece of equipment can be found based on the sum of current values that maximize profits of the future. If the subscript x is used to identify the existing piece of equipment and subscript r for its replacement, an equation can be written to relate profits to revenues and costs. The estimating of these future amounts might look ahead 15 years as explained earlier. One set of amounts is based on keeping the existing piece of equipment j years (i.e., $n = j$). The equation for profit, P, in its simplest form is

$$\sum_{n=0}^{15} P = \sum_{n=0}^{j} R_x{}^* - \sum_{n=0}^{j} C_x{}^* + \sum_{n=j}^{15} R_r{}^* - \sum_{n=j}^{15} C_r{}^*, \qquad (2\text{-}15)$$

where R_x = estimated revenue for the existing equipment in any future year it is used;

C_x = estimated cost for the existing equipment in any future year it is used;

R_r = estimated revenue for a replacement piece of equipment in a year of its use;

C_r = estimated cost for a replacement piece of equipment in a year of its use; and

* means the present worth factor at interest, i, for n years (PWF, i, n) as appropriate for each term in the series.

By varying the additional years, j, to keep the existing piece of equipment, the maximum profit from having such equipment can be found. In the next 15 or so years there might be several replacements. This is obvious if each piece is kept five to seven years.

It takes little imagination to realize that each summation (Σ) on the right-hand side of the equation has many terms in it. Furthermore, they are complex with their differing interest rates and time elements to find PWF values. So it is not hard to recognize that the determination of economic life by the current-value approach is very laborious. In fact, it might not be feasible or worthwhile without the use of a high-speed, electronic computer. The developer of this approach to the economic life of equipment, Dr. Douglas at Stanford University, has devised a computer program to ease such determinations.

This method for finding economic life seems to agree with the federal government's new emphasis on "total cost bidding." In that pricing of equipment the supplier agrees to sell a piece to the government at a bid to cover the initial purchase price plus a maximum repair cost and a minimum repurchase price. These repair and repurchase amounts are for a certain number of years which should be the economic life of the equipment.

2.4 Maintenance of Construction Equipment

In order for construction equipment to be effective, it must be working in the way it was designed to operate. A piece of equipment must be working that way to earn money to pay for the investment in it. Of course, a piece may be temporarily out of working order and in need of repair to be operated again. One must also recognize that there is more than the loss of value in the equipment when it is not working as it should. If it were working with its operator and many other workers and interdependent equipment connected with its operation, a breakdown in the operation of the piece of equipment would be very costly. For these important reasons it is essential to keep construction equipment in good working condition. This is the purpose of good maintenance of all equipment.

The economic importance of equipment maintenance can be noted by referring to Sec. 2.2(5). There it was mentioned that the major repairs and overhauling of a piece of equipment generally costs between 50 and 100% of the original investment in the equipment over its lifetime. In anticipation of such an expense, the construction equipment manager should do what he can to minimize the lifetime cost of maintenance. Manufacturers help by designing various pieces of equipment with interchangeable parts. It is widely agreed among equipment manufacturers, owners, and users that minimizing maintenance cost requires a program of regular care and maintenance for each piece of equipment. Such a program is called preventive maintenance.

2.4.1 Preventive Maintenance

A program of preventive maintenance is a plan and procedure to take the right measures for keeping equipment in good working condition. It might follow a motto such as "protect constantly, clean thoroughly, inspect frequently, lubricate efficiently, repair immediately." That seems easy to understand, but the problem is to establish the program and keep up the maintenance.

Equipment managers find that it is difficult to get the operators, mechanics, and construction supervisors to accept the concept of preventive maintenance. The reasons are that it is difficult to set up a good program, it takes some paper work and office work without obvious benefit, and the program tends to die out with time. These reasons should not prevail. A good program of preventive maintenance should be a paying proposition. It can amount to substantial savings on repair bills and a considerable increase in equipment availability.

Equipment manufacturers often stress *availability* as the important characteristic of a reliable piece of construction equipment. The term refers to the amount of time the piece is able to function effectively. Availability is usually expressed as a percentage of the total time planned for the equipment to work. A reliable piece of equipment properly maintained should have better than a 90% availability. Older equipment may have a somewhat lower availability but should not fall below 80%. Any piece with a lower availability should be replaced. An ongoing program of preventive maintenance is designed to improve equipment availability.

2.4.2 Maintenance Operations and Records

The maintenance operations to perform on equipment generally can follow the points of the motto given in the previous section. Equipment must be protected to prevent its parts from rusting or otherwise deteriorating. This can be done by covering or painting those parts that may be hurt by exposure to weather, moisture, dust, etc. However, painting is not recommended for parts that must be inspected for very small cracks or other signs of distress. Certainly, moving parts that must mesh and run in close tolerance with each other must not be painted. If exposed, they should be coated with the specified oil or lubricant. Thorough cleaning is recommended to prevent or to detect the wearing of critical parts. If dust and dirt get between close-fitting, moving parts, there will be bad wear and tear. Grease and dirt will cover up cracks and other danger signs. Cleaning might be by washing or steam blasting. In the case of electrical parts a dry air jet must be used instead of moisture. Equipment exposed to wet concrete should be washed by water as soon as possible, certainly within that day.

The need for frequent, even some daily, inspection is to detect trouble with a part of the equipment before there is a costly breakdown. Trouble might be detected in various ways. For instance, one can listen or feel for unusual vibration, knocks, or noises in the running parts. Or worn and weak spots in structural parts may be detected by visual inspection for cracks, etc. To inspect or check the need for maintenance in the engine's running parts, certain tests may have to be made. These might check fuel consumption, engine RPM speeds, or heating of the radiator system and other parts. Lab analysis of a sample of the lubricating oil can detect engine trouble by the amount of dirt and other contaminants it contains. The sample is taken from a dip-stick setup. Such an analysis can discover engine trouble before there are noises, vibration, or overheating.

A necessary repair, even though the equipment can still be operated, should be made as soon as possible. The appropriate time might be at the end of the work shift in which it is detected. If the repair is not done that early, it may be overlooked for days. Then the trouble may eventually cause a complete breakdown and work stoppage. That would be much more costly in total operating expenses. And the repair probably will cost more than it would have if done when first detected.

The need for efficient lubrication is to minimize friction and wearing of the moving parts. The lubrication requirements are generally spelled out by the manufacturer's specification for each piece of equipment. They must be done on a regular schedule.

One of the key points of a good preventive maintenance program is the timing of the maintenance operations. Timing frequently is based on (1) the mileage travelled by high-speed hauling equipment or (2) the hours of operation for all other construction equipment. However, those times for any piece of equipment will not come at regular intervals on the calendar. That may cause difficulty in setting up a program of regular maintenance operations. An easier program procedure may be to do them on a daily, weekly, or monthly basis. Of course, major overhauls may be needed only when required by a breakdown or once a year. In that case the operation will probably be done on the appropriate basis—hours of operation or mileage. And a necessary repair will be done as needed.

The paper work for a good maintenance program can help with the timing. This part should be as simple as possible to be sure it is done regularly. Forms suggested by one equipment service manager are only five in number, as follows:

1. Operator's Daily Memo—to tell what construction operation it did and for how many hours, and to report any problems observed;

2. Monthly Unit Record—made by an office clerk to summarize daily reports on each piece of equipment;

3. Preventive Maintenance Check List—to be sure that all points about the cooling, transmission, undercarriage, and other systems are checked as well as all lubrication points to be covered by the maintenance shop people at the major 100-hour (or other suitable time) inspection and overhaul;

4. Unit Repair and Maintenance Report—to tell what has been done on a piece of equipment in the shop for irregular repair or maintenance; and

5. Preventive Maintenance Control Report—for the equipment manager to account for the major inspections and repairs made, and the hours of operation between each, during the lifetime of the piece of equipment.

Of course, there could be other forms in addition to these or replacing some. Whatever is used must be simple and provide all the necessary information for a good preventive maintenance program. That information should be helpful in the determination of economic useful life for a piece of equipment.

2.5 Equipment Ownership, Leasing or Rental

The person planning for construction equipment to work on his job has many alternatives to choose from. Productively, he has to choose among the various types and sizes of equipment that could do the job. This part of the decision, and the variety from which to choose, will be understood better after studying all the following chapters. Economically, the planner who is considering owned equipment must decide whether to use existing pieces or replace them with new equipment. This was discussed in Sec. 2.3.

There are fundamental advantages and disadvantages for a contractor owning construction equipment. The main disadvantages are that (1) the money invested in equipment cannot be used as such to help finance other expenses of the construction work and (2) having the equipment on hand necessitates finding work of the right sort to keep it busy. The main advantage to owning the equipment is that it provides an economical, competitive advantage over an identical rented piece for work it is designed to do. Such bald statements overlook a number of ramifications that will be discussed and investigated in later sections of this book.

Now we look at another set of alternatives in the economical choice of equipment. These are the alternatives to owning equipment—leasing or renting it. The leasing alternative is a way of having the use of equipment on a long-term basis without having to make the initial investment for it. A leasing company purchases the equipment and leases it to the user, who makes monthly lease payments. The lease agreement is for a year or more, so it is contrasted with renting equipment by the day, week, or month. Leasing equipment is more like leasing office space.

There are several important advantages in leasing equipment. The user can have the equipment he needs for an extended period of time without paying the lifetime ownership expense. Therefore, he will conserve working capital for current expenses. The monthly payments and expenses, of course, will cover all the equipment ownership and operating costs while he has the equipment, as well as a reasonable profit for the leasing company. The user's profit should more than cover the lessor's profit. That margin might be very small. But it can be improved if the lease agreement gives the user some recovery when the equipment is resold after its use. Monthly lease payments should be less than a month's rent of the equipment.

2.5.1 Equipment Rental Rates

The rental rate for any given piece of equipment is based on anticipated length of use, whether it is rented "cold" or "hot" (with fuel, etc., to run it), coverage of extras such as transportation, and the supply and demand for that type of equipment. The basic rates follow closely the listings suggested in the Associated Equipment Distributors' "nationally-averaged rental rates for construction equipment" for any given year.[7] These are rates for the equipment rented "cold," i.e., with no allowance for the operating labor nor fuel costs in the rental agreement. The party providing the rental equipment is the owner. Obviously, the rental rates will be higher than the equipment ownership expense discussed previously. The excess charge above the ownership expense is mainly to cover handling and accounting costs, servicing between rentals and a reasonable profit on the investment. Usually the major overhauling repairs done between rental agreements are paid by the rental company while the renter pays for all lesser repairs, particularly those handled on the jobsite. Generally, the renter will pay the transportation—freight or drayage—charges from the shipping point to the place of use and return. Also the renter will pay any loading, unloading, assembling, and dismantling charges.

Rental rates are usually set on three different time bases. There is a daily, weekly, and monthly rental rate for each piece of equipment. A long-term rental rate for perhaps half a year or the duration of a job may be negotiable. This would be less than a monthly rate and approach a lease rate. It might include maintenance in the agreement. The monthly rate is comparable to the equipment ownership expense with the extra charge added for the reasons discussed in the previous paragraph. The monthly rental may be 10 to 60% higher than the ownership expense rate. The weekly and daily rates are progressively higher for a common time interval to take into account the more frequent turnover, more severe wear and tear due to a variety of operators, and more likely idle time between rentals. The weekly rental is commonly 0.25 to 0.4 of the monthly rate, which means it would be from 8 to 70% more expensive for a full month than using the monthly rental rate. Similarly, the daily rental is generally one-third the weekly rate, which means it would be 67% more expensive per week than to rent at the weekly rate.

These three different rental rates are based on a reasonable number of straight-time hours for the rental period. The rental hours will be shown as H_r. The daily rental rate is based on $H_{rd} = 8$; for the weekly rate $H_{rw} = 40$ hours, or 5 days of 8 hours each; and for the monthly rate $H_{rm} = 176$ hours, or 22 working days of 8 hours each. For two- or three-shift operation the hours beyond the first 8-hour shift may be charged at half the normal hourly rate. Thus, rental equipment used for three shifts a day would be charged for the equivalent of two shifts at the normal rate.

The basis for any rental charge should be clearly spelled out in the rental contract. Trying to be practical and fair to the parties concerned, it might be assumed that an agreement between the rental company and the user can be reached whereby the following pro-rata time limits are to be used: the daily rental will be charged for any work up to 3 days or 24 hours; then the weekly rental will be charged as a minimum and for all working days up to 15, or 120 hours prorated; and thereafter, the monthly rental will be charged as a minimum and any excess over 176 hours based on the pro-rata ratio of actual hours worked to 176. These are rather arbitrary limits but they seem logical considering the problems of both the equipment rental company and the renter.

2.5.2 Economic Comparisons

It is quite likely that an equipment planner will compare the economics of using an owned piece of equipment to renting a similar piece for the job. Unfortunately, this is not a simple comparison. One important reason is that the time considerations are quite different. Rental rates are for periods in multiples of a day, week, or month. All equipment and related expenses to include will be for the same periods. If the time of use may approach a year or more, leasing of the equipment should be considered instead of renting.

In the case of owned equipment the full year of expenses must be considered because of such expenses as insurance and taxes. Even more important is the overhead expense that goes on whether the piece of equipment is being used or not.

The first example is used to determine the necessary chargeable working time per year in order to justify owning the piece of equipment. The hourly cost to own a piece of equipment was shown in Eq. (2-14):

$$e_o = e_{ow} + e_{op} = \frac{E_{ow}}{H_{cy}} + e_{op} \qquad \text{dollars/hour,}$$

and to rent the same piece of equipment:

$$e_r = \frac{R_m}{H_{rm}} + e_{op} \qquad \text{dollars/hour,} \qquad (2\text{-}16)$$

where e_{op} = equipment operating expense per hour;
e_{ow} = equipment ownership expense per hour;
E_{ow} = equipment ownership expense for the year;
H_{cy} = chargeable hours of working time per year;
R_m = rental charge ("cold") per month; and
H_{rm} = rental hours charged per month, generally 176.

For economical ownership of the piece of equipment, $e_o < e_r$; therefore,

$$\frac{E_{ow}}{H_{cy}} < \frac{R_m}{176},$$

or the chargeable hours of probable working time,

$$H_{cy} \geqslant 176 \frac{E_{ow}}{R_m}. \qquad (2\text{-}17)$$

For example, consider a piece of equipment that cost $22,000 new ($C_o$) and has an estimated useful life (U) of 5 years. Its depreciation and major repair costs for one year amount to $8,000, and the total of interest, insurance, taxes, and storage annually run 20% of the average annual investment (A). The rental charge per month is $1400. Assuming straight-line depreciation,

$$A = \tfrac{1}{2}(6/5)\,22,000 = \$13,200;$$

from Eq. (2-4)

$$E_{ow} = \$8,000 + 0.20 \times 13,200 = \$10,640.$$

Therefore, for economical ownership the piece of equipment should have a minimum chargeable working time found as

$$H_{cy} = 176 \times \frac{10640}{1400} = 1340 \text{ hours/year.}$$

In other words, the owner of the piece of equipment would have to be able to justify charging jobs where it will be or could be used for at least 1340 hours in one year's time. Otherwise, he should be renting the piece of equipment at $1400 per month.

This comparison is based on other costs being equal; for instance, the moving "on" and "off" costs are the same. Also the comparison is valid only if the two pieces of equipment are identical and of the same age, model, etc. If the rental piece of equipment is new as against an older owned piece with an original potential use of only 2000 hours a year, the contractor might well be advised to rent on this comparison rather than have his own unit. On the average the planner has to figure that the down time of a piece of equipment increases by at least 10% per year for every year of its age, and it seems to accelerate with age.

The determination of continuity of work illustrated above to justify ownership assumes that the owned piece is the best suited for the work to be undertaken. That is not always the case. The next factor to consider then is suitability for work. Often the owned piece of equipment was purchased because of its versatility or usefulness on a variety of work. For any given operation of construction work there is probably another piece of equipment differing in size, reach, capacity, or the like, which would do the job a little better or more efficiently. Assuming that such a piece is available for rent, the comparison now between owned and rented equipment necessitates introducing their different productivities. This will also involve a difference of time to accomplish a specific job of Q total units. Therefore, the overhead cost, C_v, will differ and we will assume it varies in direct proportion to the days of work.

To illustrate the effect of this factor of suitability, we will compare the cost of equipment on an earthwork operation, where the cost of materials purchased and labor not operating the equipment is negligible. The total cost of the operation, $C = C_e + C_v$.

(1) If the equipment is owned, the total cost is

$$C_1 = (e_o H_{cd} + C_{vd}) D_{w1}, \qquad (2\text{-}18)$$

but since the total hours, $H_{cd} D_w = Q/q_a$ or $D_w = Q/q_a H_{cd}$,

$$C_1 = \left(e_o + \frac{C_{vd}}{H_{cd}}\right) \frac{Q}{q_{a1}} \qquad (2\text{-}19)$$

(2) If the equipment is rented, the total cost is

$$C_2 = (e_r H_{cd} + C_{vd}) D_{w_2} = \left(e_r + \frac{C_{vd}}{H_{cd}}\right) \frac{Q}{q_{a2}} \qquad (2\text{-}20)$$

where C_{vd} = daily overhead cost;

H_{cd} = hours charged per working day;

D_{w1} = working days to complete the job by using owned equipment;

q_{a1} = productivity rate for owned equipment;

q_{a2} = productivity rate for rented equipment; and

D_{w2} = working days to complete the job by using rented equipment.

The break-even point would occur when $C_1 = C_2$, or when

$$\left(e_o + \frac{C_{vd}}{H_{cd}}\right) \frac{Q}{q_{a1}} = \left(e_r + \frac{C_{vd}}{H_{cd}}\right) \frac{Q}{q_{a2}}. \qquad (2\text{-}21)$$

Therefore, for the cost using the owned equipment to be cheaper than rented equipment, the ratio

$$\frac{e_o + k_v}{q_{a1}} \quad \text{must be} < \frac{e_r + k_v}{q_{a2}},$$

where k_v is the constant hourly charge for overhead. However, if the owned equipment is much less suited for the work than the rented, the chances are that the average production, q_{a1}, will be considerably lower or less efficient than q_{a2} for the rented equipment. In that case the original advantage of lower cost per hour for owned equipment compared to rented will probably be overshadowed in the total economic cost analysis.

In order for the prospective owner of a piece of equipment to be more assured that the ownership will be to his advantage, he should take certain precautionary steps. First, he would do well to be dealing with standardized equipment. This is equipment that is designed as far as possible with standard parts, which generally have proven to be the best design for their purpose and are readily available for purchase and repair. Furthermore, the standardized equipment is designed in that manner because its versatility and usefulness are in great demand and its ruggedness has been tested and found to require a minimum of replacements. Second, the careful purchaser of equipment should take advantage of the general availability of rental equipment for experimental purposes. When he or his operator are unfamiliar with a piece of equipment which seems like a good one for their operations, one of its model should be rented for a while before purchasing one for ownership. The lease type of agreement provides one way of doing this. However, ordinary rental of a piece with possible change to another make or model during the experimentation may be more economical. Then, after the trial period, the purchaser and his operator can feel more assured that they are obtaining for ownership the best piece of equipment currently available for their operation.

With all the previous points in mind, we can draw some fundamental conclusions regarding the financial advisability of owning equipment:

1. Own only those pieces of equipment which experience has shown are best suited to the variety of work that the contractor will probably undertake frequently.

2. Own a minimum of equipment and yet enough to enable the owner to have a sound competitive position on the type of operations he is qualified to contract.

2.6 Cost Accounting and Control

The introduction to this chapter emphasized the importance of the costs for equipment. Discussion that followed identified the specific costs and pointed out that they are incurred at various times and in different ways. This all suggests that there must be a system for gathering equipment costs. Furthermore, the costs should be collected in a way that they can be used for the control of operations.

A cost accounting and control system is a means for recording the essential costs and data on experience for future decision-making. The procedures have been used extensively for industrialized work. An important reason is the need for pricing products accurately. The need for careful cost estimating for construction work is probably even more important. So the planning in setting up a cost accounting and control system must be well done. The need for expert advice on this is imperative.

In the broadest sense, cost accounting and control give guidance and regulation to the operation and use of equipment by current recording of cost measurements. A distinction must be noted between this cost keeping and bookkeeping. The latter is done to record debits and credits in order to "balance the books." Cost accounting is a system of measurements to show construction costs and productivities to get work done. It is used in the determination of a piece of equipment's economic life or time for replacement. Another important use is to compare actual operations with pre-work estimates. This will also help for estimating and bidding future work. In any of these uses the cost accounting does not try to record every last penny to balance the accounts. However, the system must not overlook any significant cost measurements.

The costs to be recorded are all the equipment ownership and operating expenses. In a review of the ownership costs it is noted that some may be charged only once or, at most, a few times each year. At the other extreme, the fuel and operator costs add up each hour of equipment operation. Some of the expenses, particularly those for repairs and equipment downtime, are incurred at irregular times. Along with the costs there must be other measurements in the control system. These include especially the equipment's productivity and maintenance record.

With the variable information for the cost accounting and control system there is a timing problem. How often or quickly does the information have to be processed for effective decision-making? A correct answer depends on the uses to be made of the information. Daily recording of costs and control information would cover all possible data for equipment. But that could mean much more effort and information than is necessary. Weekly, or even monthly, recording from more frequently chargeable costs and production experience may be often enough for the system of controls. Whatever timing and involvement of people is used for the cost accounting and control, it must be well planned and accepted.

2.7 Use of High-Speed Computers

The cost accounting and control system will gather a tremendous amount of data. If daily records arc made on continually working equipment, there will be hundreds and even thousands of bits of each kind of data in the lifetime of the equipment. The man-hours to record all that data would be very tedious and costly, especially for a fleet of equipment. This task can be minimized by using a high-speed electronic computer. Its effectiveness is even extended by the computer storage of the data.

The type of computer suggested is called a digital computer. This is basically a counting machine that can discriminate and work on separate numbers by following certain rules of mathematical logic. Generally, the mathematical operations are done by additions and subtractions or combinations of those. For example, multiplication is performed by successive summations.

A high-speed digital computer is an electronic data processing piece of office equipment. The large memory and storage computers are expensive, specialized equipment that may not be afforded by many users of construction equipment. However, all should be able to afford access to the service of a large electronic computer through time-sharing remote terminals.

This type of digital computer may be operated directly or through a remote terminal. When properly programmed for use, it serves its purpose by reading, sorting, comparing, calculating, and tabulating data. So it has much use in cost accounting and control. With the computer's storage or memory ability through magnetic cores or drums or on electronic tape, it can handle the cost records for the lifetime of equipment.

The procedure to follow in the computer system for the control of equipment is quite simple. When the job is estimated, each item of work and operation is coded and its expected costs and productivity are recorded in the computer's memory. As the job is carried on, daily reports and equipment records are posted in the computer by office personnel familiar with its operation. Then construction or equipment supervisors can request any specific information or desired analysis of the costs or productivities. Such an analysis should take the computer, at most, a minute or two, if the computer has a good program. This is done to help management in its decision-making role.

Large high-speed digital computers are also very helpful in checking costs against pre-work estimates. They can be used to advantage with a good program for testing alternative construction operations and equipment plans. As mentioned earlier, a computer is almost essential for determining the economic replacement time for equipment based on maximizing the money return. Another important use for such a computer is with the critical path method discussed in Chapter 1.

2.8 Statistical Analysis of Data

The differences in values for supposedly the same construction operations or equipment performances must be recognized. The cost for doing an item of work is not the same for each time it is done. The productivity of the equipment to do an operation will vary from one time to the next.

A piece of equipment does not even go through its repetitious work cycle in the same time for any two consecutive cycles. These examples and many others illustrate the need for some sound means to analyze data. That is the reason for considering a statistical analysis to find the best average value or range of values to expect for a quantity.

A statistical analysis of data is more reliable as the volume of comparable data on a given operation or piece of equipment increases. This objective is helped tremendously by the use of high-speed computers. As long as the input to the computer can be simple enough so the man-hours for recording are minimized and daily inputs can be expected, the computer will gather sufficient data for statistical analysis. The right amount depends on the variation between values gathered. This is shown by the standard deviation or coefficient of variation. Such terms will be explained briefly in the following discussion.

2.8.1 Statistical Data and Results

A brief discussion will be included here to bring out statistical concepts. The reader who plans to apply a statistical analysis to his data should refer to a complete treatment in any good book on statistics. This discussion will refer to the data gathered and other terms as these are important to several key steps of statistical procedure.

First, the data or values that are statistically equivalent will be gathered. Values are equivalent if they represent the same thing, i.e., for the same working cycle, operation, equipment, etc., and apparently under the same conditions. The data are gathered manually or by instruments and may be fed into a high-speed computer from either source. There will be an n number of readings for the statistical analysis. Then, n is known as the statistical population.

Next, the variable recorded values (X) can be placed in statistical order, generally in numerical order. If there are very many values, a computer could do this to advantage. With this ordering, statistical measures such as the mean value ($\overline{X}$), median value, and mode can be found easily. The mean is the average value found by taking the sum of all values and dividing by the number, n. The median is the middle value in the series arranged in numerical order; and the mode is the value that occurs most frequently. These measures give the experienced analyst an idea of the confidence he can put in his variable data.

With the data gathered and key measures made on it, an important statistical check can be made next. This is for the standard deviation or variation to be expected from the average value found. It is found by

$$S_x = \sqrt{\frac{\Sigma(X - \overline{X})^2}{n - 1}} = \sqrt{\frac{\Sigma X^2 - \dfrac{\overline{X}^2}{n}}{n - 1}} \qquad (2\text{-}22)$$

where S_x = standard deviation of the sample;

X = variable recorded value;

$\overline{X}$ = the mean or average value; and

n = number of values recorded, which should be 20 or more, if possible. Following this determination the coefficient of variation, V, can be found from the following equation:

$$V = \frac{S_x}{\overline{X}}(100) \text{ in } \%. \qquad (2\text{-}23)$$

This will give another measure of the consistency in the statistical data. If the V is too great, more observations (n) of data are needed or the values are inherently too variable. A value below 10% indicates reliable data, statistically, to be assured of a reasonable average and range of values.

The accumulation of recorded values can be put on a statistical plot. A "normal" plot would result in a symmetrical bell shape when the points are connected. In that case the mean and mode values will be equal. If the shape is skewed to one side or the other, these key measures for the data will differ (Fig. 2-4). The amount that they differ compared to the difference between any two numerically consecutive values, such as X_4 and X_5, is an indicator of the scattering of values.

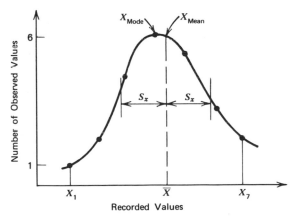

FIGURE 2-4 Statistical plot of accumulated values.

A more important measure for analysis makes use of the standard deviation (S_x). Two-thirds of the values are between the limits of the mean value plus or minus S_x. For more confidence, 95% of the values should be in the range $\overline{X} \pm 2S_x$. If values frequently fall outside this range, the operation or performance should be considered too variable for much confidence in the values obtained. A more extensive statistical analysis might be made. The expense could be justified if the results will lead to a better understanding of the variable values for successful estimating and profitable operations.

References

1. "Are Depreciation Rates Realistic?", *Construction Methods and Equipment*, McGraw-Hill Book Company (New York, N.Y., October 1958) pp. 134–144.
2. Douglas, James. "Optimum Life of Equipment for Maximum Profit," *Journal of the Construction Division*, Vol. 94, No. C01, American Society of Civil Engineers (New York, N.Y., January 1968), pp. 41–54.
3. Alltucker, J. W. "Equipment Replacement Decisions," *Civil Engineering—ASCE*, Vol. 37, No. 4 (New York, N.Y., April 1967), pp. 46–47.
4. "Planned Equipment Replacement," booklet prepared by Market Division, Caterpillar Tractor Co. (Peoria, Illinois, 1970).
5. "Equipment Maintenance" issues, *Construction Methods and Equipment*, McGraw-Hill Book Company (New York, N.Y., annually in November).
6. "Rent It or Buy It: Good Reasons for Both," *Engineering News-Record*, McGraw-Hill Publications Company (New York, N.Y., Feb. 21, 1963), pp. 70–72.
7. "Compilation of Nationally Averaged 1970 Rental Rates," 22nd edition by Associated Equipment Distributors (Oak Brook, Illinois).
8. Drevdahl, E. *Profitable Use of Excavating Equipment*, Desert Laboratories, Inc. (Tucson, Arizona, 1961).

Chapter 3

■■■■■■■■■■■■■■

■■
BASIC COMPONENTS
AND FUNDAMENTALS
■■

Introduction to the Basic Equipment

The process of construction involves equipment of all sorts. Some equip
ment is designed for specific purposes or projects and might be considered
custom made. However, the majority of construction equipment has been
designed by manufacturers to serve with some flexibility for a variety of
of projects or jobs. In either case the equipment is designed to carry out
some facet of the material handling necessities of construction operations.

The fundamentals and key factors in the selection of suitable construction equipment are discussed in this book. Generally speaking, these are
the all-important considerations for the forces and elements of nature
to be dealt with in construction; the common variables and limitations
for carrying out construction operations; and the costs and controls in
using equipment for these operations. Now that we have this framework
for thinking and calculating, the specific construction equipment parts
and units can be covered more effectively to achieve the objectives of this
book.

This chapter will launch into basic equipment components and units

67

that are common to many, if not all, construction operations. The basic components to be discussed include the power units so vital to all construction equipment. Without the power unit no equipment mentioned in this book would be fully operable. A piece of equipment could be kept in a holding position, as in the case of a crane holding its load by a friction lock or lugged brake on the drum without the power unit helping, but to move the load other than strictly by gravity would require the power unit. All pieces of construction equipment must have power units to operate as designed.

In order to understand and be able to calculate the work for a piece of equipment, it is necessary to know the nature of construction materials. The various ways material is encountered and handled for a construction operation are important. How the power in a unit of equipment can be applied to work on material and the forces of nature must be understood. These fundamentals will be covered later in this chapter. An understanding of the most effective way to apply the power and, consequently, the bases for planning an operation for the minimum power requirement is worth acquiring.

This chapter will also cover other basic equipment components common to a variety of construction uses. The supporting mechanisms of crawler- or wheel-type mountings are common variations for tractors, earth excavating units, cranes, and material handling and paving equipment. Tractor equipment is common to all construction operations involving the application of force parallel to the ground or supporting surface. So these types and components of common construction equipment will be discussed in this chapter.

3.1 Power Sources for Construction

The construction equipment of a century ago was driven by animal and steam power. The use of animal power persists to this day for construction work carried out in some of the underdeveloped countries. That form of power for construction equipment faded out of the picture in the United States and other industrialized nations with the disappearance of the Fresno scraper around the turn of the 20th century.

The evolution of the various forms of generated power for construction equipment proceeded from steam power through electrical and internal-combustion power units, compressed-air power, and hydraulic power, and may, before many more years go by, include nuclear power.

Even though animal power is no longer used on construction, the memory of it does give us a necessary understanding. The need represented by the animal is for energy to do work. The energy is a result of fuel that the beast has eaten. The animal applies the energy to move some force through a distance, thereby doing work, i.e., work = force × distance. The rate of doing some given work is then defined as power. It is from a given rate of a number of horses doing certain work that we get the term horsepower. The modern definition tells us that one horsepower (1 hp) is 33,000 foot-pounds of work per minute (written ft-lb/min), or 746 watts.

Generally, the animal applies its force in the direction of motion. The only modern form of power that has such a strictly linear relationship with the body it is working on is jet power action. The ones to be discussed for use with present construction equipment deliver their power by rotation or reciprocating action. In this way it is possible to have a compact power unit and still do the necessary work. However, to understand each power need, it is well not to lose sight of the linearity of the force being applied in the direction of motion.

The horsepower (1 hp = 33,000 ft-lbs/minute) requirement can be expressed simply as

$$P = Fv/33,000, \qquad\qquad (3\text{-}1)$$

where F is the force in pounds to be moved or to work against, and

v is the speed of motion or equivalent in feet per minute.

If there could be a one-to-one relationship, this power, P, would be the required output sometimes known as the brake horsepower of the power unit doing this work. However, slippage, heat generation, and other losses occur between the power take-off and what actually results in work done. These losses generally must be estimated in any power application and will be discussed where appropriate in this text.

Now to gain a more complete understanding of the power story, we will trace the power development from input to output. The power potential of a generating unit is the energy in the fuel or source to convert to useful power. This could be in the form of:

1. chemical energy—as in a lump of coal or a gallon of gasoline;

2. thermal energy—as in the steam stored in a boiler or the burning gases in an internal combustion engine;

3. electrical energy—as delivered at the terminals of a storage battery;

4. mechanical energy—as delivered through the drive shaft of an engine or motor; or

  5. atomic energy—as in the nuclear, high-energy, chain-reactive materials.

Most power units for construction equipment represent a combination of several of these forms of energy transfer. If the unit starts with a fuel, there is a potential for work as indicated by the following values:

Fuel	Average Heat Value	Other Data
Wood	5,000 Btu per lb	
Coal	12,500 Btu per lb	
Diesel Fuel	19,250 Btu per lb	7.3 lbs/gal
Kerosene	19,800 Btu per lb	6.8 lbs/gal
Gasoline	20,500 Btu per lb	6.2 lbs/gal

where 1 Btu equals 778 ft-lbs.

  The actual power generated at the point of conversion—at the fire jacket in a steam boiler or the piston of an internal combustion engine—is less than the fuel's potential. The differences occur because of imperfect burning or firing in the ignition chamber, heat losses to parts of the equipment not directly receiving the thermal energy to do useful work, leakage, and the like. The ratio of the effective conversion of the fuel's energy to its potential energy might be called the fuel efficiency, E_1. If this is in an internal combustion engine, the next energy transfer is from the expanding gases in the cylinder chamber to move the piston. There will be heat losses and leakage here with the resultant efficiency, E_2. Then the piston acts on the cam which rotates the drive shaft. Some loss is experienced in this transfer due to slippage and frictional heat loss leading to efficiency, E_3. There would be similar considerations at the universal joint and other gear-reduction and transfer points. The efficiencies of these transfers might be lumped together as E_n. The net result of all these efficiencies can be used to find the output of the given power unit. Thus,

$$P_o = E_1 \times E_2 \times \ldots \times E_n \times P_i, \tag{3-2}$$

where P_o is the power output, and
  P_i is the power input or potential.

  Generally, the variety of efficiencies such as described above are not known nor readily found. What has been determined by overall performance checks are the overall machine efficiencies for the average commonly used power unit. The machine efficiencies, i.e., ratio of energy output to energy input, expressed in percent, are:

Steam engines	6–10%
Gasoline engines	20–30%
Diesel engines	30–40%
Electric motors	75–95%

To see what these values might mean in operating with such power units, we will make some fuel-cost comparisons to develop 100 hp output. To do this, it will be helpful to have some power and energy equivalents:

1 hp = 33,000 ft-lb/min	1 Btu = 778 ft-lbs
1 hp = 2,545 Btu/hr	1 Kw-hr = 3413 Btu
1 hp = 746 watts	1 Kw-hr = 1.34 hp-hr

For a steam engine using coal as fuel let us assume the overall efficiency is 8%. The amount of coal, X_c, in pounds per hour to develop the 100 hp output can be found (using the table on p. 70):

$$.08\ X_c\ 12,500 = 100 \times 2545;$$

$$X_c = \frac{254,500}{.08 \times 12,500} = 254\ \text{lbs/hr};$$

and if coal costs $14 per ton, the fuel to generate this 100 hp will cost

$$C_c = \frac{254}{2000} \times 14 = \$1.78\ \text{per hr.}$$

For a gasoline engine let us assume the overall efficiency is 25%. The amount of gasoline, X_g, in gallons per hour can be found by

$$.25\ X_g\ (6.2)\ 20,500 = 100 \times 2545;$$

$$X_g = \frac{254,500}{.25(6.2)20,500} = 8.01\ \text{gal/hr};$$

and if gasoline costs 30¢ per gallon, the fuel to generate this 100 hp will cost

$$C_g = .30 \times 8.01 = \$2.40\ \text{per hr.}$$

For an electric motor let us assume the overall efficiency from power-line input to drive-shaft output is 85%. The power input from a 3-phase alternating current source would give the kilowatts (kw). The power input, X_e, in kw-hr, to develop 100 hp output is found by

$$.85\ X_e\ (1.34) = 100\ \text{hp for an hour};$$

$$X_e = \frac{100}{.85(1.34)} = 87.8\ \text{kw-hr};$$

and if this electrical power costs 2¢/kw-hr, the cost for 100 hp generated per hour is

$$C_e = .02 \times 87.8 = \$1.75 \text{ per hr.}$$

The cost of electric power would be even less if the charge were part of many thousands of kilowatt-hours by this customer.

The previous comparisons were only for the fuel or energy costs. The selection of an appropriate power unit for a piece of construction equipment must take into account much more. To understand these other factors in the selection process, the several forms of power—steam, internal combustion, electric, air, and hydraulic—will be discussed independently.

3.1.1 Steam Power

Steam power was the first form without the use of animals to transmit force mechanically to operate construction equipment. The development of such power was promoted most effectively by James Watt's innovations for the steam engine in the second half of the eighteenth century. The first patent by Watt was obtained in 1769 though the first steam engine was built by Thomas Newcomen in 1705 in England. After the turn of the nineteenth century, progress with the steam engine was rapid. The power was delivered by the linear, back-and-forth motion of pistons acted on by varying steam pressure in the piston's cylinder. The pistons moved driving rods connected eccentrically on the drive wheels like steam locomotive drivers.

Now the use of steam power for construction equipment has practically faded out of the picture. It had some fine advantages, but also some important disadvantages. Key drawbacks to steam power are the bulkiness and hazards of the system. To generate and deliver steam power for such equipment as power shovels and tractors necessitated a sizable, bulky boiler for generating the steam. Then the transmission of this energy to move wheels required sizable reciprocating parts for effective power. The hazard of steam is due to its high temperature and pressure in the boiler and in transmission to produce driving force. This presents a safety hazard to the operators who must be on the movable equipment in proximity to the steam power unit. Also, the steam loses its effective power through heat loss along the transmission line if the pipe or hose is not very well insulated. This has been difficult to provide readily. Consequently, steam cannot be transmitted very far without adding considerable expense or bulky insulation for the system.

The use of steam power in the early days of construction equipment did provide some advantages. Once the necessary head of steam, i.e., steam pressure, is obtained with the right balance of convertible water and volume of steam in the boiler with active coals and fire that can be continually fed, this form of power has ample flexibility and sustaining force. The transmission is controlled simply by valves, which allow ideal reversibility. Probably the greatest advantage of steam power is its ability to deliver impact force. This makes steam power still useful for pile-driving operations. In all other power applications in construction, steam has yielded since the 1930's to other forms that have some of the same advantages of steam without its disadvantages.

Steam Boiler Rating.   Since steam power is not used much more, it will not be discussed further, except to explain boiler ratings in connection with pile-driving hammers. These hammers may be powered by steam and call for a certain horsepower boiler at some specified pressure, say, at 100 psi steam. A boiler does not develop "horsepower" in the sense that an engine will, so a boiler horsepower rating is merely giving an index number based on a standard definition. One boiler horsepower is said to have developed with the expenditure of energy just sufficient to evaporate 34.5 pounds of water per hour into steam "from and at 212°F." Of course, this is at sea level and standard atmospheric conditions. Under these circumstances at the boiling temperature, one pound of steam is equivalent to 970 Btu. Therefore, one boiler hp = 33,479 Btu per hr. And to relate boiler hp to engine hp, use the expressions on page 71 to find one boiler hp = 13.16 hp.

The other ways of rating steam boilers are by the area of heating surface between the fire and the water convertible to steam. One such rating scheme used by the Steam Boiler Institute (SBI) stated that one square foot of radiation is equivalent to 240 Btu per hour at 212°F. Another was the "Builder's Rating" which said that 10 square feet of heating surface equals one boiler horsepower. This was an arbitrary and very imprecise rating scheme. However, it probably allowed the old-time contractors to pick a boiler that was not too large and dangerous nor too small and inadequate. With the inherent flexibility and variability possible with a steam power source, greater precision was not necessary nor justified expense-wise.

3.1.2   Internal Combustion Power

Around 1930 the internal combustion (I.C.) engine started to replace the steam engine for mobile construction equipment in the industrialized parts of the world. This engine's power is generated from the thermal energy in gasoline or diesel fuel. The first to develop was the gasoline engine patterned after the gas engine built in 1876 by a German, Nicholas August Otto. The four-cycle arrangement is known as the Otto cycle. A great improvement was made by Gottlieb Daimler, who used vapor from oil with gas to form gasoline as his fuel. This achieved not only a much lighter engine to extend the advantage over steam power, but also reached much higher revolutions per minute—800 to 1000 rpm.

The Otto cycle is a four-stroke cycle for each piston. The first stroke in one direction, called "intake," draws fuel and air into the cylinder; the next stroke, in the opposite direction, is the "compression" of this mixture; the third stroke, in the same direction as intake, is called "power" or "ignition" started by the sparking and explosion of the compressed fuel; and finally, the last stroke, again in the compression direction, is for "exhaust" of the spent fuel gases. The power stroke is the one that produces useful work. However, the compression stroke has much to do with the relative power to size or weight of the engine. Originally, the I.C. engines had compression ratios of 4 to 1; that is, the gasoline fuel was compressed into a volume one-fourth of what it was at intake. With the use of better, more compressible fuels, stronger metals for the cylinder and pistons, closer tolerances, and more efficient design, high-compression ratios better than 8 to 1 have been achieved in gasoline engines and 16 to 1 in diesel engines.

Such improvements along with more compactly designed components have led to higher and higher power-to-weight ratios. This is a big advantage to construction equipment which must be able to apply great power without needing such a large mechanism and frame as to be awkward for maneuvering. However, a limitation to increasing the power-to-weight ratio in moving equipment is imposed by the concern for traction. As explained in Sec. 3.4.1, the power required in the form of force (F) to apply at the drivers cannot exceed the weight-force perpendicular to the surface (F_n) times the coefficient of traction (C_t). If it does, the drivers will simply spin as long as the power exceeds the maximum tractive effort (TE) that can be applied under the given job conditions. This is expressed in equation form as

$$\text{maximum TE} = C_t \times F_n \geqq F, \text{ the required force.}$$

I.C. Power Generation and Output. The effectiveness of an internal combustion engine is generally measured in terms of horsepower or torque. As pointed out earlier in this chapter, power is directly related to force applied and speed of motion in a linear fashion by the expression $P = k_1 F v$, where k_1 is a constant of proportionality. In the I.C. engine, power is developed at the rotating drive shaft moved by the pistons from the cylinders. Thus, the driving power is from a torque, which can be translated to force applied tangentially at the circumference of the drive shaft. Thus, torque,

$$T = F_t r, \tag{3-3}$$

where F_t = tangential force, and
    r = radius from center of rotation to point of force application, generally, the radius of the drive shaft. In these terms the power equation is

$$P = \frac{2 \pi N T}{33,000} \text{ horsepower,} \tag{3-4}$$

where N = shafts speed in revolutions per minute, rpm, and
    T = torque in lb-ft.

The tangential force applied might be through a belt on the engine's drive shaft or a frictional force due to braking pressure applied radially on the drive shaft. If the engine were running with only the brake to work against, it would be noted that the friction varies with the brake force and the rotating speed of the shaft. Should the brake be removed entirely, the speed could become infinite during the operation of the engine. To prevent that happening, the engine is designed with a governor to limit the speed of shaft rotation at some set maximum rpm. An engine is rated in terms of its horsepower at the governed rpm. For instance, if it can develop a torque of 500 lb-ft at the governed speed of 2100 rpm, by Eq. (3-4) the engine would be rated at

$$\frac{2\pi(2100)\,500}{33,000} = 200 \text{ hp.}$$

This is known as a brake horsepower (bhp), i.e., the power that is available for work at the output or take-off end of the drive shaft. The horsepower rating of an I.C. engine is given on the specification sheet for the engine. The characteristic shape of these curves is shown in Fig. 3-1 with the values from the above example given.

Note that, even though the brake horsepower is a maximum, the delivered torque is not a maximum at the governed speed. The maximum torque is at perhaps two-thirds of the governed speed. This torque is 10 to 20% higher than that at the maximum speed. Such a feature in the engine's design provides a reserve to take on a momentary overload and keep the engine from stalling out with the overload. Of course, if the

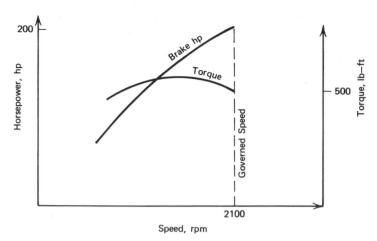

FIGURE 3-1 Relationship of power to speed of an engine.

load is still too great for the engine to overcome, it could stall. To prevent that happening, the operator must take action with the overload and shift back to a more powerful gear.

I.C. Power Transmission. How does the I.C. engine transmit the brake horsepower at its output shaft to overcome the forces of the load applied on the equipment? The mechanism for doing this is a transmission consisting of gears that reduce the speed and increase the forcefulness delivered. This translation is made by using appropriate gear ratios in the design. We note that the engine's brake horsepower is related by

$$bhp = k_2 N_g T_g = k_3 N_g F_g, \tag{3-5}$$

where N_g, T_g, and F_g are the governed speed, torque, and tangential force at the circumference of the drive shaft, with k_2 a constant of proportionality and k_3 another including the radius of the shaft. Then, with the power transmitted through gears to an output at the drive wheel, the delivered force, F_o, to overcome load is theoretically related by

$$F_o N_o = N_g F_g. \tag{3-6}$$

With this expression rearranged to produce ratios of the speeds or forces, the ratios would represent the total gear reduction ratios. The last reduction recognizes that the output torque, T_o, equals $F_o \times r$. Here r is the radius of drive wheel, on the circumference of which the force, F_o, is applied.

Unfortunately, the transmission of power from its source to the point where it must do its job is never 100% effective. Losses occur in the train of gears and elsewhere in the total mechanism. The mechanical efficiency of the system might be somewhere between 60% and 80% depending on its design and maintenance condition. This efficiency coupled with the effectiveness of converting the fuel potential to useful power in the I.C. engine gives us the overall machine efficiency given earlier in this chapter. However, the mechanical efficiency is the one of importance in calculations because we are given the brake horsepower and other helpful data on the engine's specification sheet.

An example should help to clarify these points. Consider an I.C. engine as the power unit for a tractor. Given. bhp = 280 at governed speed = 2100 rpm, and 1st gear delivers 21,200 lbs at 3.4 mph with a drive-wheel radius of 2.2 feet.

Checking, torque at governed speed,

$$T_g = \frac{33,000 \text{ bhp}}{2\pi N_g} = \frac{5,250(280)}{2100} = 700 \text{ lb-ft.}$$

The torque output to do work can be found theoretically by $T_o N_o = T_g N_g$, and where $v_o = 2\pi R N_o = 3.4 \times 88 = 298$ fpm:

$$N_o = \frac{298}{2\pi(2.2)} = 21.7 \text{ rpm};$$

from above, $T_o = (2100/21.7) \times 700 = 67,800$ lb-ft. If there were no power losses in the system between the engine brake horsepower and the tractive effort (TE) delivered at the drive wheel, then TE = 67,800 ÷ 2.2 = 30,800 lbs. But the measured output is 21,200 lbs; therefore, the mechanical efficiency,

$$\text{Eff.} = \frac{21,200}{30,800} = 0.688, \text{ or } 68.8\%.$$

To incorporate this efficiency into the application of Eq. (3-4), the output power,

$$P_o = \frac{N_g T_g(\text{eff.})}{5250}, \tag{3-7}$$

and the force available at the wheel rim,

$$TE = \frac{33,000(\text{eff.}) \text{ bhp}}{v} \text{ lbs} \tag{3-8}$$

where v = speed of vehicle in fpm.

Some engine specifications will show a family or set of horsepower vs. RPM curves. The differing values are to account for varying load requirements. These values are:

1. "maximum," or peak, horsepower, which the engine can deliver for perhaps five minutes without a drop in speed; this is an indication of the maximum power in the engine within 5%;

2. "intermittent" horsepower, which is for variable-load applications such as excavators and hoists, where the duration of sustained full power output is one hour or less; intermittent hp ≈ 90% peak hp; and

3. "rated," or continuous, horsepower, which is for load applications such as rock crushers and pumps, where the duration of sustained full power output is 8 to 24 hours per day; "rated" hp ≈ 80% peak hp.

Note that the values obtained from such curves or specification sheets are based on standard conditions. These are stated as sea level barometric pressure (29.92 inches of mercury) and a temperature of 60°F. The effect of other conditions is discussed in Sec. 4.1.3. The values generally apply for an engine equipped with an air cleaner, water pump, fuel pump, and other standard parts. If the equipment includes a fan for the cooling system, that takes about 4% of the power available.

The relationships of these various horsepowers and available torques with designed speeds will be shown next. The curves in Figs.3-2 and 3-3 are for direct-drive transmissions.

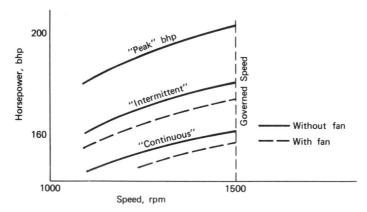

FIGURE 3-2 Speed vs. brake horsepower for various conditions of loading.

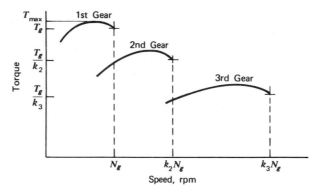

FIGURE 3-3 Speed vs. torque in various gears. (Notes: N_g is governed speed in 1st gear; k_2 is gear ratio from 1st to 2nd gear; k_3 is gear ratio from 1st to 3rd gear; T_q is torque in 1st gear at governed speed.)

Torque Converter. In the preceding discussion the problem of matching power developed to required applied load or force is of major concern. If the engine with a standard transmission is faced with an overload greater than the approximately 10% reserve above the torque at governed speed, the engine is likely to stall. The operator of the standard direct-drive engine must shift down to a gear powerful enough to overcome the extra heavy load. A torque converter makes the proper power choice automatically, "shifting" or sliding into the appropriate gear ratio to handle the applied load. It is essentially like the fluid drives or automatic transmissions introduced in automobiles after World War II.

The torque-converter type of transmission for construction equipment has some key advantages. These, which serve to describe the system, are:

1. transmitting power smoothly from the power source to the load by hydraulic means;

2. keeping the constant power source matched with the variable load at all times by multiplying the torque automatically, which in turn varies the speed of operation (refer to Eq. 3-6);

3. preventing engine lugging and stalling by allowing the engine to run independently of the load variation—the automatic torque conversion takes place between the output of the engine and the resisting load;

4. increasing engine and equipment life by cushioning shock loads; and

　　5.　reducing operator fatigue by eliminating continuous gear changing.

A period of evolutionary development of the torque-converter type of transmission has followed its introduction in 1946. It has been necessary to concentrate on designing these transmissions to match the engine output of power as closely as possible for all shaft speeds. A clutch has been developed to allow shifting to different ranges of torques and speeds for better operation under varying load conditions. Also, a direct-drive clutch is desirable to lock the converter out of the power train for certain parts of the work cycle.

The design of the gear trains for torque-converter transmissions has received major attention. The guiding principle is to achieve a good match with the working cycles typical of the different applications. It has been found that hydraulic drives in combination with simple rugged planetary gear trains can be designed into a compact and reliable variable torque transmission. Another such transmission may be described as a constant-mesh multiple counter-shaft type which shifts under power by an air-actuated multiple-disc clutch controlled by a rotary vane. These systems and others make it obvious that there are various ways of accomplishing the desired torque conversion. Each engine manufacturer has his own patented design to provide this automatic control of power delivered to overcome applied load.

The characteristic curves relating torque to speed for the engine with a torque converter are shown in Fig. 3-4. Note that four ranges are shown with two labelled "low" and two "high." The low-range pair might be obtained by a regular planetary gear train with about a 2:1 gear ratio. The high-range pair builds on the low with a gear "splitter" added to obtain an intermediate ratio. We also note that the curves are not linear, pointing to the difficulty in matching a torque converter exactly with engine power.

The variety of torque-converter type of transmissions leads to combinations with various numbers of forward and reverse gear ranges. One may provide three forward and three reverse ranges of equal gear ratios. Another may provide six forward speed ranges and one reverse speed. The one shown in Fig. 3-4 gives 4 forward (fwd) speed ranges and might have one or two in reverse. The different numbers of speed ranges are justified by the working cycles of the different construction equipment applications. Thus:

　　1.　tractors, loaders, and prime movers can best use the 3 fwd + 3 reverse combination;

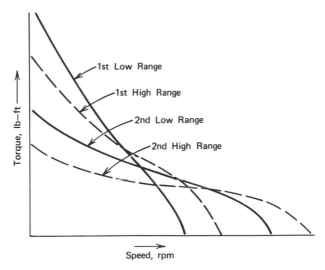

FIGURE 3-4 Speed vs. torque output for a torque converter.

2. dump trucks, scrapers, and off-highway vehicles use the 4 fwd + 1 or 2 reverse gear ranges; and

3. highway trucks for long hauling can best use the 6 fwd + 1 reverse combination.

It is important to understand why the curves are shown as gear ranges as contrasted with the speed in a given gear for the direct-drive transmission. As explained earlier, the chief purpose and characteristic of a torque converter is to automatically adjust the output power to the load requirement. Thus, the torque varies infinitely with the load within the ability of a gear range. The speed varies inversely with the required torque. If the load requirements call for higher values of torque than a range can give, the operator must shift into a lower gear range, and vice versa.

One of the real advantages of the torque-converter type of transmission, not emphasized earlier, is the ability to shift under load without hesitation. This provides a real benefit and power saver for heavy equipment because the operator can change gears without losing momentum. However, it must be noted that in general operation the equipment with a torque converter does not realize as high a power efficiency as one with a sliding-gear, direct-drive transmission.

Comparison of Direct Drive with Torque Converter. A review will call attention to a comparison of the basic transmission units and their advantages with internal combustion engines for construction equipment:

1. direct drive, sometimes called "gear drive," provides lower initial cost and higher ultimate power efficiency; and

2. torque converter, sometimes known as "power shift," provides automatic selection of right power for load, eliminates stalling and loss of momentum when shifting, reduces operator fatigue, and increases engine and equipment life.

A comparison of the output torque or rimpull force vs. speed of operation for equipment with the two basic types of transmission is shown in Fig. 3-5.

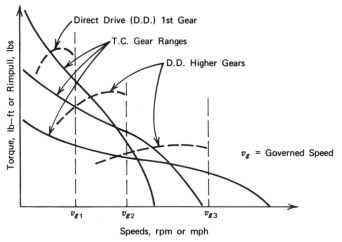

FIGURE 3-5 Comparison of direct drive to torque converter characteristics.

Summary of I.C. Engine Power. Internal combustion engine power finds a wide variety of uses with construction equipment. The power units are made in such vast numbers that their sizes range from 3-hp engines for the smallest construction pumps to mighty engines for big floating dredges with horsepower ratings approaching 10,000 hp. The variety of engines usually found in readily maneuvered moving equipment are built in a variety of sizes by many manufacturers. They range from 5- to 10-hp gasoline engines for concrete buggies and 40-hp gasoline engines for the smaller tractors to 500-hp diesel engines for large earthmovers and even more than 1500-hp diesel power units for a 200-ton dump truck hauler.

The costs of such units, of course, vary somewhat in proportion to their horsepower rating, but also with the variety of accessories and sophistications built into them. For instance, as mentioned earlier, an engine with torque converter will be more expensive than one of the same horsepower rating with a direct gear drive. Generally, the size and cost of the I.C. power unit is an integral part of the piece of equipment it powers. Its size and desirable accessories are designed to be balanced with the equipment and its expected work applications.

The construction planner should select his equipment based on the variety of work that will be done by it. He will realize that the power unit may indirectly influence his selection, but the total piece of equipment is the prime determinant. Of course, he must include the costs of fuel or energy source, lubrication, and maintenance for the engine as part of the total equipment cost. The energy cost was discussed in one way earlier in this chapter. Fuel and lubrication needs frequently are given on the specification sheet for an engine. The general approach to these parts of the equipment cost were discussed in Chap. 2.

The basis for selection of equipment with I.C. engine power will not generally be because that type of power unit is cheaper than another. More than likely it will be chosen with this form of power unit for one or more of the following key advantages of internal combustion power:

    1. self-contained power generation to give maximum equipment mobility;

    2. high power-to-weight ratio for maximum forcefulness with maneuverability; and

    3. flexibility in power application or speed of operation, especially with a torque-converter type of transmission.

The intricacies of the internal combustion power mechanism and the consequent problems of maintenance are key disadvantages. However, all selections must recognize that no worthy benefit is gained without some difficulty.

3.1.3 Electric Power

The use of electric power for construction equipment depends on the type of operations, their requirement for mobility, and the availability of electrical energy or the feasibility of its generation. Where this form of energy for power is readily available and the construction operation is

concentrated in one spot, electric power should be given prime consideration. It is the most economical form of power under such circumstances.
This is because already available electricity is the cheapest source of
energy one can purchase to power equipment. Its use should be given
first consideration for a construction operation in a populated area,
where the electrical lines are already installed, for the power requirements
of high-rise building cranes and hoists, asphalt and concrete mixing
plants, and the like. The cost of making an electrical hookup for power
in an urban or fringe area is negligible. If the selection is planned
enough in advance and the setup can be constructed as part of the permanent installation, there may be no installation charge to the contractor.
He would be charged only for the power used, the same as a regular
commercial customer. The rates charged are explained later.

In less populated areas where high-voltage cross-country lines exist,
electric power should still be given top consideration. There would be
the cost of installing transformer equipment to step down the voltage
and perhaps need for a regular electrical substation if there were several
lines to the various electric power demands on the jobsite. A large dam
construction site might have half a dozen or more electric lines from the
substation. These would power the variety of uses from cofferdam pumps,
power shovels, rock crushers, conveyers, and concrete mixers to air compressors, welders, light towers, and building lights. In this case the high
cross-country voltage would be stepped down only to 2300 volts for
electric motors of more than 50 hp, but with a further reduction for
cofferdam pumps and other equipment with a water hazard. The lines
for these exceptions and smaller horsepower motors operated out of doors
would likely be 440-volt lines. In the shops and buildings 220/110 volt
supply should be sufficient and not too dangerous.

Determining the Power Supplied. The electrical power furnished from
existing or specially installed lines is delivered in the form of amperes of
current (I) driven by the voltage (E) potential drop. Each of these quantities varies regularly in an alternating system. The power (P) available is
proportional to the product $E \times I$ and is expressed as kilovolt-amperes,
kva. The effective power in watts from this, taking into account a power
factor, is generally expressed in kilowatts, kw. The power factor with alternating current takes care of the fact that the current and voltage are
frequently not in phase with each other. That is, they do not reach their
maximums, or any other stage of their cycles, at the same instants. The
power factor will generally vary from one down to 0.7 for reasonable
operation. It is shown as "cos θ," reflecting the sinusoidal nature of alter-

nating current and voltage and involving the phase angle, θ. If all the equipment motors using electric power have lagging power factors (P.F.), the resulting cost is greater than would be the case with a resultant P.F. close to unity. It is for this reason that the selection of an electric motor with a leading P.F., such as a synchronous motor, among the various units in the system is advantageous to the economy of the total system's power demand.

Commercially available electrical energy is alternating current because of its superior transmission characteristics compared to direct current. Furthermore, it is generally three-phase as the simplest, most effective polyphase supply to eliminate the noticeable effects of the sinusoidal variations. This polyphase energy supply produces a reasonably uniform amperage and voltage which gives power as

$$P_a = \frac{EI \cos \theta \sqrt{3}}{1000} \text{ in kva,} \tag{3-9a}$$

or

$$P_a = \frac{EI \cos \theta \sqrt{3}}{746} \text{ in watts.} \tag{3-9b}$$

A single-phase supply would give maximum power of

$$P_a = \frac{EI \cos \theta}{746} \text{ in watts.}$$

The electricity input from a utility company to power motors and for other job uses is billed on two bases for metered quantities: a demand charge for the maximum kilowatts used and an energy charge for the kw-hours drawn. The demand charge is based on the highest 15-minute average kw in the given billing period, generally a month. This charge has a minimum for the first, say, 10 kw, which may be in the order of $25; then a lower amount, perhaps around $2 per kw for the next 40 kw; and successively reduced for other incremental kw amounts to perhaps $1.50 for all over 400 kw. The energy charge has a similar schedule for charges per kwhr without a minimum, as long as the demand charge provides a minimum monthly bill. Energy charges may start around two cents per kwhr for the first 3000 kwhr used and reduce in successive steps to less than 1¢/kwhr for all amounts over 300,000 kwhr.

An example will help to clarify the charging for electricity. Consider a two-ton asphalt plant which requires electrical energy for heating the asphalt and for driving the electric motor. Each has a kw demand—say, the heating requires 10 kw steadily for four hours, and the motor requires a maximum of 21 kw during the eight-hour shift. The total demand charge would be based on 31 kw, if both units operated together for any 15-minute period. The user could reduce this part of his charge to a maximum of 21 kw, if he would arrange the heating to be done before the day's shift for the motor started at 8 a.m. Furthermore, the loss of heat in the asphalt during the daytime, where ambient temperature would rise to 80°F or more, would be negligible. If there is no other electrical demand during the month, that part of the bill would be perhaps $25 (for the first 10 kw) plus 11 × $2.00 for a total of $47.00 demand charge. The energy charges per day, assuming the motor uses an average of 18 kw and if the rate is 1¢ per kwhr, would be

—for heating asphalt, $4 \times 10 \times .01 = \$0.40$
—for electric motor, $\quad 8 \times 18 \times .01 = \quad \underline{1.44}$
and for a work month of 20 days $\quad \$1.84 \times 20 = \$36.80$.

A utility company supplying electricity is always anxious to have its customers maintain a net power factor as close to unity as possible at the company's point of delivery. This helps the efficiency of electrical energy service transmitted by the company. Therefore, it encourages customers to strive for unity P.F. by discounting the demand charge for a net P.F. higher than 90% lagging, if the total kw is over some minimum—say, 100 kw. By the same token, there will likely be an extra charge if the P.F. is lower than 80% lagging.

Selection of Electric Motors. The selection of electric motors for construction use, as with other power forms, depends on such factors as (a) operating torques, (b) speed requirements—constant or variable, (c) need for reversing or nonreversing, and (d) load demand—continuous or intermittent duty. These are factors for consideration regardless of the general type—alternating current or direct current motors. As noted in the following, each type and variety within the type has its advantages and disadvantages. Furthermore, each construction operation and possible electric power use has a variety of requirements such that no one motor is ideal for the situation. However, electric motors are sufficiently advantageous under certain circumstances to make their selection desirable.

Alternating current (ac) motors are available in single-phase or polyphase variety with the latter being more frequent for heavy equipment. They generally are classified as induction or synchronous motors, which have the characteristics cited in Table 3-1.

TABLE 3-1 General Selection Factors for Alternating Current Motors

Variety	Selection Factors	Possible Use
Induction motors:		
(a) Squirrel-cage	— Constant speed — Infrequent starting — Moderate torque	— Conveyor system — Aggregate plant — Compressors, pumps for intermittent service
(b) Wound-rotor	— Frequent starting — High torque for starts and running — Low input current	— Tower hoists — Power shovels — Rock crushers
Synchronous motors:	— Constant (synchronous) speed — Leading power factor — No starting ability (need separate starter)	— Power supply for lighting — H-v-a system — Shop tools, etc. — Compressors or pumps for continuous service

Direct current (dc) motors are not used as frequently as ac motors because of the form of available energy source. However, if the electricity is to be generated specially for the construction use, direct current might be preferred. This could be based on certain advantages of dc motors. They are available in the shunt, series, or compound-wound variety. The characteristics and, consequently, the selection factors for these are shown in Table 3-2.

The characteristic selection factors for dc motors can best be shown graphically as in Fig. 3-6, relating percentages of full-load speed to full-load torque.

Full load is that maximum operating combination of speed and torque or current and voltage at which the motor could be operated continuously without burning out or damaging it. DC motors are rated in power available at the line voltage furnished for full-load conditions. The horsepower is computed by

$$P_d = \frac{e\,EI}{746},\tag{3-10}$$

TABLE 3-2 General Selection Factors for Direct Current Motors

Variety	Selection Factors	Possible Use
Shunt motor	— Constant speed regardless of load condition — Low starting torque	— Conveyor system — Compressors — Pumps
Series motor	— Highest starting torque of dc motors — Speed drops as torque increases in operation	— Crane or hoist — Power shovel — Material elevator
Compound-wound motor	— Good starting torque — Speed varies uniformly with torque	— Aggregate plant with crushers, etc. — Concrete mixer

where e is the efficiency of the motor. The efficiency will vary from 75% to 90% for full-load operation. To determine the power of the motor in usable torque, one can make use of Eq. (3-4) to find

$$T = \frac{63025\,P}{N} \text{ in lb-inches.} \tag{3-11}$$

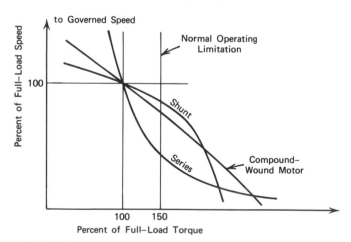

FIGURE 3-6 Characteristic selection factors for direct current motors.

An overload calling for additional current (I) may be considered up to 50% above the full load requirement for a short duration without sustaining serious damage. This 50% overload sets a normal operating limitation for dc motors.

Generation of Electric Power for Construction. There are several important construction situations that may suggest generating electricity for power. This would certainly be the case for welding equipment or a lighting system if electricity were not already available. If a construction project called for a variety of equipment and tools suitable for electric power and no electrical energy were economically available, generation in a stationary facility might also be justified. As suggested before, the construction operations would have to be close together with no need for mobility. It should also be noted that some generation of electricity is feasible and desirable within the total system of a mobile piece of equipment. This electric power situation will be discussed later.

In the generation of electrical energy for a construction project with a variety of equipment users, several key points must be held in mind. The variety and intermittent nature of many construction equipment needs introduce severe electrical demands. There may be high peaks of short duration, unavoidable surges, and short circuits. Therefore, the electrical generation should be designed with liberal capacity and adequate electrical protective features.

Small, movable electric generators are available with gasoline engine power units. These are rated, say, 15 kw at 1800 rpm, for continuous heavy-duty service. They will be designed with generous overload capacity, good electric motor starting characteristics, and for all climate conditions with a housed unit. These units are designed, generally, to deliver 115 volts to 460 volts in 60-cycle, single-phase or three phase form. Their costs are from more than $100 to several hundred dollars per kw for a power unit that weighs less than one ton.

Generation of electric power for certain parts of mobile equipment is now commonplace. It should be recognized that some power shovels, which are not particularly mobile, have used electric power for more than a half century. The use of such power was extended for construction equipment in 1922 by R. G. LeTourneau for his novel scraper. That piece of equipment, which looked nothing like present scrapers, had electric motors for the loading and dumping movements. However, the widespread use of electric power for fast-moving mobile equipment has been the case only since about 1950.

The primary use for this form of energy in mobile construction equipment is for an electric control system. The system consists of an electric generator driven by the internal combustion engine and feeding energy to the variety of electric motors strategically located around the piece of equipment. The motors on high-speed equipment handle power steering and power brakes as well as lights and other small attachments. On a bulldozer an electric motor can handle the raising and lowering of the dozer blade along with other operating adjustments. An earthmoving scraper can also have electric motors to operate the apron and tailgate in the bowl and provide the hoisting action for the whole bowl unit. It may also have an electric motor to drive each wheel as with one of LeTourneau's most recent developments.

These electric motors are strategically located to do their work with a minimum of power transmission. That is, a motor is right at its point of work application—the electricity having been transmitted to it through

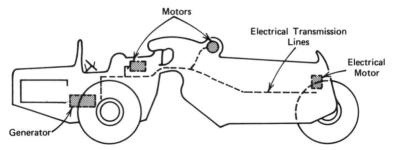

FIGURE 3-7 Electrical power system in a scraper.

electrical wires from the generator (Fig. 3-7). In this system's arrangement each motor is mounted to a gear reduction box and drives either a cable drum or an output gear.

The electric motors in fast-moving-earthmovers must be rugged, weather worthy, and simple to operate, maintain, and service. The ruggedness is to take the frequent starts and stops, high torque demands, and abusive work in an earthmoving operation. The weather worthiness is to withstand the temperature extremes, wetness or dryness, and the dust, mud, and dirt. Simplicity of operation and reversibility of mechanisms is achieved by two-way switching. These characteristics are obtained in airtight, self-lubricating components in a system of alternating current generator and motors. The type of electrical unit that seems to serve best for this purpose is a squirrel-cage, induction-type motor. The open, brushless mechanism with a ·high degree of insulation protects it against wear, weather, and foreign particles. Special design of the motor gives it high-torque characteristics not ordinarily found in this type of electric motor.

The generator will operate on the I.C. engine's flywheel moving at 1800 to 2400 rpm with full governed speed. This can produce several hundred volts for three-phase alternating current which will operate various motors up to 100-hp size. The three main electrical transmission lines provide opportunity to insert switching circuits for limit controls, holding action, and safety devices. Thus, the advantages of a mechanical system are duplicated without the power losses inherent in the mechanical transmission of power.

3.1.4 Hydraulic Power in Construction Equipment

Mobile construction equipment can use a hydraulic power system in a manner similar to the electric control system just described. The hydraulic system makes use of a specially designed hydraulic pump as the power generator near the operator and hydraulic cylinders located at the strategic points for applying the work. Separate, frequently parallel, feeder lines carry the hydraulic fluid between the pump and the cylinders. This hydraulic power is started in action by simple hand levers or electric switches at the operator's position. A lever or switch, appropriately operated, activates valves in the system to send the hydraulic fluid to do its work on the given mechanism (Fig. 3-8).

Simplicity in the valve and piston-type mechanism is one of the advantages of this form of power. Another advantage is that the hydraulic fluid

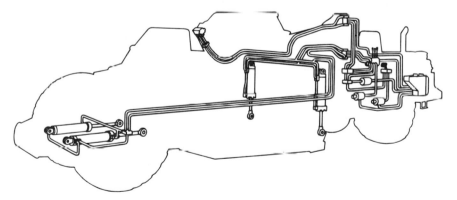

FIGURE 3-8 Hydraulic system in a piece of earthmoving construction equipment (courtesy of Caterpillar Tractor Co.).

is self-lubricating and minimizes wear and tear in the system. The hydraulic power can be used to operate the brakes and steering for a fast-moving piece of equipment. It can be used equally well to power other, various mechanisms on earthmoving equipment. For instance, it can handle the following parts on: bulldozers—blade lifting and angling; motor graders—blade angles and scarifier positioning; front-end loaders—bucket lifting, tipping, etc.; scrapers—bowl lifting and apron and ejector action.

3.1.5 Other Forms of Power for Construction Equipment

Compressed air is widely used in industry for powering hand tools because of its inherent safety features. It has been used extensively on construction operations as well because of its simplicity and its safety features. The compressed air is transmitted from a power generator through air lines similar to electricity or hydraulic fluid. The generator is an air compressor. Such a compressed-air system is not as efficient in the transmission of its energy as a similar electric system. The compressed-air power or energy cost may be three or four times more expensive than electricity to deliver the same amount of work. The air motor has no tendency to become hot when overloaded. Compressed-air energy does not carry with it the electric shock hazard to the operator. There is not the danger presented by an electric spark, steam heat, or fuel ignition in an explosive atmosphere. It is for these reasons that compressed-air power is used for hand tools, especially in the close atmosphere and damp conditions of tunnel operations. The considerations for selecting an appropriate air compressor for the necessary tools on a given construction job are covered in the next chapter.

Another form of power generation is now being introduced for use in construction. This is the gas turbine power generator. A gas turbine consists of a multistage axial flow compressor (about 6:1 compression ratio), an annular combustor for natural gas or distillate fuels, and a multistage axial flow power turbine. The turbine generator will operate at more than 20,000 rpm, which is reduced to around 2,000 rpm at the power output shaft. On mobile equipment this relatively lightweight generator can produce in the order of 800 kilowatts of power for construction use. This can be translated to 1000 kva with 480 volts and 60 cycles per second. The voltage can be varied from about 200 to over 4000 volts.

The variety of power forms for construction use is obvious from the discussion of the previous sections. The advantages of each will be better understood as the various types of equipment are discussed in following chapters.

3.2 Mobile-Base Mountings

Mobile construction equipment, such as tractors, power excavators, movable conveyors, concrete pavers, and haulers, may have some variety of base mountings. The variety will include crawler-track mountings or rubber-tired wheel carrier mountings. The wheel mountings may be single- or multiple-axle mountings, which may be only for towing behind a prime mover. If there are two or more axles, the rubber-tired unit is most generally self-propelled, such as a scraper, tractor, truck or other hauler. There are also single-engine, self-propelled, rubber-tired mountings for power excavators. These are contrasted with the two-engine, truck type mountings for many power cranes and excavators. The various mountings for construction equipment will be discussed in the following sections.

3.2.1 Crawler Mountings

A piece of mobile construction equipment which must work on rough or loose-material surfaces which cause poor footing, should be mounted on crawler tracks. This is particularly recommended if the equipment does not have to move much once it is on the jobsite. Such is frequently the case with power shovels working in an excavation pit or a quarry. The crawler mounting provides the greater amount of bearing area for work on ground and can withstand the greater abuse to the bearing surface on rough terrain.

The crawler part for a power shovel or crane unit consists of two continuous, parallel crawler belts supporting a base frame. The bearing length of the crawler belts depends on the surface and the depth of penetration into it. This length can be safely taken as the distance center-to-center of the crawler end sprockets or axles for the end tumbler wheels, on which the belts turn. The bearing width is two times the edge-to-edge width of a crawler belt. Normally, these dimensions for a standard power excavator with regular equipment parts will result in pressures of 5 to 12 psi of bearing area. The overall width of crawler belts can be increased on most excavators to provide greater stability for draglines and cranes. Changing the crawler tracks to give greater bearing area makes the equipment more stable on loose, soft ground. This gives crawler-mounted equipment an advantage over rubber-tired equipment.

The base frame supported by the crawler units houses the propelling and steering mechanism, which is driven and controlled from the revolving superstructure of a power excavator. The propel drive is either single- or multiple-speed. The manufacturer specifies the operating speeds according to certain standards. Crawler-mounted excavators generally travel at speeds from 1/2 to 2 mph on the level. To comply with the Power Crane and Shovel Association (PCSA), such equipment must be capable of climbing a 30% (30-foot vertical rise in 100-foot horizontal distance) grade on smooth, firm, dry surface free of loose material without a load.[7]

The power unit to deliver this capability, according to PCSA, is generally an internal combustion engine or an electric motor. The power outputs must be specified in terms as discussed earlier in this chapter. For an I.C. engine this will include the number of cylinders, the cubic inches they displace, whether two- or four-cycle combustion, the engine rpm speed, and other particulars. For an electric motor the manufacturer specifies whether it is alternating or direct current, the voltage and frequency, the type of motor, and its rating—continuous or intermittent. The power take-off from the prime mover (power unit) may be mechanical drive, hydrodynamic drive (i.e. fluid coupling or torque converter), hydrostatic drive, or electric drive from a generator.

3.2.2   Rubber-Tired Mountings

By far the largest variety of construction equipment is mounted on rubber tires. The tires used on towed pieces of equipment are usually ordinary ones manufactured for highway vehicles. This is the case for portable compressors, heaters, and concrete mixers. The exceptions would include the larger towed scraper units and the like. In the case of the great variety of self-propelled equipment, tires of many sizes are manufactured to match the needs for support and mobility.

Tires manufactured for commercial and equipment use are standardized in their dimensional identification.[5] The description is with two numbers, such as 14.00 x 24 (read as fourteen hundred by twenty-four). Referring to Fig. 3-9, the first number stands for the unloaded section width and the second for the rim diameter. The tire's contact area under normal load and on a solid surface has a loaded width approximately equal to the unloaded section width and a length nearly 50% greater than the loaded width.

However, this normal contact area has little meaning because the actual area varies with the total load transmitted, the tire pressure, and the maximum pressure the contact surface can take. The total load that a tire supports will be somewhat greater than the air-gauge pressure in pounds per square inch (psig) multiplied by the contact area in square inches. The tire's additional support capacity is due to the compressive strength of its sidewalls. If the air pressure or contact area is not great enough, the tire

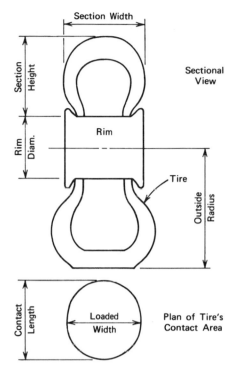

FIGURE 3-9 Dimensional identification of rubber tires.

will deflect until the load is matched. Also, if the supporting surface is not firm enough to take the tire pressure with the normal contact area, the tire will sink into it until the final contact area becomes great enough to distribute the load with a bearing pressure that the surface material can take. The number of plies, or layers of materials, in the tire's wall is a fair indication of the load-carrying capacity of the tire.

A low-pressure tire is better for equipment operating on soft ground because it will give lower rolling resistance than a high-pressure tire (see Table 3-5). This is true because the low-pressure tire is designed to have a larger bearing surface and so will not have to penetrate the surface as much to get support for a given load. It is noted that rolling resistance is proportional to the depth of penetration, recognizing that the tire has to "roll out" of the hole it sank into. At the other extreme, a high-pressure tire is better on a hard surface where there is essentially no tire penetration. This is because the roller resistance under that circumstance is predominantly due to flexing of tire walls, and a high-pressure tire flexes less.

It is important to pay particular attention to the selection and maintenance of the tires on a highly mobile piece of equipment. For one reason, the tires on equipment such as large, high-speed haulers or earth-movers are an expensive part of the total cost. The considerations to watch are:

1. percent of normal inflation or air pressure in the tire;
2. load on the tire relative to its rated capacity;
3. speed of operating tire compared to speed for expected life of tire; and
4. other sound equipment maintenance measures for good tire life.

Considering air pressure or tire inflation, it is well known that tires with lower than recommended pressure wear out faster. The relative life determinations show:

10% underinflation loses 10% of total mileage;

30% underinflation loses 40% of total mileage; and

50% underinflation loses 75% of total mileage.

Likewise, determinations for overloading tires above rated capacity show:

10% overload causes 10% loss of mileage;

30% overload causes 20% loss of mileage; and

50% overload causes 30% loss of mileage.

The effect of operating vehicular tires at higher speeds than recommended for reasonable tire life has shown:

10% greater speed causes 18% mileage loss;

20% greater speed causes 30% mileage loss; and

30% greater speed causes 40% mileage loss.

Since tires wear out sooner or later with any use, a method has been devised to relate the choice of tire for heavy-duty use to its expected loads and speeds. This is advocated by the leading manufacturers of tires for construction equipment.[6] The method makes use of a formula for the ton-miles-per-hour (TMPH). It is designed to give the distance and quantity of material-load a tire can be expected to carry without danger of early failure due to excessive heating. The basis for use of this formula is that heat is a tire's most important life-expectancy factor.

Using the TMPH formula, the planner can tell in advance if his tires should last their expected lifetime. Or if used to select the tires, it tells which ones would not wear out early by overheating. The formula can be stated as the average ton load per tire multiplied by the average miles per hour of the equipment it is on—giving ton-miles/hour. Each tire of different size on a piece of equipment needs to be checked by the TMPH formula for its selection. For all tires of the same size, the tire with the highest average load should be checked. The following is an example of how this works.

Consider the example of a 6 x 4 wheel off-highway truck (equipment discussed in Chap. 9), with each front tire carrying 7 tons of empty truck weight and 10 tons with the truck loaded, and with each of the drive-axle tires carrying 6 tons empty and 12 tons when the truck is loaded. Each front tire has an average load of $(7 + 10) \div 2 = 8.5$ tons, and each drive tire takes $(6 + 12) \div 2 = 9.0$ tons on the average. If all the tires are to be the same size, then the "average ton load" for a rear tire is used.

Next, the average truck speed, both when loaded and empty, is determined for its long-term use. This might be found by recording or estimating the miles the truck will travel during a day of a specific number of operating hours. Say, it will travel 105 miles during a 7 1/2-hour work day. It will have averaged 14.0 mph. In this example, the governing truck tire takes a loading of $(9.0) \times (14.0) = 126.0$ TMPH.

In testing tires for TMPH rating, manufacturers found that the method cannot be used for tires loaded 20% above their rated capacity. Nor can it be used if the hauls are longer than 20 miles. For less severe punishment they have been able to give all their available off-highway vehicle tire models a TMPH rating. The planner with the example calculated above should select tires that have a 126 TMPH rating, or greater.

Wheel Mountings for Crane-Type Equipment. The rubber-tired wheel mountings for power cranes and shovels may be of two categories.[7] One is the single-engine, self-propelled type of unit with the power unit in the revolving superstructure. The other is a truck-type mounting with two engines. In the latter, one engine is in the truck body and provides the power to move from one working spot or site to another. The second engine is in the revolving superstructure and powers the working operations of the crane or power excavator. On some of the truck-type mountings the operator at the controls of the second engine in the superstructure can operate the first engine in a limited fashion by remote control.

The single-engine, self-propelled type of unit may have two or more axles with power delivered at four or more wheels. These variations are indicated in the specification of the unit. Thus, it may be a 4 × 4 or a 6 × 4 crane. The first number is the number of wheels, and the second number designates the driving wheels as contrasted with the free-rolling tires. This unit requires only one operator, but its speed is limited to maximums of 10 to 20 mph. It is usually manufactured in the smaller sizes such as the 3/4-yd class.

The two-engine, truck-type carrier mounting is designed, generally, with three or more axles. Aside from the stability of operation, the greater number of axles helps satisfy on-highway movement. There are state or federal highway load limits of perhaps 18,000 lbs on one axle or 32,000 lbs for a pair of dual axles. The designation of number of wheels and drivers for a truck-type mounting is the same as for the self-propelled unit above. However, the truck unit provides higher speeds of movement with a number of gears or gear ranges and top speed approaching 50 mph.

The gradeability of rubber-tired wheel-mounted cranes and shovel is not nearly as great as with a crawler-mounted unit. And to increase its stability for lifting or carrying out its primary function on level ground, the basic unit will generally have outriggers to extend outward so as to stand firmly on the bearing surface or ground.

3.3 Construction Operations and Materials

All construction operations deal with material to be moved, processed, or placed. The selection of suitable equipment to work in an operation depends on that material. What is to be done to it, and, consequently, what physical forces must be applied to the material? The principle variations in the answers to these questions are vital to the selection of appropriate construction equipment for an operation. Therefore, these variations will be discussed in the following sections.

3.3.1 Removal of Existing Material

Many construction operations involve the removal of existing material from its natural deposits on Earth. The operation may be earthmoving, rock excavation, pavement removal, trenching, tunneling, dredging, or dewatering. Several of those may have to be done together. In any case the equipment selected for the operation must be able to take care of removing the existing material. That requires knowing the material's natural condition.

The natural state of the material may be as a solid, broken and granular, or a liquid. Material knowledge is obtained from soil or rock borings, test pits, and on-site observation. The interpretation of this information requires a knowledge of geology and soil-mechanics; without this technological background gross errors can be made. For the proper evaluation of this information, the individual, if he does not possess this background,

FIGURE 3-10 Construction equipment handling material (courtesy of the Koehring Company).

should bring in a consultant for that purpose. Most material deposits in nature are a combination of various forms. That is, a deposit may be partly solid rock with some pockets of broken rock. Another could be granular, such as sand and/or gravel, or other earthy material that is quite fluid due to a high water content. One of the states will be predominant and govern the choice of equipment. Therefore, a planner should know the principle characteristics and properties of each.

Solid material to be removed, such as rock, consolidated clay, or concrete, is a continuum with fibrous strength or cohesion. The strength is through internal tensile, compressive, and/or shearing strengths which resist pieces being divided from the continuum or mass. Equipment to deal with such solid material must be able to overcome the strengths within the mass. If it is consolidated clay, the property to hold it together is known as cohesion, which must be broken by the equipment to remove it. These strengths will range from a few hundred pounds per square inch (psi) to possibly 50,000 psi. Consolidated clay will vary from 100 psi to 1000 psi for a shale-like deposit. A plain concrete will show from a few hundred pounds tensile strength to 5,000 psi, or more, compressive strength. Solid rock masses will vary from a few thousand psi for shale to more than 20,000 psi for a sound granite.

Where the material is broken, chunky, or granular, there are somewhat different forces for the equipment to handle. The material may have internal frictional forces or apparent cohesion. These are not generally large enough to bother the equipment's operation. For that reason solid material may be broken down to this state by drilling and blasting, heating or jetting, or some more modern means. Then the material's property of concern is its weight. This is generally expressed as a density, or weight per unit volume, such as pounds per cubic foot (lbs/cu ft) or pounds per cubic yard (lbs/cu yd).

TABLE 3-3 Unit Weights or Densities of Materials Handled in Construction

	Densities, δ			% of Swell (see Chap. 5)
	In Natural Bank		Loose	
Material	lbs/cu ft	lbs/cu yd	lbs/cu yd	
Clay, natural bed	110	2960	2130	40
Clay & gravel, dry	85	2290	1940	40
Clay & gravel, wet	97	2620	2220	40
Earth, loam, dry	97	2620	2100	25
Earth, loam, wet	125	3380	2700	25
Gravel, 1/4″-2″, dry	118	3180	2840	12
Gravel, 1/4″-2″, wet	140	3790	3380	12
Sand, loose, dry	100	2690	2400	12
Sand, packed, wet	129	3490	3120	12
Sandstone	159	4300	2550	54
Limestone	163	4400	2620	67
Trap Rock	164	4420	2590	65

From "Fundamentals of Earthmoving" by Caterpillar Tractor Co.

Liquid material dealt with in construction may be as clear as fresh water or it may have many solids suspended in it. Fresh water weighs 62.4 lbs/cu ft, and salty sea water is slightly heavier. When the fluid material is perhaps 40% solid particles, its weight might be as high as 100 lbs/cu ft. This is the condition of a saturated or submerged sand or gravel deposit. For a truly liquid material, regardless of the portion of solids in it, the one common characteristic is that it will seek a horizontal top surface in an open container or holding bed.

3.3.2 Moving Material to Storage

Another common form of construction operation involves moving material to storage, where there is a minimum of careful handling. The steps of such an operation are loading the material, carrying it, and depositing it in a suitable place for storage. By storage in the broadest sense, one should think of any prepared or natural place where material can be deposited without precision. Frequent examples of this operation are the hauling of excavated natural material, or moving crushed aggregate, or trucking processed material that does not have to be deposited in one exact spot. A move to an exact point of deposit will be discussed shortly with the subject of placing finished material.

The material in this form of moving operation will most generally be in pieces or particles. A certain number or amount of them are loaded on or into the carrying part of the moving equipment. If the material is solid pieces of the end product, they will be stacked for carrying, and their sizes determine how many can be loaded into the equipment for moving. Material that is in a loose, particle, bulk form must be carried in a box-like part of the equipment. That part of the equipment for moving bulk material will be called the container. The determination of how much bulk material can be carried in a box-like container is discussed in Chap. 5 on earthmoving equipment.

The physical forces that a piece of equipment must apply in moving material should be understood by the operation planner. These are a combination of gravitational and inertial forces. Pure gravitational force due to weight of the material being moved can be figured with the help of Table 3-3. The equipment to handle such a load may be holding it with an arm or boom that does not have a vertical support. In that case the force in the equipment due to the load is greater than the weight, W. Thus,

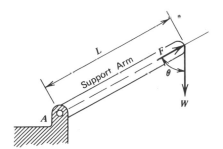

equipment's force,

$$F = \frac{W}{\cos \theta}, \qquad (3\text{-}12)$$

and the moment about the base of a supporting arm of length, L, is

$$M_A = W(L \sin \theta).$$

The arm or boom must be treated as a structural member with bending and torsion.

When the motion of the material is changed from a standstill to moving at a certain rate, there will be an inertial force. This is dependent on the weight of the material and the acceleration of that weight. Acceleration, a, is the rate of velocity.

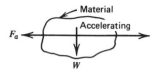

The inertial force is found as

$$F_a = \left(\frac{W}{g}\right) a = \left(\frac{W}{g}\right) \frac{\Delta v}{t}, \tag{3-13}$$

where g = gravitational pull of 32.2 feet/second/second;
Δv = change in velocity, ft/sec; and
t = time in seconds.

As shown by Eq. (3-13), the acceleration can be found from the change of velocity between that at the start of change, which would be zero at a standstill, and its maximum at full speed. If the time it takes to get up to the maximum velocity can be estimated, the acceleration is figured as the change of velocity divided by that time. The inertial force, F_a, acts in the direction opposite to the changing motion. Equipment on which it is acting must apply an equal and opposite force. Any part or all of the equipment also undergoing the same change of motion is causing an inertial force of acceleration. The same sort of determination must be applied if the weights are slowing down. That is called deceleration and produces a force in an opposite sense to that for acceleration.

A moving weight that has its direction of motion changed will exert another inertial force. This is because a body in linear motion with a constant velocity tends to continue that way unless acted upon by a force in another direction. This can be shown in relation to the curving direction of motion in the adjacent sketch. The body's force that tends to

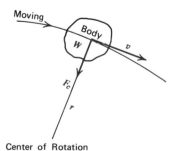

Center of Rotation

throw it out of the curving path is called centrifugal force. It must be counteracted by the equipment applying a centripetal force, F_c, found by the equation

$$F_c = \frac{Wv^2}{gr}, \tag{3-14}$$

where v = tangential velocity, ft/sec;
  r = radius of curvature, feet; and
  g = 32.2 ft/sec/sec.

3.3.3 Processing Material for Construction

A construction operation to process materials for use in the final product may have any of various actions. The processing is to take raw or refined materials and work on them to produce a more finished construction material. Examples of this type of operation include crushing rock to make aggregate; mixing aggregate, cement, water, and additives for concrete; making precast building elements; and bending metal for pipe sections.

The processing operation may involve one or more steps to produce the end product. In the production of aggregate there may be one or more steps called stages, which are the successive crushing steps to take the raw input material down to the final specified sizes. To make concrete or asphalt, the steps are the weighing out of refined materials for the mix, mixing them in a proven manner, and pouring the mixed material. The pouring may be directly into place for finished material, but more likely it will be into equipment for moving the processed material to its point of use. Generally, each step for processing material is done by a different piece or part of the equipment for the operation. This will become more obvious as the types of equipment are discussed in following chapters of this book.

The physical forces applied by the processing equipment are compressive pressure, impact force, rotary motion, and the bending force of a moment or shearing force of a load applied at right angles to the axis of the piece. These are applied by crushers, mixers, pipe benders, and construction material shears or cutters. The amount of force that is needed from the equipment has generally been designed into it. A balance has been built into the processing equipment so that it cannot take anything too big or difficult for it. So the planner merely needs to select equipment designed to accept the material which will be fed to it. The equipment will do as it is designed, applying the right actions and forces.

3.3.4 Moving Processed Material that Flows

The moving of processed material such as aggregate, asphalt, and concrete calls for special consideration. This is necessary for all flowing materials that have gained a certain specified gradation or consistency. Segregation and separation are the results if they are moved improperly.

Equipment to handle the moving of processed materials that flow must be designed to minimize segregation and separation of the material particles. This can be done by appropriate design features and operation of the equipment. In the case of concrete to be moved considerable distance, a hauling truck is used that continues to mix or turn over the material. For jobsite moves of any of these materials the use of pouring buckets, chutes, and conveyors with minimum vibration would be advised.

The physical forces for the equipment to apply are the same as those discussed in Sec.3.3.2. Inertial forces will tend to bring about separation of mixed material, so such forces should be minimized in operating the equipment. Primary concern for the effectiveness of equipment moving this processed material is with its control of the pouring action. That must be done so that the placed material is as nearly like the originally mixed materials as possible. Then the final material will have the desired homogeneous mass that is specified.

3.3.5 Placement of the Finished Material

A final type of construction operation dealing with somewhat different equipment concerns is the placing of finished material. The finished material is that which has been processed and made ready for a final constructed product. It is handled in batches of flowing material or pieces of the whole, such as blocks, columns, and pipe sections, or subassemblies, such as precast units or structural parts.

The sorts of operations using finished material are suggested by certain common ones. Steel erection, erecting precast building parts, pile driving, and pipe laying all involve solid finished material. Laying asphalt and concrete for pavement are operations for equipment to place flowing processed material. In every case the equipment handling the placement of material to finish the constructed product works in a careful, slow manner.

Equipment requirements in this type of operation are quite variable. The operations involving heavy, sometimes awkward, solid pieces or subassemblies call for equipment with adequate lifting power and reach. They generally do not have to worry about inertial forces. Those operations to place and finish flowing material require equipment that can spread and maintain its uniform consistency. The common need for equipment in all these operations is control of the placement and avoiding damage to the finished material. That leads to slower motion than with other equipment operations dealing with material.

3.4 Forces Governing Motion of Equipment

The previous sections introduced the equipment planner to the inertial forces of motion. They apply not only to moving material but also to equipment in motion. In addition, one must understand the forces that can and need to be exerted by the equipment to maintain its movement.

A moving piece of equipment is powered by a so-called prime mover. This may be the engine or motor in a self-propelled piece of equipment. Or it may be a whole tractor with its motor needed to move another piece of equipment. In either case the power unit must deliver force to the surface on which the prime moving equipment is operating. There are limits to the maximum force that can be delivered. They will be discussed in the next section. What has to be delivered to move the equipment must cover the inertial forces and resistances to motion.

3.4.1 Traction and Tractive Effort

The engine or motor of the equipment's prime mover is the source of power for motion. Power sources for construction equipment were discussed earlier in this chapter, as well as the transmission of generated power to drive the equipment. The power output or force applied at the driving wheels or tracks is known as tractive effort. This force must be enough to overcome the combined resistances to motion.

The maximum force that can be applied by a prime mover at its drivers to move equipment on the ground or prepared surface is governed by one of two limits. Either the maximum force is limited by the maximum power output delivered by the power unit to the driving contacts on the supporting surface, or the maximum force is based on the limit of traction between the contacting equipment tires or tracks and the supporting surface. The smaller of these values will be the maximum applicable force. If the traction governs, the sketch of Fig. 3-11 will show the relationship between the forces involved.

The maximum force a tractor can apply is found as

$$F_T \le (F_t = W \tan \theta) \ge F_R \qquad (3\text{-}15)$$

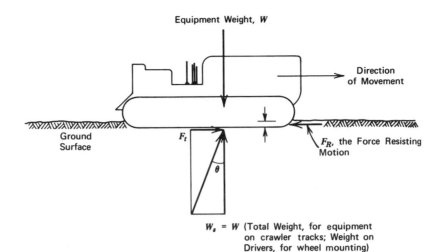

FIGURE 3-11 On traction.

The term "tan θ" is known as the coefficient of traction (C_t) and varies, depending on the type and condition of the supporting material. For most supporting surfaces an increase in the moisture on the surface will decrease the coefficient of traction. This is shown by the representative values in Table 3-4.

TABLE 3-4 Coefficient of Traction (C_t or tan θ) Values

Contact from Equipment on Surface	C_t or tan θ	
	Tracks	Rubber Tires
Dry concrete	0.45	0.90
Wet concrete	0.45	0.85
Dry macadam	—	0.70
Wet macadam	—	0.65
Dry, firm earth or clay loam	0.90	0.5–0.7
Wet, firm earth or clay loam	0.7–0.85	0.4–0.5
Dry, loose sand	0.30	0.2–0.3
Wet sand	0.35–0.5	0.35–0.4

Thus, the moisture tends to make the surface more slippery, especially for rubber-tired equipment. However, it should be noted that this is not the case on predominantly sandy soil. A certain amount of moisture gives the sand more apparent cohesion due to the surface tension of the water coating the grains and thereby adding to the stability of the sand. But if the sand takes on enough water to be essentially saturated, then this apparent cohesion is lost.

The application of the appropriate C_t value to find the maximum force, F_T, or tractive effort is shown in Eq. (3-15). Force F_T must be, at least, equal to the total of resistance to motion, F_R. Major resistances are discussed next.

3.4.2 Rolling Resistance

The primary resistance to the motion of a piece of equipment on a level surface is called rolling resistance. It is caused by two effects. One is the friction or flexing of the driving mechanism delivering power for tractive effort. There will be some of this for either a crawler-track drive system or rubber-tired drive wheels. The larger rolling resistance is due to the drivers having to "push" through or over the supporting surface. This was shown by force, F_R, in Fig. 3-11.

Rolling resistance (RR) is expressed in pounds of tractive effort required to move each gross ton over a level surface of the specified type or condition. Although it is impossible to give completely accurate values for the rolling resistances for all types of haul roads and drivers or wheels, except by field tests on a jobsite, the values given in Table 3-5 are reasonably accurate and may be used for estimating purposes.

The rolling resistance will increase about 30 lbs per ton of weight for each inch increase in penetration. Therefore, if the weather conditions keep the soil surface wet and soft, the rolling resistance may be appreciably higher than for dry, firm soil conditions. Assuming an average rolling resistance of 150 lbs per ton for an unprepared haul route on ground, one inch additional penetration due to wet weather will amount to a 20% increase in resistance to motion on level ground. It should be recognized that increased softness and so possible increased penetration may be reduced with rubber tires by deflating them to obtain greater bearing area and so better flotation. However, deflating tires will tend to increase the part of rolling resistance caused by the flexing of the tires themselves.

The actual force to overcome the rolling resistance of the support surface is found as

$$F_{RR} = RR \times W, \text{ in pounds} \qquad (3\text{-}16)$$

where W = total weight of equipment in tons.

TABLE 3-5 Representative Rolling Resistances for Various Types of Contacts and Surfaces, in Pounds per Ton of Gross Load, i.e., RR Value

Type of Surface	Crawler-Type Tractor on Tracks	Steel Tires, Plain Bearings	Rubber Tires, Antifraction Bearings	
			High Pressure	Low Pressure
Smooth concrete	55	40	35	45
Good asphalt	60–70	50–70	40–65	50–60
Earth, compacted and well maintained	60–80	60–100	40–70	50–70
Earth, poorly maintained, rutted	80–110	100–150	100–140	70–100
Earth, rutted, muddy, no maintenance	140–180	200–250	180–220	150–200
Loose sand & gravel	160–200	280–320	260–290	220–260
Earth, very muddy, rutted, soft	200–240	350–400	300–400	280–340

3.4.3   Grade Resistance Moving Up an Incline

Any equipment moving up an inclined surface works against another form of resistance. This is called grade resistance and is due to the effect of the weight of the equipment acting down the incline. The adjacent sketch shows this effect. The degree of incline is spoken of as a grade,

$$G = \frac{V}{H} \times 100, \text{ expressed in a \%.}$$

It can be found by determining the vertical rise, V, in 100 feet of horizontal distance. For example, a 5% grade ($G = 5$) means that the incline rises 5 feet vertically in a horizontal distance, $H = 100$ feet.

The grade resistance calls for a force, F_{GR}, to move the equipment up the incline against the effect of weight, W_G. By the geometry from Fig. 3-12 it can be seen that $W_G/W = V/I$. The needed force to overcome the grade resistance is found by

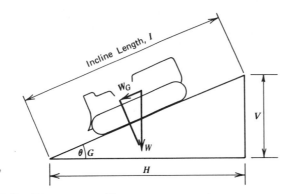

FIGURE 3-12 Showing grade resistance.

$$F_{GR} = W_G = (V/I)W;$$

but for the usual grades of less than 20%,

$$\frac{V}{I} = \sin \phi \approx \tan \phi = \frac{V}{H} = \frac{G}{100}$$

Then, by reasonable substitutions,

$$F_{GR} = W \tan \phi = W\left(\frac{G}{100}\right),$$

and for easy use with enough accuracy,

$$F_{GR} = 20(G)W, \text{ in pounds} \qquad (3\text{-}17)$$

where G = the grade in %;
W = weight of the equipment in tons; and the constant is a grade resistance factor, $GR = 20 \text{ lbs/ton/\%G}$.

One should note that moving down an incline gives a minus grade $(-G)$ and negative F_{GR} requirement, so there will be a force advantage for the equipment then.

The effective power output or maximum tractive effort for a tractor type of equipment is sometimes expressed as gradeability. This is merely a way of converting its available maximum force, F_T, to a total resistance it can overcome. The total of rolling and grade resistance can be combined:

$$F_R = F_{RR} + F_{GR} = (RR)W + (20G)W = (RR + 20G)W.$$

The term in parentheses can be called gradeability with the RR value converted to an equivalent grade simply by dividing by 20 pounds per ton per % of grade. For example, if $RR = 140$ lbs/ton, that is equivalent to a 7% grade ($= 140/20$); and if the tractor has to climb a 5% incline with that rolling resistance, it would need a 12% gradeability.

Application of the various forces discussed in the above sections will be made frequently in following chapters. These are most commonly applied in the discussions of earthmoving and hauling equipment.

3.4.4 Other Resistance to Moving Equipment

Two other significant factors use up the power generated by a prime-mover engine or motor. One is internal to the moving equipment, and the other acts on it externally.

The internal resistance is a combination of power losses due to the rotating and moving parts of the prime mover, its transmission and drive shafts, and the friction and vibrations caused by all of that motion. This power loss will be proportional to the total weight of the moving equipment and the speed (revolutions per minute, RPM) of its power unit. It will generally amount to less than 10% of the available power. Later calculations in this book will avoid a direct determination of that loss by either assuming a total internal power loss or using the power output available at the drive wheels or tracks.

The other external resistance to motion is due to air resistance. It amounts to a force against a moving piece of equipment similar to rolling resistance. At the relatively low speeds for construction equipment air resistance is generally not a major power consumer. But for high-speed hauling equipment it must be considered. Also, for any equipment moving against a strong wind the air resistance is a significant factor. The determining quantity is the relative difference between the speed of the moving equipment and the air velocity opposing or directed against that motion, i.e., the net velocity. Thus, if equipment is traveling at 10 mph against a 40-mph wind, that gives as much air resistance as a truck moving at 50 mph on a calm day.

The force of air resistance is dependent on the net velocity and the cross-sectional area of the moving equipment. That area is the equipment's surface on which the air pushes. In the case of a truck, the area is found as the overall width times height. Of course, the shape of the body can affect the air resistance due to that area. In high-speed racers streamlining helps cut down air resistance. So the use of a cross-sectional area is merely for an approximation.

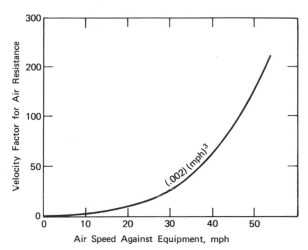

FIGURE 3-13 Air resistance relative to motion of a piece of equipment.

The differential, net velocity is the primary factor in air resistance. To give some idea of the relative effect of velocity, the curve of Fig. 3-13 will be helpful.

The power to overcome air resistance is found by the product of the velocity factor and an area factor. The area factor is proportional to the cross-sectional area of the equipment as mentioned above. That factor is only 0.155 for a piece of equipment eight feet high by eight feet wide. So, for it to move at 20 mph with no wind and at sea level will take only 2.5 hp ($= 16 \times 0.155$) to overcome the air resistance. Air resistance would seem to become significant when the net air velocity opposing motion is 50 mph or more.

3.5 Minimizing Power Requirements

In all the construction operations discussed so far there are forces at work. Forces must be applied to move equipment. The necessary force is generated primarily by the power unit of the equipment transmitting its energy to the driving mechanism. The power output can be measured by the force applied to produce a certain rate of motion. In other words, to apply a force for moving a body or weight at a given speed takes a certain amount of power. The required force is mainly dependent upon the total of rolling and grade resistances and inertial forces acting on the equipment.

Since the power generated by the prime mover for equipment is an expense of operation, the required power should be considered carefully. The cost of an engine or motor, its fuel, transmission, etc., are generally directly proportional to the maximum power required and the total power consumed. So the equipment planner does well to minimize his power requirements.

At least five key principles regarding power can be applied in planning for construction equipment. If these are followed to advantage, the equipment selected and used should result in more economical operation.

The principles for minimizing the requirement for generated power can be stated in the following way:

1. Take advantage of gravity flow or downgrade $(- G)$ travel for moving the heavier weights;

2. Plan for straight-line, constant-speed movement to minimize the power required for inertial forces;

3. Minimize lifting material as much as possible, since vertically upward movement against the force of gravity is the greatest power consumer;

4. Minimize turns or changes of direction for equipment, or its load, since each turn or change calls for power to oppose the centrifugal force;

5. Minimize starts and stops of equipment, or its load, since each one involves overcoming an inertial force which consumes energy and takes power.

The application of these principles should be instinctive to a well-versed equipment planner. Concentration on them is most beneficial when planning the construction method for an operation.

If one analyzes the principles stated above in relation to other common operations, he can recognize them as assembly-line techniques. This refers to operations perfected in industrialized manufacturing. Improvements toward optimizing such operations have been made possible by the use of time and motion studies. More will be said about them later. For the present, it can be stated that planning for construction equipment can be improved by applying assembly-line techniques. Or, if such techniques are not familiar, apply the principles stated above in this section.

Assembly-line techniques have been modernized to eliminate all unnecessary manual operation. This minimizes the stops, starts, turns, and changes of speed that all consume energy and power, if they are done by equipment. Such changes are compounded by manual control, if the operation is not systematic. If each move that is necessary can be done "automatically" by habit, there will be a minimum of hesitation and energy used. This is an important objective in eliminating manual controls.

The means for reducing manual control have been made by automating with electrical or hydraulic controls many of the steps in an operation. Automatic controls on construction equipment have been introduced, and more should be possible. Automation helps to minimize the power requirements for construction equipment.

3.6 Tractors

The use of tractor-type equipment in construction is extensive. Tractors serve as prime movers for a great variety of earthwork equipment. They are used with all variations of dozers, rippers, pusher blocks, fork lifts and loaders, and other auxiliary attachments. Tractors are also the prime movers for pieces of equipment such as towed scrapers, graders, and rollers. The rollers are for compaction, and it should be noted that the regular crawler tracks or wheel-type tractor units are frequently used as soil compaction equipment. Compaction equipment will be discussed at greater length in Chap. 12.

3.6.1 Rimpull by Wheel-Type Tractors

A tractor serves as the prime mover in any of these uses by applying its power at the rim of its drive wheels or through the crawler tracks. In the case of wheel-type tractors the power is applied as a tangential force known as rimpull. The rimpull is the tractive effort (TE) if in its application the force does not cause slippage but applies in its entirety to move the tractor. This rimpull, or TE, for the tractor can be found by using Eq. 3-8. Note that if the speed of operation is known, the radius of the drive wheel is not needed. However, if that speed is unknown, it is possible to determine the rimpull from the tractor's engine output and transmission gear ratios. Thus, from Eq. 3-8 and $v = 2 \pi r N_o$,

$$\text{TE} = \frac{33,000 \,(\text{eff.}) \, \text{bhp}}{2 \pi r N_o}, \tag{3-18}$$

where r = the wheel radius or tire's outside radius (see Fig. 3-9), and
N_o = rpm of drive axle.

The speed of the drive axle is not known any better than the speed, v. Consequently, to find the rimpull entirely from values internal to the tractor, it is necessary to work with engine output and the transmission system. We recognize that the engine output of governed speed (N_g) and torque (T_g) are related to the comparable values at the drive axle by the expression: $N_g T_g = N_o T_o$, assuming no slippage in the gears. The ratio of rpm's or torques is that of the total gears in the transmission. Then, Eq. (3-18) may be rewritten

$$TE = \frac{5250 \ (\text{eff.}) \ bhp}{r \, N_g \dfrac{T_g}{T_o}}, \tag{3-19}$$

where N_g = governed rpm of the engine and
$\quad T_g / T_o$ = total gear reduction ratio in the particular gear.

As an example, a tractor's specification sheet could give the following information:

bhp = 100 at governed speed; N_g = 2200 rpm;
Total gear reduction = 200:1 in first gear;
Total gear reduction = 20:1 in highest gear;
Speeds: 1st gear = 1.6 mph; highest = 16.0 mph;
Tire's outside radius, r = 24 1/2 in. = 2.04 ft;
Tractor weighs approximately 9700 lbs when operating.

Assume the mechanical efficiency (eff.) from engine output to drive wheel is 75%.
(a) In the highest gear with the tractor's speed known, use Eq. (3-8) to find

$$TE = \frac{33,000 \ (.75) \ 100}{16 \times 88} = 1,760 \ \text{lbs.}$$

Or this could be found from the internal power transmission using Eq. (3-19):

$$TE = \frac{5,250 \ (.75) \ 100}{2.04 \left(\dfrac{1}{20}\right) 2200} = 1,760 \ \text{lbs.}$$

This would be enough to overcome a total resistance to motion—rolling and grade resistances, acceleration, and those of trailing load—of 363 lbs/ton.

(b) In first gear, using Eq. (3–19),

$$\text{TE} = \frac{5250\,(.75)\,100}{2.04 \left(\dfrac{1}{200}\right) 2200} = 17,600 \text{ lbs.}$$

But obviously this cannot be applied in tractor motion by itself because there would not be enough traction. If the coefficient of traction (see Table 3-4), $C_t = 0.7$, the maximum tractive effort that can be applied by the tractor operating alone is $0.7 \times 9700 = 6,800$ lbs. The rimpull of a wheel-type tractor is the total resulting applicable power at the drive wheels of the prime mover. It will vary with the gear of operation for the tractor and also may be limited by traction.

3.6.2 Drawbar Pull Power by Crawler Tractors

The same sort of determinations can be made for a crawler tractor as were described above for the wheel-type tractor. However, a more directly useful value of a tractor's power has been consistently specified. It is called the drawbar pull power and abbreviated DBPP. Drawbar pull is defined as the pulling power or force available at the drawbar hitch when the tractor and its towed load are travelling on level ground. The use of a DBPP value assumes that the rolling resistance (RR) of the crawler tractor has already been covered. To make that assumption means the DBPP values specified must be based on a specific RR for the tractor. Such is the case, and the specific value is $RR_T = 110$ pounds per ton.

The universal use of DBPP and this specific rolling resistance evidently comes from the procedure established for testing tractor power. Crawler tractors have been tested for years at the University of Nebraska by a simple quantitative determination of how much each can pull horizontally. The testing site is firm earth, modestly maintained and considered to have a 110 lbs/ton rolling resistance to crawler tracks. As can be seen by the values in Table 3-5, the RR values can vary considerably. If the power of a crawler tractor is being considered, the user is advised to first decide what RR value is appropriate and adjust the DBPP if necessary.

A brief discussion for review should help to clarify the use of drawbar pull power for crawler tractors. The components of resistance for a tractor towing a load (roller, scraper, etc.) at a uniform speed are:

1. rolling resistance of the tractor, $F_{RR\text{-}T} = RR_T \times W_T$, the weight of the tractor in tons;
2. rolling resistance of towed load, $F_{RR\text{-}L} = RR_L \times W_L$, the weight of the load being towed in tons;

3. grade resistance of the tractor, $F_{\text{GR-}T} = \text{GR} \times W_T$;
4. grade resistance of towed load, $F_{\text{GR-}L} = \text{GR} \times W_L$. If the tractor with its towed load is accelerating, there is the

5. force to produce acceleration, $F_a = \dfrac{W_T + W_L}{g} \times a$, for both the

tractor and the load. The rate of acceleration, a, is in units of feet/second/second with $g = 32.2$ ft/sec/sec. These forces can best be identified and remembered by an illustration such as Fig. 3-14.

It should be noted that the rolling-resistance values may be different for a crawler tractor (RR_T) and its towed load (RR_L) if the load is on rubber-tired wheels. Also, a tractor towing a load will, generally, accelerate to operating speed before starting up a grade, so an F_a requirement will not be needed where grade forces, F_{GR}, must be overcome. Of course, the total power force that the tractor must apply cannot exceed the traction it can get on the surface. Thus,

$$\text{TE} = \Sigma F_{\text{RR}} + \Sigma F_{\text{GR}}, \tag{3-20}$$

where $\text{TE} \leq (\text{C}_T)\, W_T$. The weight of tractor only is used since drawbar hitches are designed not to transmit vertical weight from the towed load for convenience of load changes. If $\text{RR}_T = 110$ lbs/ton, then the

$$\text{DBPP} = F_{\text{RR-}L} + F_{\text{GR-}L} + F_{\text{GR-}T} \tag{3-21}$$

It is generally safer to use Eq. (3-20) to avoid the mistakes of neglecting to adjust where RR_T is not equal to 110 lbs/ton and neglecting the grade resistance of the tractor. However, tractor specifications tend to demand the use of Eq. (3-21).

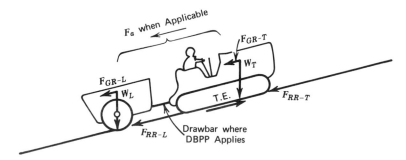

FIGURE 3-14 Components of resistance for a tractor towing a load.

References

1. "The Story of Power," booklet by Public Relations Staff, General Motors Corporation (Detroit, Michigan, 1956).

2. Schaefer, R. M. and H. C. Kirtland. "What's Behind the Trend to Torqmatic Drives," Allison Division, General Motors Corp. for Society of Automotive Engineers (New York, N.Y., 1956).

3. Richards, C. W. "Electrical Operation of Earthmoving Equipment," R. G. La-Tourneau, Inc. (Peoria, Illinois, 1952).

4. Kellogg, F. H. *Construction Methods and Machinery*, Prentice-Hall, Inc. (New York, N.Y., 1954), Chapter 4. Power, pp. 75–106.

5. "On and Off-the-Road Truck Tire Data" booklet by the Goodyear Tire & Rubber Company (Akron, Ohio, August 1969).

6. "Earthmover Tire Ton-Mile-Per-Hour Formula" pamphlet by the Goodyear Tire & Rubber Company (Akron, Ohio, January 1969).

7. "Mobile Power Crane and Excavator Standards," PCSA Standard No. 1, Construction Industry Manufacturers Association (Milwaukee, Wisconsin, 1968).

8. "Fundamentals of Earthmoving," booklet by Caterpillar Tractor Co. (Peoria, Illinois, revised 1968).

Chapter 4

■■■■■■■■■■■■■■

■■■■■■■■■■■■■■■■■■■■■■■■■■■■■■■■■■■■■■

COMPRESSORS
AND PUMPS

■■■■■■■■■■■■■■■■■■■■■■■■■■■■■■

The capabilities of air compressors and pumps and the flexibility in their use make these pieces of equipment very common for construction projects. The prime function of both types of equipment is based on the principle of the displacement of a gaseous or liquid fluid to work on or move material in the construction process. They are generally powered by internal combustion engines to achieve a high degree of mobility and, consequently, flexibility in use for construction. The types of compressors and pumps used on construction are of a relatively small size and simple, rugged design so that they need a minimum of maintenance to keep them operating.

4.1 Air Compressors for Construction

Compressors are used in construction to generate air power for a variety of hand tools, hammers, and certain atomizing and conveying equipment. Some idea of the many construction uses of air compressors is given in Table 4-1.

TABLE 4-1 Construction Uses of Compressors
for Air Power

(a) For operating hand tools to
  — cut material with circular or chain saws,
  — bore holes in timber members,
  — drill holes in rock or other crushable material,
  — dig out sticky material, such as clay,
  — break out crushable material, such as asphalt or concrete,
  — vibrate poured concrete for denser final set,
  — chip rough edges off parts to finish poured and set material,
  — drive rivets into place,
  — tighten or loosen structural bolts,
  — tamp earthfill material for improved consolidation, etc.;
(b) For many uses in tunneling, such as
  — drilling holes to load for blasting or rock bolts,
  — blowing out rock cuttings from drilled holes,
  — aerodynamic power to move drifters and jack other equipment in the tunnel, and
  — blowing out explosive, noxious fumes from the tunnel atmosphere;
(c) For mixing and atomizing to shoot fine particle material, such as paint or cement gunnite, into place;
(d) For the fluid to carry through a pipe small particle materials, such as cement, fine dry sand, or other mixing materials;
(e) For an air-operated centrifugal pump; and
(f) For an air-powered hoist drum or brake.

To select a compressor suitable for a particular construction situation *necessitates determining*:

1. *the tools and other equipment,* known as the users, to be provided compressed air from this compressor;
2. *total air requirement* in cubic feet per minute, cfm, demanded by all the users listed in (1);
3. *pressure requirements* in psi at each user of the compressed air supplied;
4. *system of piping and hoses,* including lengths for air feeder lines from the compressor, strategically located, to all the users;
5. *the user that governs* the compressor's pressure requirement, recognizing that the pressure loss in the feeder lines is directly proportional

to the equivalent length of pipe or hose and inversely proportional to the size of line, with hoses causing much greater pressure drop than pipe of comparable size and equal length;

6. *the desired air pressure in the receiver* at the compressor to provide the required pressure at the governing user and account for the pressure losses between them;

7. *the allowable pressure range* of the compressor to generate the desired pressure for the receiver;

8. *an acceptable diversity factor* for the number of users served by the compressor;

9. *the theoretical size of compressor* to generate the air capacity and pressure, taking into account the diversity of users to be supplied; and

10. *the economical compressor* that is commercially available and will meet or slightly exceed the requirements for the ideal size.

The compressed air is transmitted in simple linear movement through various mechanisms. The movement is controlled simply by valves and directing channels. The driving force causing the movement is generated by a diaphragm, a piston, or a rotating impeller in the compressor. To explain the generation in greater detail, let us first consider the simple bicycle pump, which is the most rudimentary of positive-displacement-type compressors available. A bicycle pump consists of a piston, secured by a rod to a handle, operating in a cylinder with a leather and a check valve. The cup leather is a diaphragm which insures tightness to keep the compressed air ahead of the piston in the cylinder during the compression stroke. The check valve prevents air from re-entering the cylinder from the discharge point when the piston is retracted, i.e. the valve maintains it as an "exit" just like a door so marked. The actions of this simple bicycle pump are similar to the common reciprocating compressor.

4.1.1 Principles of Air Compression[1]

The compressing of air from atmospheric or another intake level to a higher pressure depends on the measured pressure and temperature within the gaseous fluid. "Gauge pressure" is the pressure measured above atmospheric pressure, which at sea level and standard conditions of temperature and moisture is 14.7 pounds per square inch (psi). More precisely, gauge pressure is that which is above the ambient air pressure and would be marked on an air gauge. This is noted by the symbol "psig," standing for pounds per square inch gauge. "Absolute pressure" is that which is measured from a complete vacuum or zero pressure and equals gauge plus atmospheric pressure. Similarly, there are temperatures which are measured with a standard thermometer from zero degrees as the datum. The reference temperature of significance in compressed-air measurements is absolute zero, which numerically amounts to 460° below zero on the Fahrenheit scale. Thus, $0°A = 460°F$.

The laws of nature that apply to compressing air were discovered originally by scientists Boyle and Charles. If the temperature remains constant during the compression cycle, Boyle's law states

$$P_1 V_1 = P_2 V_2 = \text{a constant,} \tag{4-1}$$

where P values are absolute pressure, psia, and V values are the customary units of cubic feet or cu inches.

If the pressure is constant during the compression cycle, Charles' law applies:

$$\frac{V_1}{T_1} = \frac{V_2}{T_2} = \text{another constant,} \tag{4-2}$$

where T values are absolute temperature, A°.

These equations by Boyle and Charles can be combined into the generally applicable compressed-air expression,

$$\frac{P_1 V_1}{T_1} = \frac{P_2 V_2}{T_2} = \text{a new constant.} \tag{4-3}$$

The work that must be done to compress air depends on whether the change in volume is without a change in temperature, "isothermal," or without gaining or losing heat, "adiabatic." More commonly in the practical case such as compressors on construction, the compression is something between isothermal or adiabatic. Regardless, the work is a function of both pressure and volume and could be expressed as $W = f(P,V)$. The work can be determined with a modification of Boyle's law that would be expressed as

$$P_1 V_1{}^n = P_2 V_2{}^n = \text{constant,}$$

where $n = 1.0$ for isothermal compression and 1.4 for adiabatic compression. For isothermal compression, the relation is Boyle's law from Eq. 4-1:

$$W = P_1 V_1 \log_e \frac{P_2}{P_1},$$

for air taken in at sea level and 60°F, so $P_1 = 14.7$ psi;

$$W = 4883 \, V_1 \log_{10} \frac{P_2}{P_1},$$

where W = work in ft-lb, if V_1 is in cubic feet.

Now if we recognize as before that horsepower (hp) is a measure of work with one hp equal to 33,000 ft-lbs/minute, then

$$\text{hp} = 0.1479 \, V \log_{10} \frac{P_2}{P_1}. \tag{4-5}$$

An application of this might be to determine the prime mover needed to operate a 150-cfm compressor delivering air at 100 psig. Isothermally, at sea level and standard conditions the compressor would theoretically need

$$\text{hp} = 0.1479 \, (150) \log_{10} \left(\frac{114.7}{14.7} \right)$$

$$= 0.1479 \, (150) \, 0.892 = 19.75 \text{ minimum required.}$$

The actual requirement of a real compressor will be higher than this for several reasons. One is the compression and mechanical efficiencies of the equipment, which means that considerably more than 150 cubic feet of air is worked on each minute to get the required cfm output. And as mentioned previously, the actual compressor cannot achieve isothermal compression practically.

The determination of horsepower for adiabatic compression is more complicated. It can be estimated as 30–40% greater than that for isothermal compression. In the actual construction situation where the compression is between these two idealistic extremes, the theoretical power needed may be about 20% higher than that calculated by Eq. 4-5. The actual power output consumed will be higher than this to account for the mechanical efficiency, which can be estimated at about 90%. That is, the efficiency shows the ratio between the theoretical power and the actual power that must be delivered. The actual power required would be about 1.1 times the theoretical.

4.1.2 Types of Air Compressors[1]

The types of air compressors used for construction are of two categories —the positive-displacement type and the dynamic type of compressor. The positive-displacement type of compressor is like the bicycle pump, in which successive volumes of air are confined in the closed-space cylinder and compressed to a higher pressure before being discharged. These compressors, the ones most commonly used on construction, are of the (1) reciprocating or (2) rotary type as explained later. The dynamic type of compressor is one in which dynamic (high-speed) action of rotating vanes or impellers impart velocity and pressure to the air in a confined space. These dynamic compressors are of the centrifugal, axial-flow and mixed-flow types and find use mainly in the chemical, petroleum, and material-processing fields.

Reciprocating Air Compressors. In the reciprocating type of air compressor, compression is produced by the reciprocating, back-and-forth, motion of the compressor piston driven by a crank and connecting rod from the drive shaft of the I.C. engine. The control of the compression cycle is by simple check valves allowing the air passage in one direction only. The movement of the piston away from the valve end of the cylinder allows a suction valve to open and air to fill the cylinder. Then the movement toward the valve end opens the discharge valve when the pressure is great enough to discharge the air out of the cylinder to use or storage.

The principles of air-compression cycles as applied to the reciprocating compressor will now be discussed briefly. As mentioned previously, this could be adiabatic compression, with all the heat of compression retained, or isothermal, with constant temperature throughout the compression. Referring to Fig. 4-1, we see a sketch of the basic pressure-volume relationship in the compression cycle. Since the power to compress the air is

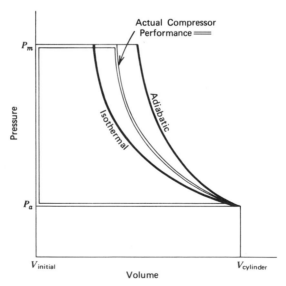

FIGURE 4-1 Idealized compression cycles.

measured by the volume under the *P-V* curve from the atmospheric, P_a, to the maximum, P_m, it is apparent that the more ideal situation power-wise would be to compress the air isothermally. This would require perfect cooling, insulation, etc. The practical or actual operation of a compressor in construction is shown between the two theoretical cases.

The actual compressor output of pressurized air accounts for the piston not moving all the way to the head block of the valve end, but leaving a piston clearance. This clearance is to provide a cushion of air, avoiding impact of the piston on the cylinder head which would lead to an engine block failure, and eliminating the hardest part of compression in the very smallest, highly compressed volume. The clearance also accounts for air pressure changes at or during the operation of the suction and discharge valves as shown in the realistic cycle sketch of Fig. 4-2.

A reciprocating compressor's theoretical air capacity is measured by the piston displacement volume, i.e., the volume swept by the piston in one minute. However, this is not the actual air compressed as can be seen by reference to Fig. 4-2. The capacity must be found as the amount of air taken in at suction conditions of the valve intake, compressed, and delivered through the discharge valve per minute. This latter amount is known as the "actual delivered capacity." The ratio of

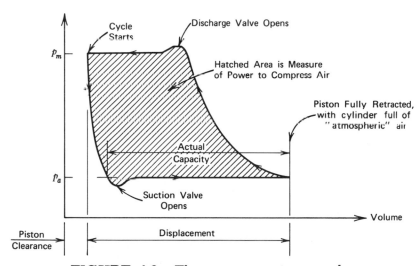

FIGURE 4-2 The compression cycle.

$$\frac{\text{actual delivered capacity}}{\text{piston displacement}} = \text{volumetric efficiency.} \qquad (4\text{-}6)$$

Some typical values of this efficiency for estimating purposes are given below for horizontal, double-acting (i.e., compression by piston in both directions with sets of valves at both ends of the cylinder), water-cooled, single-stage compressors. For those delivering at 100 psig, the volumetric efficiency is approximately 70%; and delivering at 50 psig, this efficiency is 80%. It is well to note that the volumetric efficiency is not the same as mechanical efficiency, which is the ratio of the theoretical horsepower required (given by Eq. 4-5) to the actual horsepower input necessary to power the compressor.

The system of multistage compression is one way to minimize power losses due to heat generated between the air intake and discharge at the much higher pressure. Multistaging is generally done when the pressure is to be raised in the order of tenfold, say, from atmospheric to 125–150 psi. If the compression of the air is handled in stages of a smaller pressure increase in each, then it is possible to effectively cool the compressed air at the top pressure of each stage before it goes into the next step. The power savings affected by this multistaging depends on:

1. the ratio of suction to discharge pressures;
2. the cooling medium for the cylinders; and
3. the intercooler effectiveness.

The sketch in Fig. 4-3 shows what this does to the pressure-volume curves. As explained before, the area within the curves for the cycle represents the amount of work performed on the air in the compression process. Therefore, the effect of cooling between stages and the consequent shift of volume of the air at a cooler temperature represents a saving of work and power in compressing the gas. If there is no cooling between stages, there is no power saving, but there may be improved ability to get to the higher maximum pressure by multistaging because each cylinder does not have to work on such a large differential of pressures. An intercooler between stages requires continuous circulating cool water in the order of more than one gallon per minute per 100 cfm to remove the heat from the air. The effect may be to reduce the total power required by 10–15% to compress air from the atmospheric, P_a, to the maximum of the last stage, P_{m2}, shown in Fig. 4-3.

Aftercoolers are also used on the finally compressed air to cool it to usable temperature. This also serves to remove moisture, thereby reducing condensation and losses along the pipelines carrying the compressed air.

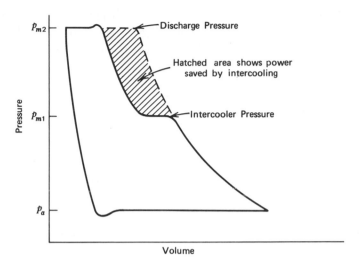

FIGURE 4-3 Effect of intercooling on pressure-volume curve.

In humid summer weather it is necessary to drain all points, such as the intercooler, aftercooler, and receiver, and any low points in the piping between the compressor inlet and nozzle tank where moisture might collect. In cool weather or when the relative humidity is below 30%, this is not a problem. If draining is not done, the condensation causes shrinkage of the volume of air and this must be compensated in calculations. These corrections are covered in Chapter 8 of CAGI's Handbook.[6] In any case it must be recognized that the displacement of multistage compressors is that of the low-pressure cylinders only, since higher ones work on the same air and no new air is introduced between stages.

The reciprocating compressor models used in construction are generally portable air compressors. A portable air compressor is a self-contained, compressed-air power plant, including engine, compressor, air receiver, and also mechanisms for starting, cooling, lubricating, and self-regulating the power plant. The engines for portable compressors up to about 300 cfm capacity are either gasoline- or diesel-powered, whereas the larger ones will generally be diesel-powered. These portable units are water- or air-cooled and, mostly, single- or two-stage compressors. They have a fairly standard discharge pressure of from 90 to 125 psig to match the requirements of the majority of portable air tools. There used to be standard model sizes for portable compressors as shown in Table 4-2.

TABLE 4-2 Portable Construction-Type Compressor Models (capacity in cubic feet per minute (cfm) at 100 psig and sea-level, standard conditions of 60°F)

Model 60	60 cfm capacity
Model 105	61–105 cfm
Model 160	106–160 cfm
Model 210	161–210 cfm
Model 315	211–315 cfm
Model 450	316–450 cfm
Model 600	451–600 cfm

Even though these standards are no longer followed by the manufacturers, they do follow the practice of having the model number for all types of portable compressors give the air capacity in cfm at 100 psig.

Rotary Compressors. The rotary compressor is a positive-displacement type with similarities to the reciprocating compressor. The main difference is that the rotary compressor functions with a rotating impeller

FIGURE 4-4 Portable air compressor of rotary screw type to power a track-mounted rock drill (courtesy of Joy Manufacturing Company).

driving the air through a confining curved chamber to compress it to higher pressure. It will operate the same variety of accessories and, consequently, has essentially the same construction uses as the reciprocating compressor. However, the rotary compressor operates at greater rpm and requires more horsepower for a given cfm delivered (Fig. 4-4).

The rotary compressor generates compressed air by several general types of mechanisms. The first developed to improve on a reciprocating compressor was the sliding-vane compression mechanism. It will compress

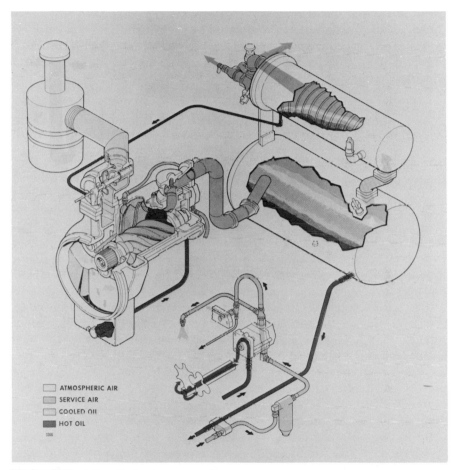

ATMOSPHERIC AIR
SERVICE AIR
COOLED OIL
HOT OIL

FIGURE 4-5 Schematic of screw-type compressor system (courtesy of WABCO Pneumatic Equipment Division of American-Standard Company).

air to around 50 psig in the first stage and 125 psig in two stages, which was not new. But this type of rotary compressor is equipped with engines developing up to 700 hp which enable the generation of much higher volumes than portable reciprocating compressors. However, the sliding-vane efficiencies are lower.

The most recent widely accepted development for rotary-type portable compressors in the United States is with the rotary screw mechanism.[2] The heart of this mechanism consists of two helical, intermeshing rotors which revolve in opposite directions in a compression chamber. They are shown in the left center of the isometric cutaway sketch in Fig. 4-5. Oil is introduced in the compressing chamber to lubricate the moving parts. It separates out of the air in the receiver tank as shown. The screw-type compressor is most effective for the larger portable units with above-300-cfm capacity.

4.1.3 Altitude and Temperature Effects on Compressors

Compressors work with air taken from the surrounding atmosphere for their first stage of compression. The previous discussions have been based on the atmospheric conditions of sea level, 36% relative humidity and 60°F, as standard. The sea-level atmosphere at standard conditions has a given volume of air under a pressure of 14.72 psi. However, the atmospheric pressure changes with weather changes as any weather watcher with a barometer well knows. There is a more predictable change with altitude, as shown in Fig. 4-6, and this should be recognized for compressors operating above several thousand feet.

The volume of air that a given compressor takes into its fixed cylinders does not change with altitude. However, the air at higher altitude is more rarefied, that is, less dense at the lower pressure shown by Fig. 4-6. The intake pressure of the fixed volume of air for the first stage of the compressor is less than expected in the compression cycle. For a given discharge pressure the higher altitude means a greater differential of pressures between discharge and intake from the surrounding atmosphere. There is, then, an increase in the compression ratio imposed on the machine.

In dealing with compressors for construction at higher elevation, one also recognizes the use to which the compressed-air power is put. The compressed air drives tools working against the surrounding rarefied atmosphere. If the tool requires 100 psig to operate effectively at sea level, its air has 114.72 psia, and at 8000-ft altitude it has 110.85 psia. In either case the air power driving the tool has a differential of 100 psi to produce

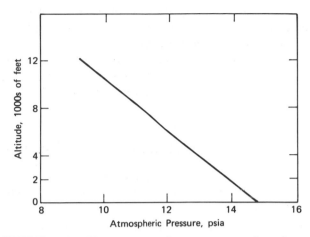

FIGURE 4-6 Pressure changes with altitude.

the tool's work. So there is no real loss in pressure effectiveness of the compressed air generated at higher altitude. This is particularly true of a multistage compressor.

However, there is a real adverse effect on an internal combustion engine that provides the power for the compressor. Since the engine takes in surrounding air for its compression stroke of the I.C. cycle, the combustion will be less effective at the higher altitude. The power effectiveness of the I.C. engine will be approximately in direct proportion to the atmospheric pressures at the different altitudes. Thus, the effect on a compressor's prime mover at an altitude of 10,000 feet will be to reduce its power from the standard engine by a

$$\frac{14.72 - 10.06}{14.72} \times 100 = 32\% \text{ reduction.}$$

This loss will be reflected directly in the volume of compressed air delivered by the compressor. Such an I.C. power loss can be minimized by the installation of high-altitude carburetor jets, turbo chargers, and other modifications to the engine.

The effect of temperature variations on the performance of a compressor is also important. As Charles' Law leading to Eq. (4-2) tells us, the compression of a given volume is inversely proportional to the absolute temperature of the air. The higher air temperature means that it has expanded and is less dense per unit of volume. So a fixed volume taken into the compressor at higher temperature will need to be compressed more to reach a specific discharge density and pressure. And conversely, a lower temperature of the intake air means that less compressive effort is needed. The standard temperature for comparison sake is 60°F (520° Absolute) as indicated in previous discussion. From this one can calculate that the effect of a 10°F variation from standard temperature results in a mere 2% change of compressor output; and the practical extreme of 60°F variation causes about 12% change.

4.1.4 Compressor Mountings and Receivers

To be most effective in the various possible construction uses, portable compressors are equipped with a variety of mountings. Where a high degree of mobility is advantageous, they will be mounted on wheels to be movable to different locations on a construction project or to different projects. For the construction use that requires stability and long-time stationary operation, such as on a barge, the mounting may be skids or a sled type.

A four-wheel mounting for portable compressors is the most frequently found in use. All sizes of portable compressors are available with this type of mounting, which may be on steel wheels for hauling on a trailer or on pneumatic wheels for towing behind a vehicle. The mounting may be with or without springs. The towing of a compressor without springs should be limited to 10 to 15 miles per hour, whereas one with pneumatic wheels and springs could be towed at 20 to 30 miles per hour.

Portable compressors of the smaller sizes may be mounted on two pneumatic tires with springs. These units can be easily towed on the highway with even a small pickup truck moving at 25 to 35 miles per hour. The two-wheel mounted compressor is best suited for a variety of small uses involving quite a bit of tow time in comparison to the total use time of the compressor.

Where there is much movement between compressor uses and essentially daily need for compressed-air power such as in utility work, the compressor usually is mounted permanently on a truck. This type of mounting, of course, limits the places of use to those accessible by the truck. Its speed of movement between uses is set by the truck's cruising speed.

The skid-mounted type of portable compressor is for uses that require little movement between compressed-air generation. This may be for construction operations where the compressor is used on a barge or on a bridge structure. The small amount of moving that is necessary for such a compressor can be made by cables operating as a winch to slide it from one spot to the next. With such little movement, generally of larger compressors, the prime mover can very conveniently be an electric motor instead of an internal combustion engine.

Most portable compressors are equipped with their own air receiver. The purpose of the air receiver is to reduce the effect of air pulsations and maintain the desired high pressure of compressed air delivered in the lines feeding the various tools powered by the compressor. An air receiver allows for a sudden draw down of air such as all tools starting up at once. It permits also the use of a large volume of air for a short period of time such as would occur in the jetting out of an open-end pile with an air water jet, a process that requires a fantastically large amount of air for a matter of minutes and then not again for perhaps hours.

The size of such a receiver generally would be from 1/6 to 1/10 of the cfm capacity for the portable compressor. Such a receiver is built to the standards of the ASME pressure vessel code. This calls for provisions to keep the receiver free of condensate by draining and the provision of a safety valve for air pop-off. The air safety valve should be checked regularly for effectiveness.

4.1.5 Air Tools and Devices Powered by Compressors

In order to select the appropriate size of compressor for construction use, it is necessary to have an understanding of the variety of tools and other construction devices that would be powered by the compressed air. The term air "tools" refers to all the hand-moved and controlled devices of great variety, the most important of which are described in the following sections. The compressed air power for these tools provides some important advantages:

1. simplicity of the mechanism which allows lightweight tools;
2. instantaneous load variations and reversibility;
3. tools that do not overheat where they are handled;
4. opportunity to use the tool in water or wet weather without danger of electrical shock; and
5. a mechanism which needs a minimum of attention and maintenance.

Demolition or Digging Tools. One type of air-power tools is the variety used for demolition or digging. These tools remove materials, such as asphalt, clay and other earth material, concrete, masonry, metal, or wood, by impact through a steel point or wedge-shaped tool end. The variety of these range from the small chipping hammer, weighing 2 to 12 pounds and handled like a rifle, to the large pavement breaker. The latter, in sizes from 35 pounds to the 80-pound class, can generally be handled by a single man. The demolition and digging action is derived directly from the reciprocating piston driven back and forth by the compressed air. Some factual information about the compressed air requirements for these tools is given in Table 4-3 with information provided by the Compressed Air and Gas Institute.[3]

TABLE 4-3 Air Requirements for Demolition and Digging Tools

Tool	Class or Size	Avg. cfm Reg'd.
Chipping hammer	Lightweight	15–25
Chipping hammer	Heavyweight	25–30
Clay digger or chisel	20-pound	20–25
Clay digger or chisel	25-pound	25–30
Clay digger or chisel	35-pound	30–35
Light paving breaker	35-pound	30–35
Medium paving breaker	60-pound	40–45
Heavy paving breaker	80-pound	50 cfm

Drilling Tools. The variety of drilling tools are for boring holes into wood, or other such material, and for drilling holes into rock and other crushable material. The drilling tools derive their applied action by the reciprocating piston going through a slight twist with each back and forth stroke. The boring tool uses a screw-fed working part driven by a vane-type rotary air-powered mechanism or a reciprocating piston motor. These tools vary in size from a small, one-hand-held drill weighing $1\frac{1}{2}$ pounds to

Table 4-4 Boring Tool Air Requirements (average air consumption of rotary air reversible wood borers according to the Compressed Air and Gas Institute[3])

Drill Diameter	Rotating Speed, rpm	Weight of Tools, lbs	Air Consumption cfm
1″	700–1000	14–16	35–40
2″	450–800	25–30	50–75
4″	250–350	25–30	50–75

the large, two-man tools weighing 175 pounds. The latter are operated by a six-horsepower unit at 20,000 rpm for boring in heavy timbers.

The largest single use of compressed air in construction is for drilling blast holes in rock material. The rock drills, in some cases called "drifters," use the reciprocating piston, driven by the compressed air within the cylinder causing rapid-fire blows. A rifle bar in the cylinder causes a partial rotation of the drill steel on every upward return stroke, giving the rotary action. Rock drills are hand held, such as sinkers and jackhammers, since they are light enough for one man to carry. Others, such as the drifter drills are mounted on tripods, crossbars, or wheels (Fig. 4-7). A single drifter is held on a guide shell or cradle which can feed its drill two to four feet at a time before changing the drill steel length. In a tunnel with large cross section, many drifters are mounted on a jumbo frame which matches the tunnel's cross section closely. Then a number of the drills can be operated at one time and all can be moved away from the tunnel face at once, with the mobility of the jumbo frame, to permit blasting of the drilled face. A drifter drill mounted on wheels is known as a wagon drill. This type of rock drill has great mobility, and its standard feed of six feet permits considerable drifting before a change of the drill steel length is necessary.

FIGURE 4-7 Self-propelled rock drills wheel-mounted for portability (courtesy of Ingersoll-Rand).

Specific information on the required air for these drills as provided by the Compressed Air and Gas Institute[3] is given in Table 4-5.
More about the advanced developments in drilling equipment is discussed in Sec. 4.1.7.

Riveters and Power Wrenches. The compressed-air-powered riveters and power wrenches are impact tools like the demolition and drill tools. A riveter is for fastening metal parts with driven rivets, so this tool has a reciprocating action like that of the demolition and digging tool. On the other hand, a power wrench is for driving lag screws and structural-steel or machine bolts. This tool, then, has a rapid series of rotary or torsional loads like the rock drills. These tools with rotary action are designed so

TABLE 4-5 Air Consumption for Rock Drills
(for operation at 90 psig and sea-level, standard conditions)

Size of Tool	Depth of Hole, ft	Air Consumption, cfm at 90 psig	Principal Uses
10-pound	up to 2	15–25	Drilling shallow
15-pound	up to 2	20–35	holes in concrete, brick, stone, etc.
25-pound	2 to 8	30–50	Anchor-bolt holes,
35-pound	8 to 12	55–75	boulders and dimension stone
45-pound	12 to 16	80–100	General rock excavation
55-pound	16 to 24	90–110	
75-pound	8 to 24	150–175	General rock excavation
Drifters:			
3″ diam. piston	(more than hand held)	150–175	Extensive blast hole drilling;
3½″ diam. piston	(more than hand held)	180–210	larger for harder rock,
4″ diam. piston	(more than hand held)	225–275	

TABLE 4-6 Air Requirements for Riveters and Power Wrenches (for operation at 90 psig and sea-level, standard conditions)

Diameter of Rivet or Bolt, inches		Net Wt. of Tool, lbs	Air Consumption, cfm
Riveter:	5/8″	15	25–30
	3/4″	18	30–35
	7/8″	20	35–40
	1 1/8″	22	40–45
	1 1/4″	25	40–45
Wrenches:	5/8″	8–12	15–20
	3/4″	15–20	30–40
	1 1/4″	25–30	60–70
	1 1/2″	35–40	70–80
	1 3/4″	60–65	80–90

that there is essentially no twisting action on the operator at his hand controls. The air requirements for the various sizes of riveters and wrenches are given in Table 4-6 with data provided by the Compressed Air and Gas Institute.[3]

Other Air Tools and Devices. Compressed air is used on construction to operate other tools and devices for doing work or moving material. The new Darda rock and concrete splitter in the place of a jackhammer or paving breaker makes use of up to 7000 psi hydraulic pressure for wedging action. The splitting force is said to be over 400 tons. Less than normal time and, so, labor can cut costs considerably. The Darda tool can be powered by air (50–80 cfm at 100 psi), gasoline, or electric motor.

Other compressed-air tools convert the air power into reciprocating or impact action delivered to the material being worked or moved. As in the previous sections, the air requirements of these tools and devices are provided in Table 4-7 for the sake of determining the needs for selecting the appropriate size of air compressors.

Air Conveying Equipment. Variety also characterizes equipment used in construction to convey material. This compressed-air-powered equipment generally moves the material based on the atomizing principle of putting material into suspension in the air and moving it with high velocity. The material moved with such equipment includes water and other fluids, as

TABLE 4-7 Air Requirements for Other Air Tools and Devices. (Provided by the Compressed Air and Gas Institute[3])

Air-Driven Device	Size-Weight	Air Consumption, cfm at 90 psig
Circular saw	12″ diam. blade (4″ max cut)	40–60
Chain saw	18″–30″ blade, 40–50 lbs	85–95
	36″ blade, 50–55 lbs	135–150
	48″ blade, 50–60 lbs	150–160
Reciprocating saw	20″ blade, 15 lbs	45–50
Concrete vibrator	2½″ tube diameter	20–30
	3″ tube diameter	40–50
	4″ tube diameter	45–55
	5″ tube diameter	75–85
Material hoist	Single drum, 2000-lb pull	200–220
	Double & triple drum, 2400-lb pull	250–260
Boom-mounted pneumatic breaker	Weighs under 1000 lbs and delivers 1000 ft-lbs per blow	250

well as dry cement particles and cement mixes. In the case of fine particle material, it is easy to see how it can be blown into suspension just like a dust storm carries fine particle material. In the case of liquids, the high-pressure air moves with great velocity and causes the material to separate into droplets which are carried in suspension. The mixed materials, such as a cement paste, are injected with the various components of water and fine aggregate into the high-velocity air stream without being allowed to stick together until deposited at the end of the air pressure line. There it is shot into place, such as gunniting on a wall or surface to be covered uniformly. The air requirements for these pieces of conveying equipment are given in Table 4-8 for information to use in the selection of appropriate size air-compressor equipment.

TABLE 4-8 Air Requirements for Conveying Equipment (These air requirements are at 90 psig, unless otherwise noted, with all information provided by Compressed Air and Gas Institute[3])

Air-Driven Equipment	Air Consumption, cfm at 90 psig unless noted	Productivity (roughly), for estimating
Cement mortar spray	210–315	area sprayed 1″ thick: 1400–4300 sq. ft./ 8 hrs
Paint spraying using 40–70 psig	8–15 14–30	production spraying heavy material spraying
Pumps:		
Low head, single-stage	80–90	200 gpm up to 40′ total head
Higher head, two-stage	160–180	150 gpm at 100′ total head
High head, single-stage	150–170	125 gpm at 150′ total head

4.1.6 Selection of an Air Compressor

The various key points in the selection of the appropriate, economical air compressor for a given construction application were covered in outline form earlier in Sec. 4.1. The detailed information necessary for each individual tool that might be powered by such an air compressor has been given in the previous sections. The correct selection necessitates knowing the variety and number of tools and equipment to be served by the compressor in a total compressed-air system. The air compressor serving a system with a number of tools is generally not required to feed the maximum requirement to all of the tools at the same time. In those few instances when this might be the case, the overload capacity of the compressor will accommodate such an exceptional system, or the tools may have to realize a lesser supply momentarily. So one of the points in the selection of the appropriate, economical air compressor with an adequate air receiver is to take into account the diversity of the tools supplied by this single piece of equipment. That is done by using a "diversity factor" based on the number of tools to be supplied by the particular air compressor. The equation for this diversity factor is as follows:

Diversity factor, DF

$$= \frac{\text{Actual cfm used for all tools}}{\text{Sum of maximum individual tool requirements}} \quad (4\text{-}7)$$

TABLE 4-9 Diversity Factor Values

Number of drills	1	4	8	12	20	30
Diversity factor	1.0	0.85	0.75	0.68	0.59	0.53

(For other numbers of tools an interpolation between these values will be reasonable.)

Feeder Lines of Hose or Pipe. The feeder lines carrying compressed air in the total system from the compressor to the tools or other devices may be hose or pipe length and connectors. Lines of pipe will be useful for transmitting air from hundreds to several thousand feet without flexibility and movement while being used. On the other hand, hose is useful for short distances with great flexibility of movement in the operation of portable hand-controlled tools. A system of compressed-air feeder lines may be a combination of piping for some distance and hose to the tools for flexibility. To insure that the system does provide the required air at each use point of a tool or other device, it is important to have receivers or air storage tanks strategically located in the pipe system to minimize surges of air along the lengthy system of feeder lines.

Compressed air fed through pipe lines will have pressure drop due to frictional resistance to the air flow. It is desirable to hold this pressure drop to a maximum of 5%, say, 1 to 5 psi, for pressure of 80–125 psig from the compressor. To achieve this nominal pressure loss in the system, the Compressed Air and Gas Institute recommends certain pipe sizes according to the volume of air transmitted and length of feeder line of pipe.[3]

Hoses carrying compressed air cause an even greater pressure drop due to their frictional resistance to the air flow. This is because of the lining material and inevitable curvatures of the hose. The selection of appropriate sizes of hose to connect with a portable tool is based on a similar consideration to that for pipe. That is, it would be desirable to hold the

TABLE 4-10 Recommended Pipe Sizes for Compressor System

Volume of Air Transmitted, cfm at 80–125 psig	Length of Feeder Line, in feet			
	50–200	200–500	500–1000	1000–5000
	Nominal Pipe Diameter, in inches			
30–60	1	1	1¼	1½
60–100	1	1¼	1¼	2
100–200	1¼	1½	2	2½
200 500	2	2½	3	3½
500–1000	2½	3	3½	4
1000–2000	2½	4	4½	5
2000–4000	3½	5	6	8
4000–8000	6	8	8	10

pressure drop to a maximum of 5% from end to end of the hose. The Compressed Air and Gas Institute recommends certain sizes of hoses to be used in transmitting air from 80 to 125 psig.

Using a larger hose than the recommended diameter will mean less pressure loss per foot of hose but it will likely be an uneconomical selection. This is true unless the hose size is already available and a smaller size would cost more. On the other hand, selecting a hose which is a smaller size than that recommended may cause too much pressure drop.

TABLE 4-11 Hose Size Recommendations (for compressed air at 80–125 psig, according to the Compressed Air and Gas Institute[3])

Volume of Air Transmitted, cfm	Recommended Hose Sizes Nominal Diameter, in inches		
	Hose to 25' long	25'–50' long	50'–200' long
up to 15	5⁄16	3⁄8	½
15–30	3⁄8	½	½
30–60	½	¾	¾
60–100	¾	¾	1
100–200	1	1	1¼

In fact, it may not even have the air capacity at the required initial pressure because the conduit is too small. The same reasoning for the alternatives can be said for pipe sizes other than the ones recommended. Since the pressure loss is much greater per foot of hose than in a comparable-sized pipe, it is well to use no more hose length than necessary for flexibility of operation with portable tools.

Leakage of air through the connecting parts of a compressed-air transmission system is inevitable. This is due to less-than-perfect pipe connections, loose valve stems, leaky hose, and loose hose connections. For estimating purposes it should be satisfactory to assume a loss of 2 cfm/100 ft of equivalent pipe or hose length.

Examples of Compressed-Air Equipment Selection. To better understand the various points made in the previous discussion and data provided on air compressors for construction, an example of compressor selection will now be given. Before doing this, it would be well to review the outline given in Sec. 4.1 covering the various steps to consider in the selection of a compressor.

(1) Given situation: the construction operation to work the first 300′ into a 40′-diameter tunnel. Figure 4-8 shows a sketch of the setup to use one 25-pound clay digger with 90 psig at tool, and five 55-pound rock drills with 100 psig at the tools.

    Select: compressor to meet the extreme setup—i.e., full 300′ of feeder pipe line in use. For 55-lb drills each with 30′ of 3/4″ hose (5 × 30′ = 150′); 5 × 100 = 500 cfm at initial 120 psig. For 25-lb digger with 50′ of 1/2″ hose; 1 × 30 = 30 cfm at initial 100 psig.

    Use: maximum of 275 ft of 2 1/2″ pipe with approximately 3 psi drop: Compressor capacity required based on 6 tools (DF =

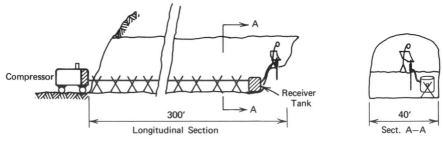

FIGURE 4-8 Construction setup for a compressor application.

0.77) and, say, 500 ft of operating line for leakage, so 5×2 = 10 cfm loss; $(500 + 30 + 10) \times 0.77 = 415$ cfm.

Referring to Table 4-2, could select:

Alternate #1: one Model 450 compressor with equipment—cost $10/hr estimated

Alternate #2: two Model 210 compressors—2 at $7.50 = $15/hr estimated

The selection may depend on operator requirements —whether one for each compressor or one for two, where an operator will add, say, $6 to $8 per hr. If only one operator is needed, there is the additional advantage of carrying on when one compressor is shut down.

(2) Given situation: a standard Model 315 compressor being used at 5000 ft altitude to erect a steel bridge with high-strength bolts tightened by 1 1/2″ impact wrenches. Setup has compressor located for a maximum of 600 feet of 3″ pipe to feed wrenches with 50 feet of 3/4″ hose for each.

Select: The number of wrenches that can be operated satisfactorily by the given compressor. At 5000′ altitude, Model 315 delivers maximum

$$315 \times \frac{12.18}{14.72} = 260 \text{ cfm.}$$

Each 1 1/2″ wrench requires, say, 75 cfm with a differential 90 psig pressure so can use a maximum of 3 to 4 tools. Try 4 tools with D.F. = 0.85:

$$0.85 \times 4 \times 75 = 255 \text{ cfm required.}$$

Note: if the operating compressor costs, say, $10/hr, the distribution of that to four tools will be $2.50/hr per tool (about 33% of the total cost); whereas for three tools it is $3.33 each. Also, the bolt tightening will take less time with four wrenches operating.

4.1.7 Compressor Costs and Trends

The portable, wheel-mounted reciprocating or rotary types of smaller air compressors used for construction are either gasoline- or diesel-powered. A diesel compressor will generally cost 7 to 12% more than the same size gasoline-powered compressor. For extensive, continuous compressor use, the lower cost of diesel fuel should eliminate this difference in the total equipment expense. Any form of modern compressor may have to conform with local anti-noise ordinances and be equipped with a suppressor or silencer.

Manufacturers make several of the smaller compressors on a self-propelled, wheel-mounted model. This is in sizes 100 to 250 cfm for small demands of short duration with need for easy moving between uses. Equipment of that ability will cost 20 to 30% more than a regular portable compressor of the same capacity.

With the escalating labor costs, there has been a continuing effort to develop bigger equipment that can do more work without increasing the labor involvement. This has encouraged the production and use of portable compressors that can deliver 900 and 1600 cfm. These air-power generators are used advantageously to drill bigger diameter holes of 8″ to 12″ in rock for blasting. In many situations the larger holes and greater amount of explosives lead to more breakage and less labor demand per cubic yard of excavation. The proportionalities, of course, depend on the rock and geological conditions. These larger compressors of the 900-cfm and 1600-cfm sizes are likely to be higher in cost per unit of air produced compared to smaller capacity units. However, when the labor to operate them is added to find the total equipment expenses per cfm, it will be less than for smaller units.

An interesting cost comparison for compressors is with the equipment or tools they power. A smaller compressor costs about the same as the total of the three or four lightweight tools that can be operated from it. For example, a Model 105 compressor costs as much as 3 or 4 clay diggers, single tampers, or lightweight paving breakers. When one selects a larger compressor, it will probably cost 25 to 50% more than the total cost of the heavy, hand-held tools that it can power.

The special heavy-duty drilling rigs now very popular and available from leading manufacturers of drilling and compressor equipment are self-contained rock-drilling units designed to spot and hold accurate holes on rough ground. The drill will be guided on a mast which is an integral part of the drilling rig as shown in Fig. 4-4. It is a variation of the older self-propelled wagon drill mounted on crawler tracks. All movements and adjustments of the drilling position are air- or hydraulic-powered.

Any position of the 6′- to 14′-long boom-mast, such as shown for the drills in Fig. 4-4 and 4-7, for different drilling angles is accomplished by one or more hydraulic pistons. The boom-mast permits horizontal drilling from about 5′ to 15′ above the ground. The crawler-track tension for a drill mounted like the one in Fig. 4-4 is easily adjusted by air; and the power for movement, including steering, is by a high-torque radial air motor with hydraulic track oscillation. The unit can pull its own compressor, which costs about the same as the mounted drilling equipment. Also, the operator can have separate controls for hammer blow, rotation speed, feed pressure, and blowing air. These self-propelled, wagon-drill-type units cost two to three times the comparable size drifter drilling unit, but should save considerably on time and labor.

Another type of special drilling rig is truck-mounted with the drill mast on the truck's tail end. Such a unit cannot get to the inaccessible spots that the crawler-mounted one can. But it has the advantage in deep vertical drilling of using 20' to 25' drill-rod lengths to a total depth of 100 feet or more, and a maximum inclination of 35° from vertical. Feed pressures can be from a gentle few ounces to more than 30,000 lbs with the drilling speed varying from 0 to 100 rpm. This allows the desired selection by the operator of the best speed and feed for the given rock to be drilled.

These and other special pieces of equipment, added to the more standardized compressed-air-power units, show the variety and versatility of this form of equipment for construction.

4.2 Water Pumps for Construction

Most construction projects require the use of a water pump at some stage of the construction. In fact, many projects have various requirements for water pumps, and a variety of pumps may be most desirable. Some pumps are used to provide clean water or water under high pressure for some construction operations. On the other hand, some may be used to remove water of various degrees of impurity which hinder the construction process or working space.

The water pumps used in construction are for a great variety of operations such as:

1. dewatering cofferdams;
2. removing seepage water from an excavated foundation, trench, or hole;
3. dewatering underground utility lines or tunnels;
4. lowering the water table around an excavation;
5. providing water for an aggregate or concrete plant or for an operating paver;
6. providing water for jetting or sluicing operations; and
7. providing water for foundation grouting.

To select a water pump suitable for a particular construction situation *necessitates knowing or estimating:*

    1. *amount of water,* in quantity per unit of time, that must be handled to fulfill the operation's requirement;

    2. *fluctuations in the amount of water* needing to be pumped—if it diminishes to nearly nothing at times, that calls for a self-priming type of pump;

    3. *foreign material in water* to be pumped—if water is muddy, sandy, full of sludge or trash, certain types of pump will be needed;

    4. *total static head* to work against between intake surface and discharge height;

    5. *height of pump above intake* water surface governing suction lift for feasible operation in the given construction situation;

    6. *velocity of discharge water* needed, if a requirement calls for discharge head for jetting or sluicing;

    7. *altitude of pump* at point of its given use;

    8. *size and lengths of hose and/or pipe* needed for transmission of water to be pumped; and

    9. *fittings and valves needed* for transmission system according to construction operation being planned.

4.2.1 Heads and Heights for Pumping

A water pump has certain limitations on its location of operation relative to the fluid it will handle. The height of the pump above the water surface is limited. Theoretically, with a perfect vacuum this height is 33.9 feet at sea level and standard conditions of 60°F temperature and atmospheric pressure of 14.7 pounds per square inch. Even the best of equipment in perfect operating condition cannot be expected to create a perfect vacuum. The practical limit to which a pump can pull up water at sea level is 25 feet.

It must also be recognized that water flowing through a pipe or hose meets certain resistance, primarily due to friction. This resistance to movement in the water line can be converted to potential force or "head" needed to cause the water to flow against it. The friction head is the equivalent height of the inlet end above the outlet to just cause the required flow. Also, if the fluid is to have some forcefulness at the outlet instead of leaving with just a trickle, it must have some pressure in it when leaving. This is known as a pressure head, which can also be thought of as an equivalent height of water level above the outlet.

A pump works against several "heads" in handling water involved in a construction operation. These are known by such terms as:

1. static suction lift;
2. friction head;
3. intake suction head;
4. velocity or pressure head at the outlet;
5. discharge head; and
6. total dynamic head.

The static suction lift, h_1, is the actual vertical height of the pump above the surface of water it is pumping. Friction head is the equivalent height advantage of water needed from the inlet end of a section of flow line to its outlet end to overcome the frictional resistance to flow. To determine the total resistance to flow in a pumping system of hose and pipe lengths with connectors and other fittings, such as an inlet strainer, it is necessary to find the equivalent length of line. The conversion of the usual connectors and fittings in an ordinary water pumping operation on construction will probably add the equivalent of 10 to 20% to the actual water line used. The total friction head, h_f, for the system is based on the sum of frictional resistances in all the equivalent lengths of water line. Intake suction head is the combination of static lift and friction head for the water flowing from the inlet of the pumping system to the pump.

A velocity head is the equivalent height of water advantage that would be necessary to give the flow from the outlet a certain velocity. This can be determined by Newton's law applied to the droplets of water considered to be solid particles. Thus, for a desired velocity, the velocity head is

$$h_v = \frac{v^2}{2g}, \text{ in feet} \tag{4-8}$$

where $g = 32.2$ feet/sec/sec at sea level.

Sometimes it is more meaningful to speak of the outlet flow in terms of pressure. This can be done by using the expression $p = wh$. Then the pressure head at the discharge outlet is found for a desired pressure as

$$h_p = p/w, \text{ in feet,} \tag{4-9}$$

where p is in pounds per square foot, and w is weight of fluid in lbs/cu ft. For fresh, clean water with $w = 62.4$ lbs/cu ft, Eq. (4-9) becomes $h = 2.31p$ with pressure in pounds per square inch (psi). The discharge head is the combination of (1) pressure or velocity head at the discharge outlet, (2) the friction head loss or resistance between the pump and the pumping system outlet, and (3) the difference of elevation between the pump and the outlet (add height, h_o, to other two if outlet is higher; otherwise, subtract it).

The total dynamic head is the sum of all heights and heads that a pump has to work with in its given operation. This can be formulated as

$$h_t = h_l + h_f + (h_v \text{ or } h_p) \pm h_o, \tag{4-10}$$

where h_t = total dynamic head;

h_l = static suction lift;

h_f = total friction head;

h_v = velocity head at the outlet;

h_p = pressure head at the outlet; and

h_o = elevation difference between pump and outlet ("+" if outlet higher and "−" if outlet lower).

In the preceding section the limit for the static suction lift was given for sea level and standard atmospheric conditions. If the temperature is above the standard 60°F, a water pump is not able to gain as effective a vacuum for its suction lift. The effectiveness is reduced directly as the temperature increases up to the practical construction extreme of 120°F. This reduction is due to the thinner, lighter air of higher temperatures. At 120°F the practical suction lift limit is about 10 feet at sea level. The effect of increased altitude is also to reduce the suction lift limit. This is because of the thinner atmosphere to support a column of water in an evacuated pipe or hose at higher altitude. The combined effect of increased temperature and altitude on the practical suction lift limit of a water pump for construction is shown in Fig. 4-9.

4.2.2 Types of Water Pumps

The consideration of pumps must take into account that water is an incompressible fluid so that its movement is the direct effect of a piston or impeller acting behind the body of water to be moved. The types of water pumps used generally for construction can be categorized as (a) the positive-displacement type, which operates with a piston like that for steam engines and similar displacement compressors, and (b) the dynamic type, similar to comparable dynamic compressors. The positive-displacement pumps that have been used are known as (1) reciprocating pumps and (2) diaphragm pumps. The common dynamic-action pumps used for construction are known as centrifugal pumps. Another type of pump used in certain conditions is the air or water lift pump; the principle being the use of high-pressure air or water issuing from a Venturi at the base of an

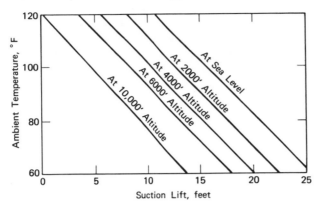

FIGURE 4-9 Effect of temperature and altitude on the suction lift of a water pump.

open pipe set into the water. The high-velocity jet starts a movement of water throughout the pipe. It is used a lot in pile work.

The commonly used pumps will be discussed next in more detail for an understanding of their operations, capabilities, and performances. With such understanding it is more likely that the best possible water pump will be selected for a given construction situation.

Reciprocating Pumps. A reciprocating pump operates basically with the water moved by a piston in a rigid cylinder with valves for intake and discharge. The piston may move the water only when it is moving in one direction and is then said to be single-acting. However, by appropriate additional valves acting, the pump may take in more water behind the piston as it discharges the previous volume of water. Then the piston will discharge this new water when it moves in the other direction. This case is known as a double-acting pump. Such a pump handles essentially twice as much water per full back-and-forth stroke of the piston as a single acting pump cylinder.

A reciprocating pump with just one cylinder is said to be a "simplex pump," whereas one with two cylinders is a "duplex," and one with three cylinders is a "triplex" pump. The quantity of water handled by a reciprocating pump is almost equal to the volume swept in the total piston displacement discharging water. A slight difference is due to water that passes in the reverse direction as the valves open or close. This loss is known as "slip" and usually amounts to 3 to 5% of the displaced volume.

A straightforward equation is available to compute the maximum quantity of water that a reciprocating pump can deliver. This is based on the total cylinder displacement, taking into account the slip and points of the pump design mentioned previously. It recognizes that, unlike the compressor's volumetric capacity, a pump is working with an incompressible fluid. Thus, the water pump's maximum capacity is found by

$$g_w = N(c)\frac{\pi d^2 lan}{924}, \text{ in gpm} \qquad (4\text{-}11)$$

where d = diameter of a cylinder, inches;
   l = length of piston stroke, inches;
   a = 1 for a single-acting cylinder or 2 for a double-acting cylinder;
   n = number of full back-and-forth strokes of a piston per minute;
   N = number of cylinders;
   c = slip factor, estimated as 0.95 to 0.97;
   π = 3.14; and the 924 includes the fact that one U.S. gallon of water equals 231 cubic inches.

To pump the water at this rate requires an engine or prime mover of certain power output. That power requirement can be calculated for a given head to pump against, by

$$P = g_w wh, \qquad (4\text{-}12)$$

where g_w = the pump's capacity, in gpm;
   w = weight in pounds of a gallon of water, 8.33 lbs for clear water;
   h = total head at pump outlet, in feet; and
   P = pump's power output, ft-lbs/minute.

An example of applying these formulas will be given next. Consider the case of pumping clear water with a duplex ($N = 2$), double-acting ($a = 2$), reciprocating pump with cylinder's bore, $d = 3''$, and length, $l = 6''$; if the pump engine shaft operates at 100 rpm, $n = 2 \times 100 = 200$ strokes/min; assuming 5% slip, so $c = 0.95$, by using Eq. (4-11),

$$g_w = 2(0.95) \times \frac{\pi(3)^2 \times 6 \times 2 \times 200}{924} = 139.5 \text{ gallons/minute.}$$

the pump's capacity. If this is being pumped with a total dynamic head of 44 feet, the power required will be found by using Eq. (4-12):
$P_o = 139.5 \ (8.33) \ 44 = 51,000$ ft-lbs/min. This is the required power output from the pump's engine. It could be expressed in output horsepower as $51,100 \div 33,000 = 1.55$ hp. A 2-hp engine should be able to handle this situation.

Construction in the past used the sort of reciprocating pump described above extensively. It was known as a road pump, probably because of its use in supplying fresh water for aggregate and concrete plants on highway construction. The road pump delivered water under high pressure at the pump to move it considerable distance, primarily horizontal, to the point of use. The layout of pipe for thousands of feet was quite a labor requirement. With the escalating labor costs of recent decades, a method of delivering water by truck instead of a long pipe line proved more economical. Also, the operation and maintenance of the road pump with rigid cylinder walls handling water that was not always clear presented a problem for efficient construction. For these reasons this type of reciprocating pump was replaced by the diaphragm type for construction operations.

Diaphragm Pumps. A diaphragm pump is a positive-displacement type of pump. Unlike the reciprocating-piston type, the diaphragm pump has flexibility in its displacement chamber so that is can handle solids in the water. The flexibility is in its main operating part, the circular disk or diaphragm. A flexible disk is alternately expanded and squeezed in a

FIGURE 4-10 A diaphragm pump with a plate disk diaphragm (courtesy of the Gorman-Rupp Company).

hydraulic pump by the rod connecting it to the rotating shaft of the pump's engine. A stiff plate diaphragm is moved back and forth by a connecting shaft in an air-powered pump (Fig. 4-10).

Moving the diaphragm forward creates a vacuum in the displacement chamber drawing liquid or air, or a combination, in through the intake valve and line. The alternate squeezing in the diaphragm's backward movement forces the fluid out of the chamber via the discharge valve and opening. The suction effect makes the diaphragm pump self-priming and allows it to operate with less than a full capacity of water. Therefore, it will operate effectively with a low or variable flow of water to be removed. Even though it is self-priming, the pump with its own internal combustion power unit can start pumping much quicker if it is filled initially with water. An air-operated diaphragm pump will need a compressor, generally of the 60-cfm size, for its power source.

The diaphragm pump is the positive-displacement type regulated in design and capacities by the Contractor's Pump Bureau of the Associated General Contractors of America. The standards and capacities established by the Bureau[4] will be explained subsequently. The Bureau suggests that the diaphragm pump is best suited for :

1. dewatering jobs where the liquid carries a high proportion of trash, mud, or sand;

2. handling a variable seepage of water, as in trench work; and

3. pumping the combination of water and air from a small well-point system.

The A.G.C.-rated diaphragm pumps are made with 2-inch up to 4-inch inlets and outlets as standard. In a situation of low flow the inlet line might be reduced to 2″ hose to decrease the initial priming and repriming time. These pumps have a maximum capacity of about 9,000 gallons per hour (150 gpm). They are single- or double-acting pumps with the latter for larger capacity in the 4″ size. The largest of these A.G.C.-rated pumps is powered by a nine-horsepower engine. The capacities of these pumps are in the order of a thousand (M) gallons per hour (approximately 17 gpm) for each horsepower of the engine with a suction lift of 10 feet. This is a much lower ratio of pumping capacity per engine horsepower than was found in the road pump example previously. A major reason for this difference in the positive-displacement pumps is that the diaphragm pump has looseness and larger openings to handle solids in its fluid and needs the extra power to insure that the mixture is moved.

The positive-displacement pumps, such as a diaphragm type, have several key advantages to their operation:

1. For a given operating speed and regardless of head, the rate of discharge is practically constant.

2. With the relatively low operating speeds and velocities of the fluids passing through it, this type of pump is well suited to handle viscous fluids.

3. This type of pump is particularly good for use in pumping small volumes against a high head.

There are also certain important disadvantages of the diaphragm pump compared to the next to be discussed:

1. It is generally a heavy and cumbersome pump.

2. This pump is rather inflexible in its pumping capacity.

3. The diaphragm pump has a smaller capacity compared to the same investment in a self-priming centrifugal pump to be discussed next.

Centrifugal Pumps. A centrifugal pump is one that has a rotating part which moves the fluid by its centrifugal force. This rotating part is called the impeller, which comes in contact with enough of the water passing through the constricted chamber to give all the water velocity for motion. Therefore, the water gains energy for motion as shown in its velocity imparted by the centrifugal force. This is kinetic energy, which is directly proportional to the mass of the fluid and its velocity squared ($KE = 1/2mv^2$). It can be equated to the potential energy of a particle or drop of water at rest in some position of height or "head" advantage. This is the same relationship expressed in Sec. 4.2.1, where the velocity head (h_v) was discussed. As can be noted by Newton's formula, $v^2 = 2gh$, the potential head would be four times as great if the velocity were doubled.

It is easy to accommodate variations in the head of water and vary the capacity for pumping water with a centrifugal pump. This is done by varying the speed or size of the impeller, which is governed by the pump's engine and design. The capacity will vary directly with the speed or the diameter of the impeller. The head that the pump can handle will vary as the square of the speed (v^2) or the square of the impeller's diameter. The power required to operate a centrifugal pump will vary with the cube of the speed (v^3) or the impeller's diameter cubed.

Another means for increasing the performance to develop a higher head or greater pressure is with the use of a multistage centrifugal pump. This is a pump with two or more impellers where the discharge from one impeller flows into the suction of another. Each stage increases the velocity or pressure head of the fluid. Thus, a multistage centrifugal pump can work against high head or produce a great pressure in the water at final discharge. Such a pump is particularly suitable for water jetting or sluicing work.

The standard centrifugal pump is not designed to handle air. As such it has to be operated where the intake to the suction pipe can always be covered by the fluid it is pumping. If it becomes uncovered, the pump will get air-bound and not operate until it is re-primed. Since most construction uses of pumps have the equipment operating above the water to be pumped, the necessity for re-priming could be a problem. A self-priming pump was developed to avoid this constant concern.

The self-priming centrifugal pump for construction purposes uses a recirculation method.[5] It is designed with a water reservoir contained within the pump and needing to be filled only at the start of its operation. As the impeller rotates, it draws water from the reservoir and air from the suction line. They do not mix, and the air is discharged and water recirculated to pick up more air from the suction line. This process continues until all air is out of the suction line and water is flowing through it. A

FIGURE 4-11 Self-priming centrifugal pump in action.

trap valve keeps the reservoir filled with water when the pump stops (Fig. 4-11).

Manufacturers of centrifugal pumps build a special model known as a trash pump. The "trash" is any foreign material in the liquid, such as stones, sticks, leaves, and other solids. To handle such material, the pump is designed with larger openings and chambers. This is to allow pumping of fluid containing solids up to 25% by volume as compared to a maximum 10% for the standard self-priming centrifugal pumps. Trash pumps are made in the same sizes as the more common standard pumps.

Submersible Pumps. The newest form of pump for construction use is the submersible pump. It is powered by a waterproof electric motor in the common housing with the pump to operate submerged in the water. A waterproof electric cable connects the motor to electrical power controls and lines on a dry surface above the water (Fig. 4-12).

The submersible pump has several important and obvious advantages over the other types discussed previously. It eliminates the suction lift limitations, loss of prime, need for suction hose, and the noise and fumes from an internal combustion engine. In spite of these differences, this pump handles water in practically the same way that a centrifugal pump does its pumping.

For construction jobs submersible pumps are ideally used where the suction lift is greated than 25 feet and a cheaper, standard centrifugal pump could not be considered. This would certainly be the case in a deep cofferdam where there is no platform or work level below the top. There are other construction situations with similar problems that suggest the use of a submersible pump.

4.2.3 Standards of the Contractors Pump Bureau

The Contractors Pump Bureau, originally in the Associated General Contractors of America, consists of a number of manufacturers who have established certain standards for pumps made for construction use. The standards are drawn up for self-priming centrifugal, diaphragm, submersible, and trash pumps. One important standard is that the pump (except submersibles) will handle suction lifts up to 25 feet at sea level, when equipped with one elbow and 30 feet of suction line of the same nominal inside diameter as the pump inlet.

A rated self-priming centrifugal pump is guaranteed to prime itself with the above suction conditions. It should be able to do this in less than two minutes with a 15-foot static suction lift. It is also guaranteed to pass solids amounting to 10% of the flowing cross section and spherical solids

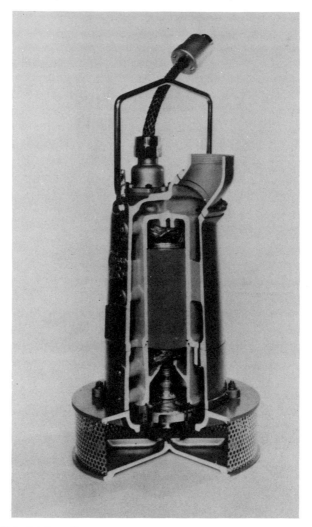

FIGURE 4-12 Cutaway view of a submersible pump (courtesy of the Gorman-Rupp Company).

equal to 25% of the nominal inlet diameter of the pump. The different sizes of these pumps are rated by the Contractors Pump Bureau (CPB) based on suction and discharge openings of identical dimensions for each size of pump. More details can be found in the Bureau's "Contractors Pump Manual."[4]

TABLE 4-12 Self-Priming Centrifugal Pumps
(rated by the Contractors Pump Bureau[4])

Pump Size and Rating (gpm)			Minimum Continuous Duty Power at Recommended Speed
1½″	5M	(83)	1.7 hp
2″	8M	(133)	3.3 hp
2″	10M	(167)	5.6 hp
3″	17M	(283)	6.6 hp
3″	20M	(333)	10.0 hp
4″	30M	(500)	17.0 hp
4″	40M	(667)	24.0 hp
6″	90M	(1,500)	35.0 hp
8″	125M	(2,080)	60.0 hp
10″	200M	(3,333)	64.0 hp

The CPB-rated size of pump represents its capacity in thousands (M) of gallons of water per hour under the stated conditions of operation. The standard pumps are given in the following tables.

Capacity tables for different heights of the pumps above the water surface and varying total dynamic heads are given in the "Contractors Pump Manual." The trash pumps are given in the standard-size models: 5MT(1 1/2″), 10MT(2″), 18MT(3″), 33MT or 35MT(4″), and 70MT(6″). These are very similar and have capacities comparable to the self-priming centrifugal pumps in Table 4-12. The power requirements for a trash pump are somewhat higher than for the more standard centrifugal pump.

The standard diaphragm pumps used to be rated for both a 10-foot and a 20-foot suction lift with the suction line five feet longer than the lift in

TABLE 4-13 Closed Diaphragm Pumps
(rated by the Contractors Pump Bureau)

Pump Size and Action	Rating at Suction		Engine, minimum
	10′ Lift	20′ lift	
2″ Single	2M	a new entry for CPB	
3″ Single	3M	(2M)	1½ hp
4″ Single	6M	(4M)	3 hp
4″ Double	9M	(7M)	4 hp

each case. Now these pumps can meet the standards by satisfying the 10-ft lift.

The electric-driven submersible pumps are now rated by the Contractors Pump Bureau. The Bureau has not set capacities yet, but they have standard line sizes: 1 1/2″, 2″, 3″, 4″, 6″, and 8″. These sizes provide capacities ranging from 50 to 300 gpm (3M to 180M), which can be compared with the self-priming centrifugal pumps in Table 4-12. For these pumps of comparable sizes and capacities there are major differences in maximum head for the horsepower provided, as shown on the curves of Fig. 4-13.

4.2.4 Pump Costs and Comparisons

All of the water pumps rated for construction use by the Contractors Pump Bureau have one design feature in common. Their power units to operate the pump are integral parts of the equipment. This is contrasted with the compressed-air-powered sump pump listed earlier in this chapter. A comparison between the cost of such a pump and its counterpart in the CPB-rated pumps will be mentioned.

The diaphragm pump is made in a limited number of sizes and has capacities only up to 150 gpm. It is designed for handling low, variable seepage flow with the chance of much air in its lines. All of this means the diaphragm pump is frequently used as a sump pump, but it operates a

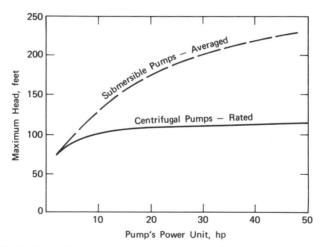

FIGURE 4-13 Comparing operating heads for centrifugal and submersible pumps.

suction line to keep the pump with its power unit out of the water. The only alternate to avoid the suction limitations used to be the compressed-air-powered sump pump. However, using a heavy duty one that can handle 125 gpm (see Table 4-8) with a Model 160 compressor would cost several hundred dollars a month for the equipment. A diaphragm pump of similar capacity with an internal combustion engine built-in will cost only one-third as much.

If there is a suction lift problem that prevents the selection of a diaphragm pump, there is now the choice of a submersible pump available. A submersible pump with the same capacity as a diaphragm type will cost about the same initially. Of course, the electric power must be available to tap, otherwise this form of pump would be expensive to use.

Submersible pumps are made in a great range of sizes for a variety of job conditions. In fact, this type is as versatile as the more common self-priming centrifugal pump. The submersibles generally rent for about two times as much as a centrifugal pump with equal capacity. Of course, a share of that differential may be due to a short supply of the newer, submersible pumps.

The self-priming centrifugal pump is the most versatile and economical of all the pumps discussed in this chapter. In the larger sizes they are provided with either gasoline or diesel engine power. The diesel-powered pumps are 50 to 100% more expensive than a gasoline-powered unit of the same size and capacity. For many water-pumping needs on construction operations, the pump will not run continuously. In that case the cheaper gasoline-powered pump will operate as efficiently as a diesel unit. The most economical pump for handling large quantities of fluid is the large 200M gasoline-powered unit.

If the pumping equipment is rented, the amount of hose becomes a significant expense. A 50-foot length of discharge hose and couplings will add 10 to 50% to the charge of a pump, and 25 feet of suction hose with couplings will add a similar expense. Such a relatively high cost is charged for renting hose because it must be replaced often with the bad wear and tear by many users. The owner-user of a pump and its accessories should have a lower relative cost for his hoses with care and attention to their maintenance.

In the final analysis for the selection of economical pumping equipment, the job conditions will hold the deciding factors. The site conditions of the amount of fluid, its location, consistency and rate of input, and where the fluid is to go, all are important facts. When these facts are accounted for, the best type and size of pump can be selected. Then the cost of getting that pump or a reasonable substitute on the job can be figured.

References

1. "Compressed Air Power," booklet prepared for Compressed Air and Gas Institute (Cleveland, Ohio, circa 1955).

2. "The Reasons Why," pamphlet by Pneumatic Equipment Division, WABCO (Sydney, Ohio, 1970).

3. "Compressed Air Power in Construction," booklet prepared for Compressed Air and Gas Institute, New York, N.Y. (circa 1960).

4. "Contractors Pump Manual," booklet by the Contractors Pump Bureau (Washington, D.C., 1970).

5. "Gorman-Rupp Contractors' Pump Manual," booklet by the Gorman-Rupp Company (Mansfield, Ohio, 1966).

6. *Compressed Air and Gas Handbook,* 3rd edition revised, Compressed Air and Gas Institute (New York, N.Y., 1966).

Chapter 5

■■■■■■■■■■■■■■

■■■■■■■■■■■■■■■■■■■■■■■■■■■■■■
EQUIPMENT
FOR EARTHWORK
■■■

Brief History of Earthworking

In ancient times the Babylonians worked along the Tigris and Euphrates Rivers, and the Egyptians performed earthmoving operations in digging irrigation channels from the Nile River. These works were followed in history by the Roman aqueducts and roadways, which necessitated much earthwork. This work was undoubtedly done by hand tools and animal-carried or -drawn devices. The American Indians in the early centuries of life on the Western Hemisphere used their hands and baskets for scoops and carriers.

The plow as the first real earthmoving device was introduced in 1819 by Metcalf and Telford on the British Isles. There followed a gradual evolution to the early Mormon buckboard scraper drawn by animals in the mid-nineteenth century. This led to the well-known horse-drawn Fresno used extensively until the turn of this century in the United States.

During the second half of the nineteenth century the steam shovel was being developed and put to extensive use. Steam power as a prime mover was also being used for tractors. Their use in the early development was primarily in farming. These early excavators and earthmovers powered by steam prevailed until the 1920's. Then Holt and Best developed their gasoline-powered tractor. Their efforts marked the founding of a tremendous earthmoving equipment enterprise which adopted the name of Caterpillar. One of its first developments of the tractor as an earthmover involved attaching a front-mounted "bulldozer" blade. In their unsophisticated beginning models the bulldozer was popular for moving earth a distance of up to 150 feet.

In 1922 a great innovator of earthmoving equipment, R. G. LeTourneau, developed the first machine-powered scraper. It scooped up loose material for carrying within itself. Other than that similarity, the difference between LeTourneau's first model and the present scraper was as much as between the original automobile and modern ones. The most significant generation of LeTourneau's scraper development was the self-propelled Tournapull introduced in 1938. Present scrapers built by many manufacturers are patterned after that innovation brought into the earthmoving scene more than three decades ago.

During the period of recent history in the United States following the great depression of the early 1930's, there has been a tremendous amount of road building and construction of dams and waterways. This construction has involved an amount of earthwork unprecedented in history and not likely to be duplicated in the foreseeable future. The need was for an unbelievable capacity of earthmoving equipment. This led to making units that had more and greater capacity and could move faster hauling their loads. This increasing demand, resulting in bigger units, was a fortunate part of the evolutionary process. It meant that the operators of the equipment were handling larger and larger amounts of earth material each trip, during a period of rapidly increasing wage rates. There was also great competition in the manufacture of the earthmoving equipment. So the total result, with these two factors paramount, was the biggest bargain of the period from 1930 to 1960. The price for moving a cubic yard of earth with a given consistency and condition has been about the same from the beginning to the end of those three decades.

A new emphasis has now been placed on construction safety in the United States. This is to include roll-over protection systems for earthworking equipment. The additions required for extra safety may add several thousand dollars to the original cost of the equipment, but it should be worth that for the owners' and operators' welfare. Major equipment to have such systems after 1972 are tractors, graders, loaders, and scrapers, which will be discussed in detail later in this chapter.

5.1 Fundamentals of Earthwork Operations

The proper planning of earthwork operations necessitates knowing the nature of the material and how to prepare it for handling and final placement. Also, the planner must be able to determine the amount of material that can be loaded in one cycle, i.e., the load capacity of an earthmoving piece of equipment. To plan an operation satisfactorily, one must be knowledgeable of the basic operations for earthwork and the components of an earthmoving cycle. The fundamentals of these important facets of earthwork operations will now be discussed in more detail.

5.1.1 Nature of Earthwork Material

The material that will be encountered in earthwork operations may vary from one extreme of sticky, spongy clay through loose or compacted granular material to solid rock. For the purpose of construction equipment planning, it is important to know how the material is to be handled and what equipment should be used for this purpose. This brings up the question of the state or condition of the material. Compared to its evolved natural condition in the earth surface, the material will "swell" during excavation and "shrink" during compaction in the earthwork operations. In its natural condition earth is measured as the in-place yards or "bank" yards. The term "bank measure" is applied to the natural state of the earth's material.

When the material is disturbed, as when rock is blasted or earth dug out of the ground by some equipment, it swells and is then referred to as "loose" material. This swelling of the ground may be partly an actual increase in volume due to the release of compressive stress due to many years of consolidation of the material. However, the majority, if not all, of the increased volume of the mass of material over its bank measure is due to the increase in the volume of the void spaces in the loose material. The increase of volume, or swell, is expressed as a percentage increase over that of the bank-measure volume. Some representative values of swell are given in Table 5-1. Not all soils, however, swell upon excavation. Volcanic ash or pumice and some alluvial soils will actually shrink in volume when removed from the natural condition.

Material that has been removed from its natural condition, then placed and compacted carefully, is called "compacted" yards. This may be said of any material placed in an embankment or fill for any purpose. Most earth so compacted will show a shrinkage, or decreased volume, from its bank measure. This is due to consolidation by equipment causing the elimination of most of the voids that existed in the loose and natural material. The one exception is broken rock, such as is used for rubble or a rockfill, which cannot be squeezed into a lesser volume than it had in its solid state. The shrinkage of earth material is expressed as a percentage decrease from the bank-measure volume. Some representative values of shrinkage are also given in Table 5-1. Not all soils will compact to less than their original volume. Certain dense sands and hard clays will have a larger volume after compaction than they had in the bank measure.

An example to show the application of this data is seen by following the changes of wet, clean sand from a borrow pit. The volume of material taken out of the borrow pit is bank measure with a conversion factor of 1.00 applied to that. The material in its disturbed loose condition will occupy a volume estimated as 1.16 times what it did in the borrow pit. And when it is compacted on a subgrade fill, it could occupy 0.86 as much space as it did in the borrow pit. Note that the weight of the total material in any of these conditions is the same as it was in the borrow pit, assuming no material has been lost in transit, which is seldom the case.

The data about the several conditions of earth materials is signicant for certain equipment-selection determinations. It will be useful in calculating the volume of material to be loaded and hauled from a given bank measure or borrow pit. Thus, the loose volume to load and haul will be

$$V_l = s_w \times V_b, \tag{5-1}$$

TABLE 5-1 Representative Values of Swell and Shrinkage for Earth Materials

(Add or subtract these percentage values in decimal form to or from 1.00 for the factor to find the loose or compacted volume from the given bank measurement.)

Material	% Swell	% Shrinkage
Dry, clean sand or gravel	+12 to +14	−12
Wet, clean sand or gravel	+12 to +16	−14
Loam and loamy sand	+15 to +20	−17
Common earth	+25	−20
Dense clay	+33 to +40	−25

where s_w = swell factor or $\left(1 + \dfrac{\% \text{ swell}}{100}\right)$, and

V_b = total bank-measure volume.

Or the operation requirement might, more likely, give the volume of fill or embankment to construct. In that case the volume of borrow pit needed to build the compacted fill is

$$V_b = \frac{V_c}{s_h},\qquad(5\text{-}2)$$

where s_h = shrinkage factor or $\left(1 - \dfrac{\% \text{ shrinkage}}{100}\right)$, and

V_c = total volume of compacted fill needed.

These equations may be combined to find the loose volume of bank or borrow material to load for constructing a given size of embankment. Thus, with the same definitions as above,

$$V_l = \frac{s_w V_c}{s_h}.\qquad(5\text{-}3)$$

5.1.2 Load Capacity of an Earthwork Container

The capacity or maximum volume of material that can be loaded into a hauling container depends on several factors. Of course, the inside dimensions of the container are of prime consideration. The inside volume up to the top edges of the container is known as the "struck" capacity. It is equivalent to the volume of water that the container could hold if it were watertight. Buckets or loading beds of construction equipment generally have an open top. Then the chance of heaping the load above the top edges of the container becomes a question for practical judgment. If the load is heaped, then what free-standing slope can the material assume regularly?

The calculations for volume of material heap loaded into a container can be found as

$$V_l = V_s + V_h,\qquad(5\text{-}4)$$

where V_l is the total loose volume, as before;

V_s is the struck volume; and

V_h is the portion of volume heaped above the container's top edges.

The components of a heaped volume are shown in the sketches of Fig. 5-1. The heaped portion is idealized by assuming a smooth inclined surface at some convenient angle, α', with the horizontal. Actually, with the material dumped into the container, the upper part will have very irregular dimensions and surfaces. An average angle to use would be a conservative value of the angle of repose for the material, say $15°$ to $20°$. Then the total would be found as (a) the struck volume, $V_s = LBH_s$, and (b) the heaped portion,

$$V_h = 2(\tfrac{1}{3} BHL_1) + \tfrac{1}{2} BHL_2 = \frac{BH}{6}(4L_1 + 3L_2),$$

where $H = B/2 \tan \alpha'$ and $L_1 = B/2$ and $L_2 = L - B$;

$$V_h = \frac{B^2 \tan \alpha'}{12}(4L_1 + 3L_2) = \frac{B^2 \tan \alpha'}{12}(3L - B);$$

$$\text{total } V_t = LBH_s + \frac{B^2 \tan \alpha'}{12}(3L - B). \tag{5-5}$$

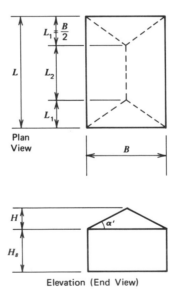

Plan View

B

Elevation (End View)

FIGURE 5-1 Theoretical dimensions of heaped container.

Note that the equipment manufacturers generally give the load capacity only by struck volume. This is conservative, but it can be measured directly without regard to the material loaded. The user who plans to use heaped loads must check the weight of his maximum load to make sure the equipment's capacity by weight is not exceeded. This may be a weight limit of the supporting pins or mechanism for the container, or it may be the axle limit for a mobile piece of equipment as given by the manufacturer.

Determination of the weight in a capacity load requires knowing the material's density in addition to expected swell and total volume. The density of the material is taken as measured in the natural condition. It is generally expressed in pounds per cubic foot (pcf) and is shown as δ in this book. Representative values for earth material densities were given in Table 3-3.

To see how the volume, swell, and density are used in earthmoving determinations, a few applications will be worked for examples.

Example # 1: to find the heaped volume and weight on a loaded wagon.

Given: a loamy sand from a borrow pit loaded into wagons specified as having a heaped capacity of 11.2 cu yds.

Assume: from Table 3-3, $\delta = 105$ pcf and from Table 5-1, swell is 20%.

Find: (a) maximum volume of actual material loaded, using Eq. (5-1):

$$V_b = \frac{V_l}{s_w} = \frac{11.2}{1.20} = 9.34 \text{ cu yds.}$$

(b) weight of maximum, capacity load:

$$W = \delta V_b = 105 \times 27 \times 9.34 = 26,500 \text{ lbs}$$
$$\div 2000 = 13.25 \text{ tons.}$$

Example # 2: to find the heaped volume of maximum weight of load on a haul unit with given bed dimensions.

Given: wet sand loaded into a hauling container $6' \times 10' \times 4'$ high; for Eq. (5-5) $L = 10$ ft, $B = 6$ ft, and $H_s = 4$ ft.

Assume: the sand's density, $\delta = 3100$ lbs/cu yd; swell $= 12\%$; and angle, $\alpha' = 20°$.

Find: (a) the volume from Eq. (5-5):

$$V_l = \frac{6 \times 10 \times 4}{27} + \frac{(6)^2 \tan 20°}{12 \times 27}(30 - 6)$$

$$V_l = 8.89 + 0.97 = 9.86 \text{ cu yds.}$$

(b) the weight of this capacity load is

$$W = V_b \delta = \frac{V_l}{s_w} \delta$$

$$= \frac{9.86}{1.12} (3100) = 27,100 \text{ lbs, or } 13.55 \text{ tons.}$$

5.1.3 Preparation of Material for Earthwork

The earth material to be handled or worked in frequently requires some preparation. The understanding needed here is partly of the material's nature and partly in preparation for the construction operations. If it is a tough, hardened material, the nature of its bonding for solidification must be broken. Strong bonds in all directions, as in solid rock, will probably have to be broken by blasting. The methods and techniques of blasting are both an art and a science requiring specialized knowledge. Some degree of understanding can be gained from the "Blaster's Handbook" published by the duPont Company, a leading producer of explosives for the blasting operation.

An operation requiring the removal and wasting of originally solid rock, as in tunneling, necessitates accurate blasting. Of course, the limits of the tunnel's dimension must be reached as closely as possible for an economic operation. Any amount of rock protruding into the tunnel and so inside the "neat" or pay line of the tunnel will not be tolerated as usually stated in the contract plans and specification. On the other hand, blasting or digging out too much of the rock surrounding the tunnel or for a foundation is uneconomical. This is because excess excavation must be filled in with concrete as an added expense to the contractor.

The need for blasting to given, specified dimensions is not the only reason for accuracy in blasting rock. Another economical reason for careful blasting in breaking the rock bonds is that the extent of breakage will be reasonably proportional to the energy of explosives used; also, the cost of explosives will be directly proportional to the explosive power. So the planner wants to use just enough explosive in his blasting to reduce the rock to manageable size for removal, if it is not needed elsewhere in a smaller specified size. The desired size depends on the equipment container size planned for removing it after blasting. Excessive breakage will mean too much expense in explosive. Not quite enough blasting will likely mean difficulty in loading for removal.

The blasting process for loosening and removing rock is further complicated by the fact that each type of rock in its particular situation has a special blasting requirement. More will be discussed on this for tunneling operations in Chap. 6. Rock removal for an open cut is not quite so difficult but has most of the same considerations.

Some solidified earth material may have a predominance of bonding in parallel planes and very little perpendicular to these. This is the case with crusted material or material that has been deposited naturally in layers. Shale or slate rock is of this nature. Such friable material cannot easily be dug out and removed without some preparation. However, it does not need to be blasted, generally. It can be ripped or scarified to loosen it for earthmoving.

There are certain lightweight, fine granular materials that appear to be solidified in their natural state. Loess is the most prevalent of this type, which will stand with an almost vertical unsupported face. However, its bond and stability can be broken down quite readily by a water jet from a high-pressure pump. The earthmoving method planned for removing such a material will govern how it should be prepared for the operation. If a power shovel is to be used for digging this material out, the vertical face should be maintained. On the other hand, a bulldozer or scraper would need to have the material spread out in a more horizontal fashion.

One other aspect of the earth material's nature should be mentioned at this time. This is the matter of water within the material being handled and adjacent to it. A wet material will be heavier than when it is dry because water is filling the voids instead of air. This not only makes heavier earthmoving loads but may lead to other disadvantages. When earth is dug out of a naturally saturated earth deposit with some slope to it, the earth's stability may be undermined. In such a situation the water would provide a lubricant, especially in a uniform silt, and an earth slide might result. Unfortunately, it may not be possible to get rid of the water in a soil. This would be the case in predominantly clayey earth material. Excavation of a saturated clay deposit can be an extremely tricky operation. In other cases of mainly granular material with the right topography the excavation may be done by hydraulicking, using water under high pressure to move the earth material.

In coarse-grained soil, such as sandy soil, it may be economical to lower the water content in the earth to be moved, especially if work is to be done in the excavated space. The high water content could be due to a high water table with its high point above the working level. Such a water table could be lowered by strategically located sand drains, a deep well, or a well-point system.

Also, excess water may exist in the earth material of a borrow pit. If this is being dug by a rubber-tired earthmoving piece of equipment, it may have a traction problem. This could be helped by lowering the water table in the borrow pit area to dry the surface somewhat. On the other hand, if this material is to be used in a fill with a moisture content of 10 to 15% for optimum compacted density, the high water content should be maintained in the material to be moved.

These are but a few of the variety of considerations for the nature of earthwork material to be faced by the construction equipment planner. He would be well-advised to consult a soils engineer or blasting expert if his problem is serious enough.

5.2 Basic Earthmoving Operation

Now if the nature of the earth material is reasonably understood, it is appropriate to identify the key parts of an earthmoving operation. The first efforts are made to prepare the material for moving. As discussed previously, this may involve loosening the material by blasting, ripping, etc. It also may involve removing excess water content. Another possibility in the preparation may be removing top soil or brush to expose the mass of earth to be moved. These preparatory operations, known as stripping or clearing, will be discussed later in this chapter.

Digging of the earthwork material follows the basic operation of loosening or other preparation. In fact, some construction situations or equipment combine the loosening and digging operations in one integrated step. The digging is the first step of moving the material from its natural location, though the loosening may move it a little. The material must be in manageable form or sized properly, in the case of blasted rock, to fit the digging part of the earthmoving equipment. Following the digging is the step of actually moving the material from its original, natural location to the place for its deposit. The distance involved in this moving may be a matter of only a few feet as in the earthmoving for an irrigation ditch. Here the material may be cast from the ditch to a bank alongside. In that case the steps of digging and moving the earthwork are combined in the operation of a single piece of equipment.

At the other extreme of distance in an earthmoving operation, the moving may be a matter of miles. For a highway or other construction projects there is commonly the need for moving large quantities of earth materials on and off the location or along the route. In this book the term "cut" will mean moving material away from its original location. And, conversely, "fill" will mean moving material into place. Sometimes these terms are used together as "cut and fill," which agrees with the given definition.

The next basic operation in earthmoving is that of dumping the material. It is dumped in the general area selected for its final deposit or use. If this is a place to get rid of the material as waste because it is not needed, it probably will be dumped somewhat haphazardly and not touched again during that construction. On the other hand, the plan may call for it to be used as fill or embankment. In that case it will be dumped in an orderly manner to be spread evenly and compacted by other equipment.

The final step in an earthmoving operation is the work done on the dumped material to be used for an embankment, fill, or subbase. The dumped material will need to be spread in a uniform layer. It may need to have water added to bring it up to the specified moisture content for a density in the vicinity of 95% optimum or higher. Finally, it will be compacted in about a 4″ to 10″ layer and brought up to final appropriate grade by adding successive layers in a similar way.

To summarize, certain distinct basic steps characterize an earthmoving operation. These have just been described as:

1. loosening the material to put it into a workable state;
2. digging the material to start the move from its original location;
3. moving the material from its original location to deposit point;
4. dumping the material at its place for deposit; and
5. working on the material to put it into final specified condition at its place of use.

Some earthwork operations will not include the last step listed. And the material may be ready for digging without any special loosening. The knowledgeable planner also knows that certain equipment used for some earthwork operations will do several of these steps without much obvious distinction between them. Such is the case when a dragline is digging, swinging, and casting its load. Regardless of the combining of steps that may be possible with a given piece of equipment, the planner must be sure that all the necessary steps in his earthmoving operation will be done.

5.2.1 Work Cycle for an Earthwork Operation

The work cycle for the equipment in an operation refers to the repetitive steps or components of work that the selected equipment does over and over again to accomplish the job. In the case of an earthwork operation, the primary work cycle is the digging-hauling (moving)-dumping-returning. This may all be done by one selected type of equipment, or it may be done by a selection of two or more types working as a team. A secondary cycle would be developed if the earth material is moved for use as the fill and needs water for proper compaction. The work cycle for this part of the earthwork operation is spreading-watering-compacting. In such a case it is likely that each step of the cycle is done by a different piece or set of equipment. This suggests that each piece or set has its own work cycle which is interdependent with the preceding or following equipment. The interrelationships of these cycles can be shown schematically for clarification (Fig. 5-2).

Now let us concentrate on a more detailed analysis of the primary earthmoving work cycle. Once an appropriate set or combination of equipment has been selected for the earthmoving, the planner must be able to make an economic analysis of its work. This is done by being able to study key parts of the cycle to decide whether they are being carried out as efficiently and economically as possible.

The most convenient basic common denominator for analyzing a cycle is time. This is true for the economic analysis of an earthwork operation because the cost elements—manpower and equipment costs—are mainly time related. Therefore, the earthmoving cycle will be studied on a time basis. This cycle reduces into a loading time, moving time loaded (haul), dumping time, and moving time empty (return).

The loading time (LT) is the total time it takes to dig and fill the earthmover to its planned capacity. It is the time required to get a full blade in front of the dozer or to get a full bowl in the scraper, wagon, etc. This time will depend on the condition of the earth material, the size of the haul container and the means for loading it, and the operating efficiencies controlling all the equipment. The loading time is controllable and should be very much less than half of the total earthmoving cycle, generally, to result in an economical operation. The exception to this may be in the hauling blade-type operation—bulldozer, etc.

The moving time to haul a load between the completion of the loading

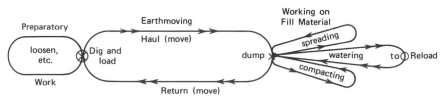

FIGURE 5-2 Schematic of earthwork cycles,

and the start of dumping is most dependent on the distance moved. It will vary from the short distance moved by a bulldozer or the swinging distance of a dragline to the several miles that a truck or other hauler may haul its load. For equipment that is mobile during its operation, the moving time will vary with the gears used, which depends on the job-governed conditions as well as the natural condition and topography of the haul route. Therefore, moving time is generally called the variable time. In this book the variable moving time loaded will be shown as VTL. The return moving time with the equipment empty is likely to be approximately the same distance as with it loaded. However, the topography may change and speeds of operation empty may be quite different. Therefore, this moving time with the equipment empty is another variable time which will be shown as VTE.

The fourth time component of the earthmoving cycle, as shown in Fig. 5-2, is the dumping time (DT). This time will depend on the conditions of the hauled material—is it free flowing, sticky, or in big chunks? It will also be influenced by the method of dumping. Is it dumped all in one spot, roughly spread, or to be carefully spread? In any case the DT amounts to only a small fraction of the total cycle. It should be less than the loading time in all earthmoving equipment selections.

The total earthmoving cycle time (CT) is basically the sum of the four components discussed above. Thus,

$$CT = LT + VTL + DT + VTE. \qquad (5\text{-}6)$$

The unit generally used for construction equipment determinations is the minute. Sometimes the times to load and dump material are called "fixed times" because they are fairly constant for any given operation regardless of the travel distances. Parts of the VTL and VTE for accelerating, decelerating, braking, and turning are also called fixed times, shown as FT, by some planners. This will be shown in discussing specific equipment later.

5.3   Types and Coverage of Earthwork Equipment

The general types of equipment designed to perform or help with one or more of the basic earthworking operations include:

1. tractor-mounted dozers, pusher, or rippers;
2. motor grader, possibly adding scarifier;

3. scraper, self-propelled or tractor-drawn;
4. front-end loaders;
5. belt loaders and wheel excavators;
6. power excavators (to be covered in Chap. 7);
7. haulers (to be covered in Chap. 9); and
8. compactors (to be covered in Chap. 12).

Each type of equipment in this chapter will have the following sub-headings: (1) equipment design features, (2) construction operations where used, (3) productivities figured for the equipment, and (4) costs of equipment on such operations. The first two subtopics on design and uses and the last two on productivities and costs might be combined and still be discussed with enough understanding.

5.4 Tractors for Earthwork

Several earthworking pieces of equipment are mounted on tractors to do their job. These include dozer blades, rippers, and pusher blocks. The tractor for this type of earthworking operation is designed to have a low center of gravity. This will concentrate the power delivered by the moving tractor at a height nearly that of the pulling winch or drawbar. The significance of the drawbar was explained in Sec. 3.6.2. If there is much difference in the height of force transmission from the tractor to its load, the tractor will tend to dig into the ground unnecessarily. Such action results in poor power effectiveness; also, the steering control of the tractor will be hampered.

Practically all tractors used for earthwork operations are powered by internal combustion engines. It is suggested that the reader review the discussion on I.C. power in Sec. 3.1.2 and particularly the sections on direct-drive and torque-converter transmissions. Each of these primary forms of I.C. power transmission has its advantages for tractor-mounted earthwork equipment. A brief tabulation is given in Table 5-2 for reference.

A number of different arrangements exist for these two basic power transmission systems called "trains." One that is particularly desirable for tractor-mounted earthwork equipment operation is called the power shift. This new transmission scheme combines the advantages of direct-drive and torque-converter systems. It has the impact, shock load ability of direct drive for the tree clearing and ripping type of work. On the other hand, it can automatically adjust to a variable load like the torque converter

TABLE 5-2 Tractor Variations with Primary Power Transmissions

Variation in Operation	Transmission Move with	
	Direct Drive	Torque Converter
Speed selection	Shifting gears to best speed for conditions	Select gear range, then speed automatically adjusts to conditions
Variable load	Shift gears to one to handle load; can get impact force	Load ability varies inversely with speed; shift to lower range for greater load
Operating in reverse, unloaded	Shift to top reverse gear	Maximum speed obtained by pushing throttle to limit
Operator efficiency	Shifting requires skill, results in wear and tear, operator fatigue	Less skill needed, less wear and tear, less operator fatigue

does for pusher-type operations. Now some of the key tractor-mounted equipment for earthwork operations will be discussed.

5.4.1 Design and Operations of Tractor-Mounted Equipment

The dozer-type attachment on a tractor may be a straight blade, an angle blade, or a U-shaped blade. There are other special blades mentioned with clearing operations. Each of these has its use for certain operations, as will be discussed. The design features of the common types are covered briefly.

An angle dozer blade has no curvature in its length but may be curved in its vertical dimension. It can be adjusted on the tractor to a maximum angle of about 25° from the perpendicular to the tractor's axis of movement. This and the tipping angle are shown on the sketches of Fig. 5-3. There may be no tip of the blade or it may be set at the other extreme for digging down to two feet below the track base. The angle blade, generally, cannot be tilted from the vertical plumb the way a straight dozer blade can.

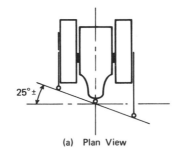

(a) Plan View

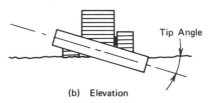

(b) Elevation

FIGURE 5-3 Angle dozer blade angles.

The straight dozer blade is similar to the angle blade in shape but is not pivoted at the center for the angular rotation. This will insure more rigidity for applying great force. It can be tipped for digging a V-cut in the ground. And this blade may be tilted about a horizontal axis to give it more of a scoop action. Since the straight blade has essentially nothing to catch material at its ends there is spillage that may not be wanted.

The U-shaped dozer blade is designed so that looking down on it from above it appears like a shallow, spread U as the name suggests. This blade is "curved" in its length by several angular breaks in the blade plates. Its main difference from the straight blade is that there are end walls to cut down spillage from the U-shaped blade. Its shape is designed for the greatest load-carrying capacity of all the dozer blades. Generally, there would not be any provision for tilting nor angling the U-shaped dozer blade.

A pusher block is merely a short, sometimes pivoted, load transfer bumper mounted on front of the tractor. It is at the right height to contact the rear of a scraper (described later) earthmoving unit. Then the pusher tractor can help give more power for the loading of a scraper. This pushing operation can be done with a straight dozer blade. It must be done with care to push on the best-supported portion of the blade near its center. There is also a pulling attachment for tractors to help load scrapers. This is a controllable hitching frame mounted low on the back of the tractor at the drawbar level. It has an eye to catch a hook at the bottom of the radiator frame on the front of a scraper.

A ripping attachment is a plow-like piece of equipment generally mounted on the rear end of a crawler tractor. It is designed to loosen material at the ground surface by penetrating into and breaking up the ground material as the ripper is pulled along. The construction type of ripper is generally either of a hinged or parallelogram design. These are shown in the sketches of Fig. 5-4.

The hinge-type of design pivots about the supporting beam through an arc of up to 30°. This may cause problems in getting the ripper shank to penetrate some tough materials. The design does allow the ripper attachment to move laterally a small amount to go around particularly hard spots or rocks. The parallelogram-type of design has a better penetration ability since the ripper shank is maintained in a vertical position and extra force can be applied to it through the dual supporting arms. Either design should give a penetration of 2 to 4 feet in order to effectively rip surface material.

A tractor equipped with rippers is shown in Fig. 5-5. It shows an assembly of a set of ripper attachments on the supporting beam.

The effectiveness of a ripper attachment depends on (1) down pressure at the tip, (2) the tractor's power to advance the tip through the material to be ripped, and (3) the tractor's weight to develop enough traction as well as (4) properties of the surface material. The number of ripper attachments that can be used together on one tractor will depend on the availability of the four above factors. Down pressure is governed by design of

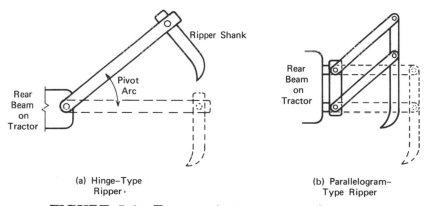

(a) Hinge–Type Ripper ·

(b) Parallelogram– Type Ripper

FIGURE 5-4 Types of ripper attachments.

FIGURE 5-5 Tractor-mounted ripper equipment (courtesy of Caterpillar Tractor Co.).

the mounting beam and supports as well as the hydraulic control mechanism to apply the penetrating force. The other three factors need no further explanation. Some designs increase their rippability by including a vibratory action to the ripper assembly, like that for compaction equipment discussed later in this chapter and in Chap. 12. A special ripper, also called a scarifier, may be mounted behind the front or rear wheels of a motor grader. In that case the requirements discussed for the tractor above must apply to the motor grader. More will be discussed about a ripping operation in the next section.

5.4.2 Dozing and Ripping Operations

A dozing operation is done with a push blade, as described in Sec. 5.4.1, mounted on the front end of the tractor. The loading conditions for the tractor will vary from fairly light and uniform to an impact force needed through the blade on a boulder, root, or tree. For the impact-force requirement a direct-drive transmission is preferred. Otherwise, for the lighter, more gradual variations of dozer loading, a tractor with torque converter would be better. The power-shift drive is best for all-around use.

A specific operation which will often call for a tractor mounted with both a special angle dozer blade and a ripper attachment is land clearing. Brush is cleared by ripping it out or by a V-shaped shearing blade cutting at the roots and piling it, frequently with a tractor-mounted clearing rake, for burning or hauling away. Trees up to 12 inches in diameter can be removed by a shearing angle dozer. Roots are cut by the blade, and the tree is pushed over by the dozer (Fig. 5-6).

The V-notch to cut roots generally requires a sudden impact force applied at an edge of the blade. The effort is made with a high concentrated force or repeated shearing cuts applied by the lowest forward gear. To weaken or bowl the tree over calls for repeated impact forces applied by a stinger-like projection on the leading edge of the shearing blade. A special high-contact "tree pusher" can be used above the dozer blade to gain more leverage to push over trees. For instance, if the regular dozer blade can push on the tree five feet up with a force of 15,000 pounds, the forcefulness is expressed as 75,000 foot-pounds (ft-lbs). With the special pushing blade that might contact the tree 15 feet up the trunk with the same force, the pushover effort would be around 200,000 ft-lbs. This advantage will mean the rate of clearing the land can be increased by a third or more, saving that much time to do the operation. A thorough coverage of all aspects of "land clearing" is found in a Caterpillar Tractor Company book with that title.[3]

FIGURE 5-6 Tractor dozing for tree removal (courtesy of Caterpillar Tractor Co.).

The operation to strip the ground cover or topsoil requires special dozer operation, or it may be done with a scraper. If a bulldozer—straight or U-shaped blade—is used, the operator will have to show special skill. This is because the natural tendency is for the digging blade to cut deeper as it digs moving forward. However, stripping the topsoil means cutting off only the top 3″ to 5″ with roots and vegetation in it. The stripped topsoil, which may be used later for covering backfilled ground surface, is frequently stored in a mound on the building site. In that case, the bulldozer has an advantage in a stripping operation. The dozing for this operation may mean cutting and pushing up to several hundred feet in one direction. This is generally easy loading for the bulldozer and either basic type of power transmission should be satisfactory. If the return to the original position for the next load is several hundred feet away, using the top reverse gear would make an efficient operation.

Ripping Operations. Solid or highly consolidated materials may have to be loosened either by a ripping operation or a drill-and-blasting operation. Igneous rocks, such as granites, basalts, and trap rocks, are the most difficult, if not impossible, to rip. This is because they lack the stratification and cleavage planes necessary for ripping hard rock. Sedimentary rocks are highly stratified, i.e., they are naturally made in layers. So sandstone, limestone, shale, caliche, and conglomerate rocks are the most easily ripped solid materials. The metamorphic rocks, such as gneiss, quartzite, schist, and slate, vary in their rippability.

The physical characteristics of material that favor ripping are fractures and planes of weakness, generally found in weathered rock, brittleness and crystalline nature with much stratification or lamination, and material with large grains and low compressive strength. The harder a material is, the less rippable it may be. The relative hardness of a rock may be found by using the fingernail and pocket knife test. All rocks can be compared on Mohs' scale of relative hardness (discussed in Chap. 6 on Tunneling; see Tables 6-3 and 6-4). A fingernail has a Mohs' hardness value over 2 and can scratch mica, gypsum, or talc. A copper penny with hardness of 3 will scratch calcite, shale, or slate. The pocket knife should have a hardness of at least 5 and scratch gneiss, schist, and granite and will mark any sedimentary rock.

A more positive and quick way to determine rippability was developed by the Caterpillar Tractor Company in 1958 and is described in their booklet "Handbook of Ripping."[4] The method makes use of the refraction seismograph which indicates the degree of consolidation at and near the ground surface. The seismic analysis principle makes use of the time it takes a seismic wave to travel through different kinds of subsurface materials. The speed in hard, tight rock is fast—up to 20,000 fps; whereas in loose soil the speed may be as slow as 1000 fps.

The seismic waves are produced by a sledge hammer striking a steel plate on the ground surface. They are made at points of various distances, giving 10-foot intervals, from a geophone receiver on the ground. The geophone picks up only the first wave to reach it from each blow on the steel plate. The time in seconds taken by that wave is plotted against the horizontal distance on the ground between the plate and geophone. From this graph one can determine the seismic wave speeds and so hardness of materials under the surface. These results are then compared to rippability charts developed for each type of ripper equipment in various materials.

An example of a rippability chart developed by the Caterpillar Tractor Company is shown in Fig. 5-7. As can be seen from the chart, there is a level of seismic wave speed for each rocky material above which the drill-

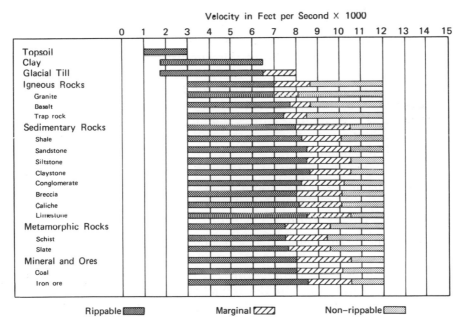

FIGURE 5-7 Ripper performance for the D9G tractor with No. 9 Series B ripper (courtesy of Caterpillar Tractor Co.).

ing and blasting method should or must be used. In the transition range—for instance, in sedimentary rocks with seismic wave velocities between 8,000 and 10,000 fps—there is chance for debating between a ripping operation or a drill-and-blasting operation. The Caterpillar booklet[4] concludes that (1) it is generally cheaper to rip than to drill and shoot, (2) greater production can be obtained by ripping as compared to drilling and blasting, and (3) increased convenience and versatility in equipment and safety in operation result when a material can be ripped rather than drilled and blasted.

When the job conditions call for a ripping operation, several other preliminary determinations must be made. The end use of the material is important. Is it to be wasted or used as fill or graded material? If it is to be crushed for graded material, the maximum size of feed to the crusher is a key determination. Likewise, a fill material is specified by its size and condition. Size of the largest chunks ripped out of the ground surface can be varied. By controlling the distance between passes or ripper attachments, the size of shattered material is fairly controlled. Closer ripper paths will result in smaller material.

The method for moving the ripped material also may dictate the size of the largest chunks. If a power shovel and hauling units are used to load out the material, it must be small enough to pass through the shovel's bucket. A dozer moving the material has no particular requirement on the largest size of its loaded material, so ripping may be done more economically. When scrapers are used to move the ripped material, no piece can be more than a couple of feet in diameter. Smaller sizes would be preferred and have less wear and tear on this equipment and its tires. If the ripped material is to be moved out on conveyors, it must be less than 12″ in maximum dimension.

There is one other special point about ripping material to be moved. If it is friable material that disintegrates too easily, the planner may find it better not to do much ripping. The reason is that the material may be reduced to a powder which will be harder to load than in its natural state.

Earthmoving by Bulldozing. The regular bulldozing operation to move earth is practically a straight-forward-and-return set of movements. This is called shuttling or shuttle dozing. As the tractor moves forward, the dozer blade in front cuts into the ground to get its load. The dozer has to travel tens of feet to get its full load. For instance, a typical blade cutting a 6″ depth will take more than 40 feet of forward motion to gain a full load. The full load will be like a moving wedge of earth pushed ahead of the dozer blade. Figure 5-8 with dimensions sketched on it shows how this load can be estimated for production purposes.

The dimensions to use for estimating purposes can be assumed:

FIGURE 5-8 Shape of bulldozer's load (courtesy of TEREX, Division of General Motors).

$$\frac{W}{H} = \frac{1.5 \text{ to } 1.67}{1} \text{ and } \alpha = 30° \text{ to } 33°.$$

The actual load wedge is slightly rounded, as can be noted in the picture. The maximum slope and spillage will depend on the material being moved. A well-graded granular material or loamy and sticky earth will remain on steeper slopes than will a uniform (one-size) gravel. To estimate the load volume, the dimensions as shown in Fig. 5-8 can be used. Thus, the dozer's load in cubic yards of loose measure (as in Sec. 5.1.2),

$$V_l - \frac{WHL}{54}, \tag{5-7}$$

where $W = (1.5 \text{ to } 1.67)H$, and all the dimensions—W, H, and L—are in feet.

The dozer is best suited for earthmoving in short hauls of perhaps a minimum of 50 feet up to 300 feet. The track powered dozer operation is ideal for the high power requirements of deep cutting, in tough material, or where there are steep grades to move material not requiring special handling.

The several forms of dozer blades described previously have more specific uses. An angle dozer would be most productive for earthwork operations requiring:

1. shallow cutting or stripping with the material being pushed into a windrow along the trailing end of the blade;
2. cutting a V-shaped ditch with the wider slope not too steep for the tractor to cut; and
3. backfilling a trench with material from a windrow parallel to it.

The U-shaped dozer blade is designed to carry the greatest amount of material by any dozer. It would be most useful for an earthmoving operation to cut and move a lot of material several hundred feet. Also, this material may be tough to cut or load, and it does not need to be picked up before dumping. A straight dozer blade is the most versatile one, but it lacks the special advantages of the others.

The previous discussion of bulldozing has emphasized a shuttling type of operation. Also, it was suggested that the ideal use of a dozer occurs in tough digging, calling for the great power from a crawler tractor.

For many dozing operations a wheel-mounted tractor may be preferred. A rubber-tired unit often can deliver enough power to carry out a satisfactory shuttle dozing operation. Certainly, this type of unit is preferred if there is much need for maneuvering or changing direction. The latest rubber-tired tractors are articulated units. That is, they have a pivot joint between the front and rear axles to allow sharp turning. An articulated rubber-tired dozer can turn 180°—an about-face—in a space no more than its total length. Thus, it can operate in as little space as a crawler tractor with less strain on the operator, if much maneuvering is required.

5.4.3 Productivities of Tractor Equipment

The production that can be expected or estimated for a tractor-mounted type of equipment is extremely variable. This is so because of the great variety of operations, such as land clearing, topsoil removal, and regular excavation. If the tractor is used as a pusher, its production will be secondary to that of the scrapers. The tractor's cycle time to contact a scraper, push it through the loading step, disengage, and reverse to return for the next scraper can be estimated at 1.5 to 2.5 minutes. Such an operation should be planned, generally, so that scrapers will not have to wait for the pusher. In other words, the pusher's secondary roll means it does not govern the productivity for the earthmoving operation. Any waiting that is likely to happen should be by the pushing or pulling tractor. This will be discussed more fully in Sec. 5.6.3.

The cycle time of tractors used as pushers or pullers for earthmoving scrapers can be studied in the field. This is done with stop watches handled by careful observers or with time-lapse photography. If enough cycles are watched, a statistical analysis can be made of the results for better accuracy.

The Highway Research Board, headquartered in Washington, D.C., sponsored such a time study of pusher operations by nine large crawler tractors on seven jobs in the eastern and southern United States. The results of these field observations showed that the average total cycle time for a pusher, including time waiting for scrapers, was 2.5 minutes. The breakdown and variations on the cycles are shown in Table 5-3.

The meaning of productivity in this book is the amount of construction material handled in an hour's time. The productivity of a ripper opera- tion is difficult to estimate and frequently not necessary. The toughness or difficulty in ripping layered rock compared to crusted earth suggests some of the variety of effort needed. Also, the ripping requirement is generally not a continuous need. It is frequently done along with loading of the material being excavated. This is the reason that a tractor outfitted with attachments for a large earthwork project will frequently have both a front-end blade and a ripper mounted behind. In that situation, without a continuous ripping requirement, the productivity is secondary, very inaccurate, and not worth estimating.

However, for a major ripping operation, it is economically desirable to know what productivity to expect for estimating purposes. Determining ripper production may be done in one of three recommended ways. In any case it will be a highly variable determination because of nonuniform- ity of the material properties. So the first two methods for determining

TABLE 5-3 Cycle Times for Pusher or Puller Crawler Tractors Working with Rubber-Tired Tractor-Scrapers

Time Element per Cycle	Time, in minutes	
	Range	Average
Assist in loading scraper	0.9–1.4	1.1
Maneuvering: backing, turning, and engaging scraper	0.5–1.2	0.8
Waiting for next scraper	0.0–1.0	0.4
Minor delays	0.0–0.8	0.2
Total cycle	1.5–3.2	2.5

productivity are based on the actual experience of the given job. The third might be used to make a logical estimate for a future ripping operation.

The first method is said to be the most accurate. It is based on recording the original and final topographic cross sections of the area ripped along with the time spent ripping. This gives volume of material, which is divided by the actual working time to find a bank-measure yards per hour. The second is similar, but rather than using the cross-sectional volume, it is based on counting the material loads moved out by scrapers or hauling units. That method requires knowing the bank-measure load carried each time and assuming it is the same for all loads.

The method that can be used to estimate a future ripping job is based on assuming the average speed a ripper will move for each trip over the area to be ripped. This speed will be less than the maximum speed for the power gear needed to do the ripping. In fact, it may be closer to half that speed. Knowing the distance to cover for each pass, the travel time per cycle can be figured. Add the time to raise the ripper, pivot or turn, and lower to rip on the return, to find the total cycle time. This time will allow one to figure the ripper passes per hour. The volume to be ripped is based on the width between passes, the average depth of penetration, and the length of area covered.

The estimate for a ripping operation using that method might look like the following:

Given: crawler tractor with one tooth, to rip in first gear, max. $v = 1.6$ mph; 2 1/2 feet between passes, each with 2-foot penetration; the area to rip is 300 feet long.

Estimating: average speed = 1 mph (88 fpm) with time to maneuver between passes, i.e., so-called fixed time, FT = 0.25 min; assuming work hour is 45 minutes.

$$\text{Cycle time} = \frac{300}{88} + FT = 3.41 + 0.25 = 3.66 \text{ min,}$$

$$\text{and for 45 m/hr;} \frac{45}{3.66} = 12.3 \text{ passes/hr.}$$

$$\text{Volume ripped} = \frac{300 \times 2\frac{1}{2} \times 2}{27} = 55.5 \text{ cu yds/pass.}$$

Estimated productivity = $55.5 \times 12.3 \approx 683$ cu yds/hr.

Note: by comparing estimates by this method with actual field results, it is found that generally the estimate is 10 to 20% higher than actual production in the field.

Bulldozing Productivity. Productivity for a bulldozing operation is an important determination, particularly where the earthmoving by dozer must be done before other operations can proceed. In other words, the case is particularly important where the dozer takes some of the valuable overall time for the project all by itself.

To estimate dozer productivity, one must break down its work cycle into significant parts, as explained in Sec. 5.2.1. The dozer will be loading during part of its travel so it is not necessary to separate the load time for this operation. There is the dozer's variable time traveling forward with its load. And the time to return in reverse for the next load is with the blade raised and empty. Each of these variable times can be determined simply by dividing the distance traveled by the travel speed in feet per minute (fpm) for the gear used.

Variable times found that way do not account for the extra time taken to get from a standstill to the governed traveling speed, or vice versa. This additional time is known as the time for acceleration or deceleration. It is called "fixed" time (FT) because of its consistency. If the travel in either direction is in one gear requiring only the shift from forward to reverse, the dozer's fixed time can be reasonably estimated as 0.10 to 0.15 minute. If there is one additional shift to a higher gear in either direction, then the estimate might be 0.20 to 0.30 minute.

The dozer's total cycle time is found by a modification of Eq. (5-6):

$$CT = FT + VTL' + VTE'. \tag{5-6b}$$

For fairly tough digging to hard digging and not too long a haul, the dozer will travel forward in its lowest gear for maximum power. This means VTL' will be based on 1.5 to 2.5 miles per hour (132 to 220 fpm). The return travel will be somewhat faster since power is not a major concern, unless the tractor is backing up a steep grade.

The peak productivity, i.e., the expected production in a sixty-minute hour, can be determined using Eqs. (5-6b), (5-1) and (5-7). These will combine with a swell factor from Table 5-1 to find for the dozer

$$q_p = \frac{60\,V_l}{(CT)s_w} = \frac{60\,V_b}{CT}, \tag{5-8}$$

where q_p = peak production, cubic yards of bank measure per 60 minutes;
  V_l = dozer load in loose measure, cu yds;
  s_w = swell factor expressed as a decimal; and
  CT = total cycle time, minutes.

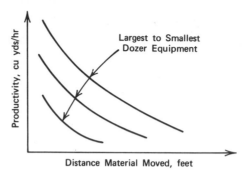

FIGURE 5-9 Comparison of dozer productivities.

Productivities for dozing operations will depend primarily on the size of blade and tractor to push it, as well as the distance the material is moved. A set of curves could be developed for different sizes of dozer blades if frequent estimates are going to be required (Fig. 5-9).

Example of Dozer Productivity. A shuttle dozing operation for earth-moving is very common. The material to be moved, distances and slopes of the work area, and equipment to use can all be varied and determined. Therefore, the productivity should be calculated for each operation. An example will be given here to show how this is done.

Given: a straight blade 13 feet long, 4 feet high attached to a 190-hp (direct-drive) crawler tractor to move dry sandy soil 100 feet horizontally.

Find: an estimate for the normal production to expect in a 50-minute working hour.

Solution: (a) to find the load carried by this blade, estimate the loose material using Eq. (5-7):

$$V_l = \frac{(1.5 \times 4) \times 4 \times 13}{54} = 5.78 \text{ cu yds,}$$

and from Table 5-1, $s_w = 1.14$, so

$$V_b = \frac{5.78}{1.14} = 5.06 \text{ cu yds, bank measure.}$$

(b) for cycle times it is assumed that the dozer travels forward in 1st gear (1.5 mph) and returns in 2nd reverse (2 mph):

$$\text{VTL} + \text{VTE} = \frac{100}{1.5 \times 88} + \frac{100}{2.0 \times 88};$$

(travel time) $= 0.76 + 0.57 = 1.33$ minutes; assuming only one shift, FT $= 0.3$, so CT $= 1.33 + 0.3 = 1.63$ minutes. (c) to find productivity, use Eq. (5-8):

$$q_p = \frac{60 \times 5.06}{1.63} = 186 \text{ cu yds/hr},$$

and for a 50-minute working hour estimate,

$$q_n = \frac{50}{60} \times 186 = 155 \text{ cu yds/hr}.$$

5.4.4 Costs of Tractor-Mounted Equipment

The basic considerations for the costs of tractor-mounted equipment are the same as for all construction equipment. Therefore, it will be assumed that Chap. 2 is understood as background for this section.

The most important and largest single part of the total cost for operating this equipment is the cost of the tractor itself. Operating the smallest of tractors (30 hp or less) would be an exception, where the wages for the operator might be larger. Such small tractors are generally not used for dozing, ripping, and other operations discussed in this chapter.

With the tractor such an important part of the cost, the planner must give attention to the basic components of the tractor. The differences between crawler-mounted or wheel-mounted and the type of power unit are important in the tractor's cost. Also, the regularity of use for the tractor is as significant as with any construction equipment. The comparative variations of costs for the main differences in tractor equipment are shown in the curves of Fig. 5-10. The cost in actual dollars for a particular tractor would be found from a dealer or current listings. Such facts should be checked each year, or more often if economic conditions require it.

If a direct, gear-drive power transmission is desired instead of the power-shift, torque-converter type, there will generally be a lower cost. This amounts to 5 to 10% less. The cost for the standardized attachments must be added to the bare tractor cost. These can be estimated by adding for:

dozers on crawler tractors	5 to 10%;
dozers on wheel tractors	8 to 12%;
ripper with hydraulic controls	15 to 20%; and
winches with power controls	16 to 18%.

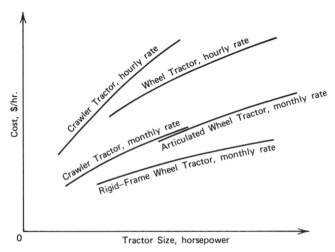

FIGURE 5-10 Comparison of cost per hour for various tractors.

An expense that must be watched carefully by the responsible user is the expense of major repairs. In the case of tractors this is generally assumed to be about one-third of the total ownership expense. It will vary from a much smaller portion when the equipment is practically new to much more when it is nearing the end of its useful life. The useful life of a crawler tractor might be assumed to be six years or 12,000 working hours, if operating conditions are excellent for the majority of its life. At the other extreme, with severe operating conditions, the tractor's useful life may be only four years of 2000 working hours each.

Example of Cost for a Dozer. Let us consider an example to illustrate these key expenses for a tractor. Referring to the second chapter, the necessary calculations can be made for a particular piece of equipment. Assume that we are dealing with a 150-horsepower-drawbar crawler tractor with a dozer blade attachment, all costing $40,000 new. It is also assumed that this equipment will be operated under average conditions for a useful life of 10,000 working hours over five years. The average annual investment by Eq. (2-2) is

$$A = \frac{1}{2} \frac{(5 + 1)}{5} \times 40{,}000 = \$24{,}000.$$

The ownership expenses for insurance, taxes, storage, and interest to be charged to each hour of operation are figured from this investment value.

The charges to recover the depreciated value of the tractor will be calculated from the original cost less any estimated salvage value when it will be sold. If we estimate 10% or $4,000, salvage value and straight-line depreciation for the five years (10,000 working hours), the expense to charge for depreciation will be

$$e_d = \frac{40,000 - 4,000}{10,000} = \$3.60 \text{ per hr.}$$

The major repairs over the lifetime of the tractor might be estimated as 75% of the original investment, less salvage value. Therefore, each of the expected working hours should be charged

$$e_j = \frac{0.75 \times 36,000}{10,000} = \$2.70 \text{ per hr}$$

to recover this expense. As suggested above, this ownership expense can vary a lot during the life of the piece of equipment. The major repair expense might be less than $1 per hour for the first year or so of operating time. On the other hand, it will probably exceed $3 per hour in the fourth or fifth year of operation. And, of course, the 75% is an estimate.

The only way to make a better estimate of the major repair expense to be recovered in the ownership of this tractor is by keeping good cost records. Then allowance for this large expense in next year's use can be based on the past year's record with some estimated increase. The record may show that the major repair expenses for one piece of equipment are twice what they are for a similar one. That difference must be taken into account when charging for the use of the different pieces of equipment. The reason for such a difference should be studied carefully. Of course, a costly tractor might be sold earlier than the end of its estimated useful life. Or there may be other reasons, such as a newer model having some advantageous features, to sell the tractor earlier than expected. This all suggests that careful cost records be kept on each piece of equipment owned. Then the charges for each hour of its use can be based on as sound an estimate of actual expense as possible.

Continuing with the example of cost for the dozer, the total ownership expense can be calculated. The costs of interest, insurance, taxes, and any possible storage expense can be found as explained in Sec. 2.2. The interest to charge for the investment is on the original cost of the equipment ($C_o = \$40,000$ for this example). The other items of expense mentioned above are figured on the average annual investment ($A = \$24,000$ for this example). With reasonable assumptions the amounts to charge for this dozer's ownership expenses are:

$$\text{Interest, } E_i \qquad = .06 \ (40{,}000) \qquad = \$ \ \ 2400 \text{ per year}$$

$$\text{Insurance, } E_p \qquad = .02 \ (24{,}000) \qquad = \ \ \ \ 480 \text{ per year}$$

$$\text{Taxes, } E_x \qquad = .015 \ (24{,}000) \qquad = \ \ \ \ 360 \text{ per year}$$

$$\text{Storage, } E_s \qquad = .005 \ (24{,}000) \qquad = \ \ \ \ 120 \text{ per year}$$

$$\text{Depreciation, } E_d \quad = \frac{(40{,}000 - 4000)}{5} = \ \ \ 7200 \text{ per year}$$

$$\text{Major Repairs, } E_j = 0.75 \ (7200) \qquad = \ \ \ 5400 \text{ per year}$$

$$\text{Total expense to charge for use} \qquad = \$15{,}960 \text{ per year}$$

With the previous assumption of working hours per year, the hourly charge for ownership, $e_w \approx \$8$ per hour for this dozer. Without much doubt we can note the importance of expenses for the depreciation and major repairs. Together they amount to more than 75% of the total cost of ownership. So a great deal of attention must be given to these two key expense items. The suggestion for a running cost accounting system for maintenance and repair costs might be given serious thought. Careful thought should also be given to the discussion on depreciation in Sec. 2.2.6 of the chapter on costs.

The operating expenses can be figured by the separate items of fuel, lubricants, minor repairs and adjustments, and wages with fringe benefits. These items are discussed in Sec. 2.1. The safest way to determine all but the wage items is by cost accounting records and experience. If the user has not had experience with the particular type of equipment, he should ask the dealer or manufacturer for representative values of fuel consumed per hour, etc. From these he can figure operating expenses at the beginning. Then he can keep his own records during the operating time of the piece of equipment. Future charges for operating time can be based on these fresh records.

The expense of operating wages has to be figured from the prevailing local rates. In recent years the wage rates for operators of construction equipment have risen practically every year. They have, generally, exceeded the rate of increase in the original cost of new equipment. The wages amount to two-thirds to three-fourths of the total operating expense.

In the example of this section, let it be assumed that the operating expense is 10% greater than the ownership expense. With this reasonable comparison the total cost to charge per hour of use for the dozer is $8 \times 2.1 = \$16.80$. If the productivity of this dozer is that of the example in Sec. 5.4.3, then the basic cost to move a yard of earth with this equipment will be $1680 \div 155$ cu yds per normal hour = 10.8 cents.

Of course, it is unreasonable to expect this "normal" 50-minute-per-hour production for long. The average productivity of the dozer over several days or weeks may be 50% of the normal, i.e., $q_a \approx 0.5\ q_n$. Then the basic cost of this earthwork by the dozer would be more nearly 20 cents per yard. To this cost must be added the overhead expenses and profit to find the cost of moving a cubic yard of earth with the given dozer equipment in this operation.

5.5 Motor Graders for Earthwork

The motor grader is a piece of equipment used to move earth or other loose material. Its function is generally to plane, mold, or grade the material it is working to a given line or contour. It is especially useful because of its blade that can be held in various positions. This grading blade is also called a moldboard, which reminds one of the origin of this type of equipment before the twentieth century. Many advances have been made since the original moldboard appeared, so that now the motor grader is a very versatile piece of earthmoving equipment. Its standard blade is 10 to 14 feet in length.

5.5.1 Design Features of Motor Graders

The blade of the motor grader can be held in positions like those of the angle dozer, bulldozer, or tilting dozer blade—all with greater flexibility of movement than the dozer. This is shown by Fig. 5-11.

The versatility of blade position does not mean that the motor grader can replace the various forms of dozer. It cannot apply the earthmoving or cutting power of a bulldozer, which applies its force through a lower center of gravity. The motor grader sacrifices forcefulness partly because it transfers the machine's power through the pivot point high above the supported blade. However, the motor grader can cut the surface material at many more angles and with much easier adjustment than a dozer has.

The extreme angles do present a problem of force applied through the motor grader's wheels. For this reason the wheel positions on the modern motor grader are extremely adjustable. That is, the front wheels are tilted for leaning when the blade is at considerable angle from the horizontal and the rear axles are "full-floating" to give full bearing of all wheels for all ground shapes. The modern motor grader is also articulated, or hinged for horizontal rotation of alignment, between the front and rear wheels. The reason for this adjustibility is shown in Fig. 5-12, a picture of angles

FIGURE 5-11   The blade angles for a motor grader
(courtesy of Deere & Company).

and forces applied in motor grader operation. It must be recognized that the material is always moving toward the open, or trailing, end of the blade. That means the cutting force at the ground's surface is applied toward the leading end and along the length of the blade. The sketches show these forces as well as their effect on the variation of angles for the operation of a motor grader.

The leaning of the front wheels and articulation to give the rear wheels an angle with the direction of movement lead to better steering stability. These design features also allow the motor grader to turn in a short radius. It is possible to turn a unit in the width of a standard two-lane highway with shoulders. The turning radius is about 20 feet.

5.5.2   Operations for a Motor Grader

The motor-grader type of equipment is used for a great variety of construction operations. This versatility is due to its flexible actions. Its

FIGURE 5-12 External forces on a motor grader (courtesy of Deere & Company).

usefulness is extended by attachments that the motor grader can handle, such as scarifier teeth, a pavement widener, and an elevating grader unit.

A basic use of the motor grader is, as its name suggests, for shaping and final grading of the total roadway width. This includes not only the base for the road surface but also the shoulders, side slopes, and back slopes. The grader can cut the V-shaped drainage ditches along the roadway, or wherever one is needed. With the attachment of a short blade, extending below the standard one, a shallow box-shaped trench can be dug by the motor grader.

Another set of roadway operations done by motor graders are of a non-finishing variety. One such operation is the maintaining of a haul road for other earthmoving equipment or trucks. The motor grader serves to move and compact the earth for a reasonably smooth traveling surface, eliminating ruts that will form on soft ground under traffic. Another operation in this category calls for the attachment of a scarifier, generally on the front of the motor grader. With this the equipment can be used to break up an old, flexible pavement roadway surface for regrading or preparing for a better surface.

The motor grader with its standard blade is also very good for mixing and spreading surface materials. The grader works to mix the materials previously placed on the roadbed in windrows. This may be done for a compacted earthfill or an on-the-site paving mix operation. In the latter the mixing may be of asphaltic or bituminous materials with aggregate for flexible pavement. Or it may be mixing materials for a soil-cement surface. In this case a disc-harrow attachment will work very well.

The great maneuverability of a motor grader makes it useful for grading airports or large building sites. This is not limited to the fine grading on a uniform plane grade but includes contour grading. To do such work more precisely, modern motor graders are equipped with automatic blade controls. These allow the operator (1) to set a desired slope of the blade and (2) to meet an established grade line. The system involves a control console to set targets, servo valves to convert electric signals from sensors into hydraulic action, and a hydraulic system to operate cylinders for moving the blade automatically. An operator can then set a desired slope from zero to 20%, which in turn sets a pendulum gauge. If the blade's slope does not agree with the setting, the pendulum signals the servo valves to correct the blade. The grade control works in a similar fashion but follows a grade string line or tracer wheel just beyond one end of the blade. If the blade end moves vertically from the established grade, a potentiometer mounted on the blade and supporting the other end of the guide rod signals the servo valves to correct the blade's position.

The motor grader can serve in a variety of other operations with special attachments. The special short blade to dig a shallow box-shaped trench has been mentioned. A similar attachment on the end of the blade is designed to use for narrow pavement widening to add a few extra feet beside the existing surface. This attachment is like a box with a front, cutting edge. When pulled along the pavement edge, the motor grader with this cutting box will be excavating the old shoulder material to make way for new pavement material. The box-like attachment may also be used as a slipform-type paver. Or it can be used to spread graded gravel uniformly for new shoulder along the existing pavement.

Another attachment to extend the usefulness of a motor grader is an elevating conveyor. This is rigged to take the loose material flowing from the trailing end of the blade, and elevate that material, casting it to the side or into a following piece of equipment to be hauled away. Such attachments as this conveyor, the shallow trenching blade, and a pavement widener all cause an eccentric load on the motor grader. The design of this equipment is able to take care of this eccentricity because of the leaning front wheels and the articulated pivot arrangement. These design features are shown in Figs. 5-11 and 5-12.

One other use of a motor grader with an appropriate attachment is worth mentioning. Since this versatile piece of equipment is often in the fleet carrying out an earthmoving operation, it can do double duty at times. The motor grader generally helps the fleet of haul units by maintaining the haul route in good condition. However, with a bumper block mounted on its front end, it can serve occasionally as a pusher to help load scrapers. Of course, if there is a constant need for a pusher, it would be better planning to have a tractor on the job specifically for that purpose.

5.5.3 Productivities of a Motor Grader

The productivity of a motor grader in its basic operation of grading is figured on the basis of time to do its work. This is different from productivity for a bulldozer or other earthmoving equipment. They are estimated on the basis of cubic yards moved per hour. In the case of a motor grader the actual volume of material moved is too variable and not of prime importance. What is more significant for this piece of equipment is the number of passes that will be needed to grade an area. That is, how many times will the motor grader have to go over an area to completely grade it? This will be several times, or many, depending on the initial condition of the grade and how precise the finish must be.

Generally, the speed of forward travel is fairly slow and constant to allow the operator full control of his grading. Experience with the operator under various grading conditions will make it possible to estimate the number of passes needed to do the grading operation. Estimates of these values and others can be used to find a motor grader's productivity. The formula to find the time required to get a grading operation done is

$$T = \left(\frac{d_f}{v_f} + \frac{d_r}{v_r} \right) \frac{N}{E}, \text{ in minutes} \qquad (5\text{-}9a)$$

where d_f = distance in linear feet for the motor grader to travel forward in one direction per cycle;

d_r = distance traveled returning to begin next grading cycle;

v_f = average speed of forward travel in feet per minute;

v_r = average speed of the return travel in feet per minute;

N = number of forward passes the motor grader must make past a given point in the length of grading operation; and

E = efficiency of operation for the motor grader.

If the operation is short enough, the return may be in reverse gear over the same distance as the forward travel. In that case the speed of travel can be taken as the average of forward and reverse speeds, v_a. Then the above equation can be changed to

$$T = \frac{2dN}{v_a E}, \text{ in minutes} \tag{5-9b}$$

where the terms are as defined above with d being the one-way distance in feet.

Efficiency in operating a motor grader depends on a variety of factors. These include the operator's ability and what guidelines he has to go by, i.e., to what stakes or lines is he grading? Another key factor is the extent of uniformity, regularity, or straightforwardness of the operation. The grading for a football field should be much more efficient than grading the side slopes and drainage ditches for a curved stretch of road. If the motor grader is working by itself and the operator controls its productive time, the efficiency, E, may vary between 70% and 90%. Frequently, a motor grader is working in a fleet of earthmoving equipment. Under those circumstances it probably is not the piece controlling the production for the total operation. The efficiency of the motor grader operation, then, would be reduced below the range given above.

Example of Motor Grader Production. An example will be used to show how to determine the productive time for a motor grader. Consider the grading for a football field on which loose material has been dumped fairly uniformly. This material is granular enough to promote drainage while raising the level of the existing surface a half foot, compacted. The plan view of the field area in Fig. 5-13A shows two approaches to the grading operation. The first method shown, starting at the upper side of the field, as a shuttling back-and-forth pattern. A pass is counted each time the grader goes forward from one end of the field to the other. It will be assumed that each pass of the motor grader with an 11-ft blade covers an eight-foot-wide strip of the field. Furthermore, each such strip requires four passes to knock down the windrow and grade the material satisfactorily. Then $N = 80/8 \times 4 = 40$ passes.

This approach to the grading will make use of Eq. (5-9b) with an average forward and reverse speed. If the grader travels at maximums of 4 mph forward and returns at 12 mph, it will have an average speed with time for acceleration, etc., $v_a \approx 6$ miles per hour. The efficiency of the motor grader operating independently of other equipment will be assumed at 80%. Then the required time for the operation will be

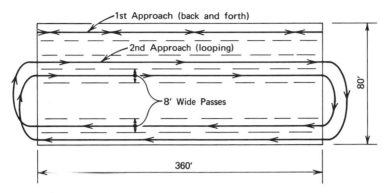

FIGURE 5-13A Grading patterns on field.

$$T = \frac{2 \times 360 \times 40}{6(88) \times 0.80} = 68 \text{ minutes.}$$

The other approach is travel continually in forward gear at an average speed of, say, 3 mph, making loops beyond the area of the field. This means that each trip from one end of the field to the other in either direction is a pass. The length of each pass must be increased by the semi-circular turn at each end. Then, using Eq. (5-9a) modified,

$$T = \frac{(360 + 3.14 \times 20)}{3(88)} \times \frac{40}{0.80} = 80 \text{ minutes.}$$

This type of operation is shown in Fig. 5-13B.

5.5.4 Costs of a Motor Grader

The cost of operating a motor grader must include the ownership and operating expenses. These are covered just as previously discussed for other construction equipment, and specifically tractor equipment in Sec. 5.4.4. Of course, if the motor grader is rented, the total rate will include the ownership and operating expenses as well as the addition of a per-centage for profit.

The variations in cost of a motor grader are generally based on the total weight of the equipment. One recognizes that the blade lengths vary only from about ten feet to fourteen feet because of the basic grading performed. However, the weight of motor graders varies considerably more, from the lightest under 10,000 pounds to the heaviest over 100,000 pounds total. The heavier unit has the advantage of being more effective with

FIGURE 5-13B Grading a field with a motor grader (courtesy of Deere & Company).

tough, variable bank cutting and other operations where maximum force must be applied. Of course, the heavier units need larger engine power to operate. The horsepower rating of engines in motor graders is practically proportional to the total weights. So the cost of a heavier motor grader has not only a higher ownership expense but also higher operating expenses. These expenses are practically in direct proportion to the total weight of the equipment. The smaller motor graders cost about the same or slightly less than a tractor of comparable weight and engine power. However, larger motor graders are more expensive than tractors of equal weight and power. This is due to the greater complexity and desire for more precise control in delivering power through the motor grader. The addition of a scarifier attachment increases the cost by only 5 to 8%.

With the above comparatives, we can assume that the motor grader doing the grading of the football field in Sec. 5.5.3 has a cost fairly close to that of the dozer in the cost example earlier in this chapter. Let us assume that the grader costs $16 per hour of operation. Then the cost for grading the field of $(80 \times 360) \div 9 = 3200$ square yards figures to be approximately 0.7 cent per sq. yd. However, with such a short operation the expense of getting the motor grader on the job site and other expenses of setting up for this operation will be relatively high. Also, we must recognize that the grading is not the total cost, nor even a major part of the cost for this earthwork. The cost of getting the loose material on the field distributed for grading is a major part of the cost. This might be done by scrapers or trucks hauling in from some source off the field site. The costs for such equipment will be discussed later.

5.6 Scrapers for Earthmoving

The earthmoving scraper was invented and developed as a piece of equipment that can load, haul, and dump loose material. The part of the equipment which handles the material is called the scraper. The other necessary part provides the power and is known as the prime mover. So the equipment may be called a tractor-scraper. For simplicity, the entire combination is generally just called a scraper.

The scraper with its prime mover can work independently of any other equipment. With this capability the equipment is known as a self-loading earthmover. Scraper equipment is widely used in earthwork operations. For this reason it will be discussed in more detail than most other equipment in this book.

There are several possible basic designs for scraper equipment. Most scrapers have a single-engine prime mover. In that case the power unit is in the front end of the tractor part and serves to pull the scraper part. There are also dual-engine scrapers, such as the one pictured in Fig. 5-14, with the second power unit behind the scraper bowl. It serves to push the whole equipment, and must be coordinated with the front, pulling power unit.

Another basic difference between scrapers is in their tractor support mechanism. A few scraper bowls are pulled by track-type tractors. Most scraper combinations move entirely on rubber-tired wheels. Then, the distinguishing feature for a scraper is whether the tractor part has one or two axles. With the same weight and scraper size, a single-axle prime mover can count on more traction than the two-axle tractor, such as the one shown in Fig. 5-15. The single-axle tractor is more maneuverable, but is also more difficult and dangerous to operate.

A third basic design difference is between the conventional type and an elevating type of scraper. This distinction can be made by studying the scraper parts in Figs. 5-15 and 5-17. Each of the main design features and their benefits will be discussed in the following sections.

5.6.1 Design Features of Scrapers

The overall dimensions of a scraper are governed by the bowl, which is the container to hold the material loaded, hauled, and dumped. At the

FIGURE 5-14 Dual-engine scraper in action (courtesy of TEREX, Division of General Motors).

bottom, front edge of the bowl is the cutting blade. The front wall of the scraper is a movable gate called the apron. The scraper loads material when the cutting edge is down in the material with the apron raised and the bowl moving forward.

The basic features of a scraper's design are shown in the dimensioned plan and elevation views of this type of earthmover (Fig. 5-15). It also shows the ejector gate in its back-loaded position. This part helps in both the loading and dumping actions of the operation. The ejector moves back and forth in practically a vertical position extending from one side to the other inside the bowl. At the start of loading, this gate is in a forward position near the apron and cutting edge. It moves back as the load is increased, serving to help the loose material boil up while being loaded in the bowl. The material moving against the ejector gate is deflected like an ocean wave against a breakwater. The action is shown in the cutaway cross section of a scraper's bowl (Fig. 5-16).

This boiling action promotes greater consolidation and larger total loads in a given scraper. The increased density is from 15–25% more than possible by merely dropping material into the open top of a hauling con-

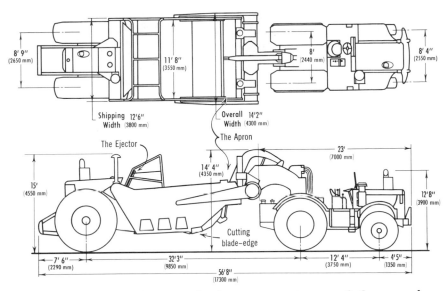

FIGURE 5-15 Plan and elevation views of three-axle
tractor-scraper combination (courtesy of
Caterpillar Tractor Co.).

tainer. For example, if the normal swell, $s_w = 30\%$, then the payload in a
12.5-cubic-yard container would be $12.5 \div 1.30$, or 9.6 cu yds bank
measure. With a scraper bowl of that size and greater natural compaction
so that $s_w = 25\%$, the payload would be 10.0 cu yds. Therefore, the

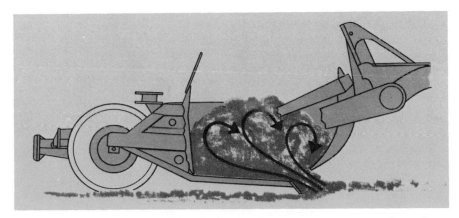

FIGURE 5-16 Sketch to show boiling in scraper bowl.

scraper can get a payload 4% greater due to the benefit of an ejector gate in the scraper.

Also, with the ejector working to concentrate the weight of material toward the front of the scraper, a greater amount of the weight is carried by the driving wheels of a single-engine total scraper unit. This helps with the traction of the moving equipment. The real significance of this will be seen in following sections.

Another scraper design to get a full load is called the elevating scraper. The special elevator feature of this type of scraper is shown in Fig. 5-17.

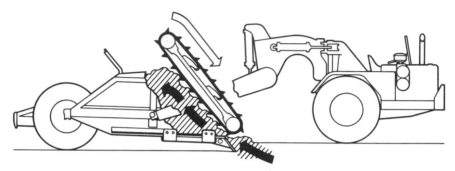

FIGURE 5-17 Cutaway showing loading action by an elevating scraper (courtesy of Caterpillar Tractor Co.).

The elevator works to help load the material by moving it in reverse direction from the boiling action in front of the ejector blade or tailgate of a conventional scraper. That is, the elevator lifts the loose material over itself along the underside of the elevator until it is dumped on top of the load.

The complete scraper equipment with the material-carrying bowl part and a prime mover may have one of several axle variations. The total combination may have 2, 3, or 4 axles for an all-wheel unit. Or it may consist of a two-axle scraper part with a crawler-mounted tractor unit. The load distribution for the all-wheel equipment is of great importance. This is because of the need for enough weight on the driving wheels where power is applied for loading and moving the equipment.

The best wheel arrangement for maximum traction and power application is with a two-axle all-wheel scraper combination. This is the type that has one axle for the scraper part and one for the prime mover. It might be called a 4 × 2 (wheels) unit, as a truck is identified. Such a combination cannot be separated and stand or operate independently of its counter part. In this usual scraper combination with a single engine, the approximate weight distribution to the drive wheels is 55 to 65% when empty and 50% when fully loaded. At least one manufacturer has a twin-engine scraper combination on two axles. The second engine is mounted at the rear in back of the scraper bowl. This engine puts power into the wheels on the rear axle. In this case all four wheels of what would be identified as a 4 × 4 scraper unit are drive wheels, giving the potential for a maximum of power based on the total weight of the equipment and its load.

The other common arrangement is a three-axle scraper combination. In this case the tractor has two axles so that it can operate independently of a trailing unit. If there is only one drive axle for such a tractor, it will be the rear axle of the prime mover's two or middle axle of the three-axle scraper combination. The scraper part attached with only one axle puts more weight on the drivers. The usual design of the 4-wheel tractor unit with 2-wheel scraper attached has a different load distribution than the 4 × 2. The tractor with scraper attached will have approximately 45% of the total empty weight on the drive wheels. If the scraper is full with a normal load, the load will be distributed with approximately 40% of the full load on the drivers. Some scraper tractors of this variety are designed with a 4-wheel-drive tractor. That insures that the total power can be used up to the limit of traction under the tractor's weight plus some from the scraper. These concerns are shown in the following discussion.

Self-Loading a Scraper. In order to have a scraper load itself fully, certain limiting forces must be satisfied. These include the force to cut the earth being loaded and pack the material into the scraper bowl, the required tractive effort from the tractor, and the necessary traction at the drive wheels.

An axiom for scraper operations that has proven reasonably accurate expresses the first force requirement. It takes one pound of available tractive effort (TE) to load each pound of material in a full scraper bowl. This might simply be stated "a scraper needs a pound per pound to load." Using this fact, the amount of load that a scraper can get by itself can be calculated. This determination will also help the understanding of the controlling forces in scraper operation.

The following calculations are based on information and understanding from Sec. 3.4 on forces and resistances to motion and Sec. 3.6.1 on rimpull of a wheel-type tractor. The limit of material obtained by a self-loading scraper depends on:

TE_L = tractive effort for load;

TE_R = tractive effort to overcome the resistances (rolling and grade) to motion;

$\text{TE}_{\max}$ = maximum force or effort engine can deliver to the drive wheel;

W_{dr} = scraper's weight on drive wheels;

C_t = coefficient of traction; and

TE_a = tractive effort actually applied.

Then, $\text{TE}_{\max} \geqslant \text{TE}_a \leqslant C_t \times W_{dr}$, and

$\quad \text{TE}_a = \text{TE}_R + \text{TE}_L.$

From this we find that the maximum effort that a self-loading scraper can use for loading, assuming enough engine power, is

$$\text{TE}_L = C_t\,(W_{dr}) - \text{TE}_R, \tag{5-10}$$

where $\text{TE}_R = W_T\,(f_{RR} + .01g)$;

$\quad W_T$ = total weight of scraper;

$\quad f_{RR}$ = rolling resistance factor, lbs/ton ÷ 2000; and

$\quad g$ = % grade (up is positive, down is negative).

Now we can find the limit of load that a scraper can get by self-loading. If it is a single-engine, two-axle scraper with weight distribution of 50% on the drivers when loaded, the maximum effort to get its load is

$$\text{TE}_L = C_t\,(0.5W_T) - W_T\,(f_{RR} + .01g).$$

Next, we apply the "pound for pound" axiom, so the weight of load, $W_L = \text{TE}_L$, and note that $W_T = W_L + W_E$. The term W_E is the empty weight of the scraper equipment. Then, the limit of material by self-loading is

$$W_L = [C_t\,(0.5) - (f_{RR} + .01g)]\,W_T, \text{ or} \tag{5-11a}$$

$$W_L = \frac{(0.5C_t - f_{RR} - .01g)}{1 - (0.5C_t - f_{RR} - .01g)}\,W_E. \tag{5-11b}$$

To show how this can be applied and to see how this limit compares with other weights of a single-engine scraper, let us work an example. Assume the job conditions are loading down a 2% grade ($g = -2$) on material with a rolling resistance of 80 lbs/ton and a 0.6 coefficient of traction.

$W_L = (0.30 - 0.04 + 0.02)\,W_T = 0.28\,W_T$, and

$W_L = 0.28\,(W_E + W_L)$;

$(1 - 0.28)\,W_L = 0.28\,W_E$;

∴ max $W_L = 0.39\,W_E$ by self-loading.

The twin-engine scraper was developed to improve on the self-loading ability of this type of earth mover. It is generally a two-axle arrangement where all four wheels are drivers. Then equation (5-11a) becomes $W_L = (C_t - f_{RR} - .01g)\ W_T$.

The example with a twin-engined scraper would show

$$W_L = (0.60 - 0.04 + 0.02)\ W_T = 0.58\ W_T, \text{ and}$$
$$(1 - 0.58)\ W_L = 0.58\ W_E;$$

$$\max W_L = \frac{0.58\ W_E}{0.42} = 1.38\ W_E.$$

This is more than a three-fold increase in the self-loading ability of the scraper piece of earthmoving equipment.

Comparison of Scraper-Tractors. As suggested earlier, the tractor serving as prime mover for scraper equipment can be a crawler or wheel-type. A crawler tractor would be used with a two-axled, free-standing scraper unit with a simple drawbar hitch so that the scraper is easily detached from the tractor. The ease in detaching is one advantage of this type of equipment. There is no attempt to use weight from the loaded scraper to the tractor's advantage. A crawler tractor is designed for maximum power applied with its total weight on the driving tracks. It sacrifices speed for the power it can apply. So this type of scraper equipment would be most useful for a short haul over rough terrain where wheel traction is poor or grades are adverse.

A wheel-type tractor is generally used with a single-axled scraper unit. This puts a significant amount of the scraper unit's weight on the tractor for better wheel traction. An explanation of this benefit was covered in the previous section. The prime mover tractor may be a single-axle or a two-axle unit. Generally, the rear wheels of a two-axle tractor are the power wheels and the front ones turn for steering. In this case only part, though it is the majority, of the tractor's weight is useful for traction at the power wheels. That weight is increased by a fair share of the scraper's weight supported on the tractor's rear axle.

The previous section, discussing the self-loading of a scraper, concentrated on a single-axle tractor with the scraper. The same sort of determination could have been made for a two-axle tractor unit. However, it is obvious that for self-loading, as much weight as possible must be on the power drive wheels to get the full benefit of the engine's power in loading. This is done with a single-axle prime mover or with 4-wheel drive in a two-axle tractor.

A comparison of single-axle and two-axle tractor units for scraper equipment will serve to summarize this discussion. The advantages of a single-axle tractor for scraper equipment are:

1. it can develop more tractive effort than a two-axle tractor with the same size engine, unless the latter has 4-wheel drive;
2. it is more maneuverable because the two wheels of the single axle can practically pivot about their mid-point; and
3. the single-axle tractor seems to "float" over rough terrain with the extra freedom of movement about the tractor-scraper hitch.

In contrast, a two-axle tractor has the advantages of being:

1. safer to operate in case of tire blow-out or trouble with the hitch or other working parts; and
2. more versatile for a variety of uses as a tractor when it is detached from the scraper bowl unit.

5.6.2 Operations for a Scraper

The scraper combination of equipment is very effective for operations where loose earth material must be picked up and moved some distance. It picks the material up a minimum height, saving energy, and dumps it in a uniform layer for final placement. These abilities are desirable in such operations as (1) stripping top soil, (2) contour grading around a building site, (3) cutting for a drainage or irrigation ditch, and (4) cut-and-fill earthwork for a highway or the like. The factors to recognize in these operations for scraper work will be discussed next.

A scraper is an ideal piece of equipment to cut and remove topsoil since this is a thin layer on the ground surface. The top layer to be removed is usually from 4″ to a foot in thickness. Scraper equipment can generally remove this with a single pass over the area. The planning for an economical operation will require knowing the shape and extent of the area to be stripped and what is to be done with the cut topsoil. If the area is modest in size, measured in acres, a smaller scraper with self-loading features and perhaps a crawler tractor may be the best unit to use. However, for a large-area site, such as an airfield, with the long dimension approaching a mile or more, large high-speed, wheel-type scrapers with pusher tractors to help load probably will be more economical. If the topsoil is to be stockpiled in mounds at the ends or sides of the large area, it might be well to compare an elevating grader with hauling units to the choice of other scraper units.

The scraper type of earthmoving equipment is ideally suited for contour grading around building sites. Its ability to cut material and deposit it in variable, controllable layers makes the scraper the appropriate equipment for this work. The grading for a building site usually means short haul distances. A small, two-axle scraper combination will have the maneuverability needed for such an operation, and it can work independently by loading itself. If the average haul distance is around 100 feet or less, the economics of using a scraper for this operation should be compared with the alternative selections of a dozer or a motor grader.

The cutting of a drainage or irrigation ditch can be done successfully by various pieces of earthmoving equipment. If the bottom width of the ditch equals the wheel clearances of the scraper, and if the side slopes are moderate, scraper equipment can be used for this operation. A bottom width equal to the scraper's cutting blade width or slightly less than an even number of blade widths would be ideal. Thus, if the scraper's wheel clearance is 13 feet wide, ditch bottoms of 13 or more feet are necessary, and side slopes should not be steeper than 1 on 3 for successful operation. Steeper slopes could be cut best by a motor grader. The teamwork of scraper equipment with a motor grader could be ideal for this type of operation. If the material cut for the ditch can be left along the sides, instead of hauled away, the motor grader alone, an elevating grader, or a dragline excavator (discussed in Chap. 7) may be more economical. The advantages of a scraper for ditching are realized where the contours can use the full width of the scraper's cutting blade and the material is to be hauled away.

The cut-and-fill operation is one involving excavating earth material from some places and building up or filling in other locations on construction sites. This is done on every highway project and, generally, on airfields, dams, building sites, etc. A scraper can do this type of operation well because of its designed features and capabilities. It is generally the best and most economical equipment for cut-and-fill operation, if the one-way haul distance is between 300 feet and 3000 feet. Of course, it is assumed that the right type and size of scraper will be selected to be economical for these haul distance extremes. This will be explained shortly.

The economical scraper equipment to use for a cut-and-fill operation depends on the (a) material to be cut and handled, (b) length of haul route, (c) haul route conditions, and (d) equipment to help with scraper operation. For a short haul of 300 feet or less, one way, the equipment planner must compare the choice of scraper to a bulldozer. This is done in the example of Sec. 5.6.4 (see p. 222). At the other extreme, a long haul distance of 3000 feet or more, the comparison for most economical equipment will have to include a power excavator or loader combined with haul units.

Between these extremes of economical haul distances for scrapers the selection becomes one of the type of scraper equipment. The Highway Research Board (HRB) made a series of studies on the use of the scraper combinations—crawler-tractor and scraper vs. rubber-tired tractor and scraper equipment. These studies can be summarized with the curves of Fig. 5-18. The modern high-speed wheel combination has been added as it might show with updated studies.

As these curves show, the HRB studies found that the majority of scrapers used for the crawler-tractor combination was on one-way haul distances between 150 and 700 feet. Similarly, the all-rubber-tired scrapers were used for distances from 400 to 1400 feet. It must be noted that these studies were done before the existence of the present, large, higher speed scrapers with wheel-type tractors. This is the reason for introducing the third curve in Fig. 5-18.

It will be helpful to summarize the use of the various scraper equipment combinations. Table 5-4 is based on the types of units described in the previous sections and is only for general variations.

An economical determination for a specific job situation must be calculated as shown in Sec. 5.6.4.

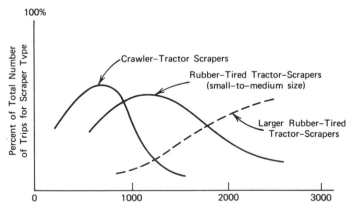

FIGURE 5-18   Comparison of scraper types, based on haul distance.

TABLE 5-4 Scraper Use Variations

Haul Length (300'–3000')	Material and Route Conditions	Type of Prime Mover, Tractor for Scraper Combination
Short	Rugged	Crawler-tractor
Short	Easy	2-wheel tractor
Medium	Tough	Crawler-tractor or twin-engine wheel unit
Medium	Average	Any wheel tractor with pusher, if needed
Long	Average or tough	2- or 4-wheel tractor with pusher and/or puller tractors helping

5.6.3 Productivity of Scraper Equipment

The production, or output that can be expected in an hour's time, from a scraper depends on several factors. Of course, the design features and capabilities of the scraper equipment, discussed previously, are most important. In addition, the productivity will depend on (1) the nature of the material dug and loaded; (2) the power available for loading; (3) haul routes—their grades, alignments, and conditions; (4) the travel speeds possible on continuous stretches of haul route; and (5) the efficiency of the operator driving the scraper equipment.

The material being handled affects the scraper's productivity by its reaction to being cut or dug and loaded into the scraper bowl. The properties of concern were discussed in Sec. 3.3.1 and 5.1.3. Material that is workable, i.e., without too much binder or cohesion, such as topsoil or a dry or moist clay or sandy loam, is best. A material that breaks out of the ground mass into chunks or lumps, such as wet clay, is poorest for high scraper yardage output because it will cause large voids in the hauled loads.

The scraper's ability to get a compact load is somewhat different from dumping a loose load into an open container. Consequently, the time it takes to get the payload in a scraper bowl must allow for a loading factor called "loadability" or digging efficiency factor, E_L. Values that can be used for this are given in Table 5-5; their use will be shown shortly.

TABLE 5-5 Scraper Loading Efficiency Factors for a Variety of Materials

Material Handled	Loading Efficiency, E_L, in %
Topsoil	100
Clay loam (low moisture)	97
Sandy loam	95
Clay, dry	90
Clay, heavy, moist	80
Sand, loose	75
Gravel, loose	67
Dense clay and shale	60
Glacial, bulky stone	50
Friable shale or rock	33

Loading Time for a Scraper. The power or tractive effort available to load material in the scraper is another key factor. As indicated in Sec. 5.6.1, a scraper needs a pound of tractive effort for each pound of material loaded. This should be recognized as a rule of thumb. It is extended with a time element by stating "it takes a scraper pound for pound to be loaded in one minute." This is based on field tests and observations by manufacturers of scraper equipment. The outcome can be graphically shown as in Fig. 5-19.

The curve shows that with sufficient tractive effort and average material

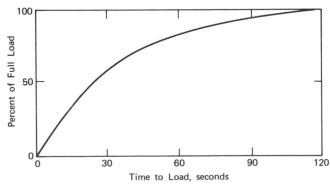

FIGURE 5-19 Scraper load-growth time.

conditions the majority of a full load is obtained in one minute of loading time. A completely full load could be packed in if 2 minutes are used, but doubling the time does not result in anything near twice the load. Also, it must be recognized that with more power available, such as by a pusher tractor, greater load can be obtained and in a shorter time. Obviously, the planner cannot reduce his loading time to zero by using the most powerful tractors available for loading. There must be an economic trade-off which appears to result from a scraper loading time of somewhat less than one minute with today's powerful equipment.

With the help of pusher or puller tractors it is assumed that 100% of the load capacity of the scraper can be achieved. A newer scheme for extra effort to load a scraper has been called the push-pull or "helpmate" method. Two scrapers work end-to-end in the loading area. To load the first, a pusher block is used, so that the second scraper serves as a pusher.

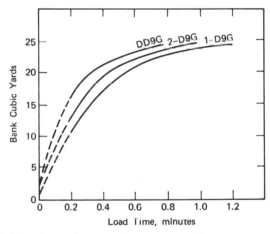

FIGURE 5-20 Load-growth curves for various scraper-tractors (courtesy of Caterpillar Tractor Co.).

Load Time	D9G	2-D9G	DD9G
.2	10.4	12.2	16.0
.4	17.0	19.9	21.5
.6	20.6	22.0	23.1
.8	22.5	23.3	24.0
1.0	23.7	24.3	
1.2	24.4		

When the first is loaded, it helps the second scraper load by pulling with a hook engaged at the block.

It should be noted that any load-growth curve, such as shown in Fig. 5-19, is for a given scraper working under a specific set of job conditions. A family of curves can be drawn for any scraper's job where it may be helped in loading by various tractor power equipment (Fig. 5-20).

To optimize the scraper-pusher production team, the planner is interested in the lowest total cost for the combined equipment. This objective will generally be reached for a scraper-loading team with the load time (LT) that results in the highest productivity. The Caterpillar Tractor Company has presented a graphical method for finding that LT in its publication "Optimum Load Time."[6] This booklet is recommended reading on the subject of scraper loading times. The graphical method is summarized in the next paragraphs.

The scraper's times for hauling, dumping, and return are constant for a given job layout. The total of those times can be shown as OA horizontally on the same scale with the load time in the load-growth curve. OA is plotted starting from the origin (O) and extending to the left opposite the LT. The vertical scale is the payload of material to load in the scraper. These are shown on the graph of Fig. 5-21, which is a summary of this graphical method.

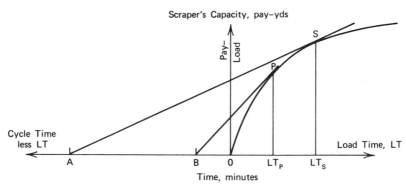

FIGURE 5-21 Graphical method to optimize scraper-pusher load time.

The optimum load time, LT_S, for the scraper can be found by drawing a tangential line from point A to the load-growth curve. It is tangent to the curve at point S. The slope of line AS gives the optimum productivity as far as the scraper is concerned. The value is found as follows:

$$\text{Slope of AS} = \frac{\text{vertical side}}{\text{horizontal side}}$$

$$= \frac{\text{pay yards/trip}}{\text{cycle time/trip}}.$$

Slope = pay yds/minute = productivity.

Any other line from point A to the curve will give a smaller productivity, so AS does give the optimum LT_S.

The same procedure can be followed for the pusher, which has a shorter cycle time between load times. Its cycle time less LT will vary with the length of the loading path and time. But the difference for the pusher can be shown with reasonable certainty by a value such as OB. The optimum productivity, as far as the pusher is concerned, is found by the tangential line BP. And the best loading time for the pusher is LT_P.

The load time to use, LT, must be between LT_S and LT_P to be an economical operation. Since the pusher's total cycle time, CT_P, is much less than the scraper's, it can help with the loading of several scrapers. The number of other scrapers the pusher may handle can be judged by dividing $(OA + LT)$ by CT_P. To optimize his fleet, the planner should select LT based on whether the scraper or pusher is working to capacity. If there are more scrapers than the pusher can normally handle, then the loading time should be shortened to LT_P. On the other hand, with excess pusher capacity LT_S would be a more economical load time.

The Caterpillar publication continues to make a comparison of productivities for one to four scrapers and a pusher with load times varying up to 1.4 minutes. These are translated into costs per pay yard. Conclusions that were reached can be stated as:

1. it is generally not economical to push-load until the scraper is filled to capacity;
2. maintain a reasonable balance between the scrapers and pusher; and
3. adjust loading time between the limits LT_P and LT_S to meet changing haul conditions so that the waiting time for the scrapers and pusher are minimum.

A way to find a reasonable load time without plotting a load-growth curve follows. Whatever equipment plan or scheme is used, it is assumed that normally close to 100% bowl capacity can be loaded. This is the basis for the following equation:

$$LT = \frac{W_L \times 100}{TE_L \times E_L},\qquad\qquad (5\text{-}12)$$

where LT = loading time in minutes;

W_L = weight of material to load, pounds;

TE_L = tractive effort available for loading from all sources, pounds/minute; and

E_L = loading efficiency factor in % from Table 5-5.

If two or more power units are working together and the actual tractive effort for loading (TE_L) is not known, it should be safe to assume that $TE_L = W_L$.

To see how Eq. (5-12) can be applied, an example will be shown. Assume that a scraper can get a full load of dry clay which weighs 50,000 pounds. Using a tractor and pusher with net available (deducting for rolling resistance, etc.) tractive effort for loading is figured to be 56,000 pounds. Therefore, the loading time is determined as

$$LT = \frac{50,000 \times 100}{56,000 \times 90} = 0.89 \text{ min.}$$

If the same scraper and tractors were used to load bulky stone, it would take, $LT = 1.79$ minutes. The reason loading bulky stone takes so much longer is that this material does not consolidate readily the way loam or dry clay will. In such a case, the guidance of the Caterpillar publication should be followed.

Effect of Haul Route. The plan and conditions of the haul route that scraper equipment must travel are key factors in its productivity. Planning the route must take into account the grades and turns. Each percent of uphill grade calls upon the prime mover to exert 20 pounds per ton of moving weight. This is the weight of scraper equipment and material load it is carrying. Traveling on a downhill grade gives a power advantage of 20 lbs/ton/% grade. Therefore, it is beneficial to haul a load on downhill grades and travel uphill with the equipment empty. Of course, this choice is not often available. The choice of hauling up a short steep grade that can be made in the lowest power gear or taking a longer, less steep grade in a higher gear with greater speed is possible in the planning. The one to select will depend on the speeds and distances for all stretches of the route and, so, the total time for a round trip. Examples of this sort of analysis are given in a following section of this chapter.

The turns that a scraper makes on its haul route take extra time. An about-face turn of 180° consumes about one-quarter minute. Therefore, a complete cycle for a scraper from loading to dump and return will have a minimum of two such turns (360°) and, so, 0.5 minutes/cycle. Additional turning in the haul route plan will add to the cycle time 0.25 minute/180° turned. Obviously, unnecessary turns should be avoided.

The condition of an earthen haul route may vary from "compacted and well-maintained" to "rutted, muddy, with no maintenance." These surface conditions were discussed in Sec. 3.4.2 and included in Table 3-5 on rolling resistance for various types of contacts and surfaces. For rubber-tired equipment traveling on compacted, well-maintained earth, the resistance to motion is 40 to 70 pounds per ton of moving weight. If the earth haul route surface is rutted, muddy, and not maintained, this resistance will be 150-220 lbs/ton. The rolling resistance calls for an equivalent tractive effort from the power unit to propel the moving scraper equipment. The greater this resistance is, the more power will be needed. For a given power unit, a greater power requirement will mean less speed of travel. In the comparison of possible haul surface conditions given above, there would be a difference of 2 to 4 times in the travel speeds if they were directly proportional to the resistances.

Of course, the actual speed of travel is not directly proportional to power requirements. But there is a close enough relationship to suggest that the construction equipment planner pay attention to the condition of his haul routes. The reason is that the planned cycle time for scraper equipment hauling its load should be as short as possible for an economical operation. If the rolling resistance of the haul route can be reduced to permit travel at higher speed, this should be done. For a large earthmoving operation with many scraper units required, it may be economical to put a motor grader on the job. This piece of equipment would be used to maintain the haul route and keep the rolling resistance for the scrapers to a minimum. An example of this economic determination is given in a following section.

The travel speeds possible for scraper equipment depend on the resistances to motion, primarily grade and rolling resistances, and the prime mover to overcome these. The layout, composition and condition of the haul route as just explained are key factors. The type of power unit in the prime mover is also significant. If it is a torque-converter or power-shift type of unit, the operator will merely have to choose the appropriate gear range to use for different parts of the scraper's cycle. This may mean only one shift from the low range for loading to a higher range for traveling. The power unit will adjust automatically with varying power demands and, consequently, speeds for the different travel conditions on the haul route.

If the scraper is equipped with a direct-drive-type power unit, the operator will have to shift to the most appropriate gear for each part of his cycle. The loading part will probably be in a low gear. An uphill grade with full load will perhaps be in a second gear, whereas travel on a fairly level surface will be in a high forward gear. The normal speeds for covering the distances of each stretch will be the governed speed for the chosen gear, as explained in Sec. 3.1.2. However, time is taken to shift and accelerate up to the new speed of a higher gear. Likewise, some time is consumed in decelerating and braking when reducing to a lower speed.

The time given in an earthmoving cycle for acceleration, deceleration, and braking can be determined, based on such factors as the type of power unit and the gears used for the specific haul cycle. When these are known, the so-called acceleration-deceleration-braking (ADB) time can be considered a fixed time. The definition of a fixed time (FT) is one that does not depend on the distance moved in the equipment's cycle of movement.

Values of the ADB fixed time have been determined by careful time studies made on equipment operating in the field. Representative values are given in Table 5-6.

TABLE 5-6 Acceleration-Deceleration-Braking Times for Tractor Operation

Tractor Unit, Gear	ADB Time, minutes
Crawler tractor, all gears	0.5
Wheel-type, power shift	
2nd range (0–10 mph)	0.4
3rd range (0–25 mph)	0.7
Wheel-type, direct drive	
1st to 2nd gear (10 mph max.)	1.0
to 3rd or 4th gear (to 20 mph)	1.5
to 5th gear (to 30± mph)	2.0

The operating efficiency is an important factor in determining the hourly production or output from an earthmoving scraper. The ability and skill of the operator running the equipment is obviously important. However, his efficiency is also governed by such factors as (a) the design and capability of the scraper-tractor combination, (b) the planning for its coordination with other equipment in the loading step and at the dumping site, and (c) travel conditions on the haul route. Some points on these have been discussed in the previous sections.

The effect of the various work conditions on the operator of the equipment and their operation is shown in efficiency factors. During certain ideal times, a planner can look for maximum, peak production realizing the full capacity of the equipment. This productivity is said to have 100% efficiency and, if carried on for a whole hour, would amount to a 60-minute productive hour. The peak productivity should be used for determining the pusher tractors or any other secondary equipment to work with the earthmoving scrapers. To plan for such primary equipment's production for the duration of an operation, a reasonable, lower efficiency should be used.

The normal operating efficiency of scraper equipment over several hours or more will differ somewhat with the type of equipment. Crawler-tractor scrapers should be easier on the operator because there is less bumping and vibration with slower transitions in the moves compared to an all-rubber-tired unit. Therefore, it is reasonable to figure the normal operating hour for crawler-tractor scrapers as 50 minutes. That is, the normal working efficiency, $f_w = 0.83$ ($= 50/60$). In the case of rubber-tired scraper equipment with faster moves, quicker turns, and more floatation and vibration, the normal operating hour should be 45 minutes. This means a normal working efficiency of 75% ($f_w = 0.75$).

These values are comparable to ones found in time studies made by the Highway Research Board with observations of more than 6,000 available working hours on more than a dozen earthmoving jobs. Such studies were extended to find the overall operational efficiencies. They took into account all delays in the operation of scrapers, including weather. The results showed the overall or average operational efficiencies (f_a) to be about 60% for crawler-tractor scrapers and about 30% for rubber-tired tractors which are much more subject to delays caused by wet weather.

Calculating Scraper Productivity. The calculation for the production, or hourly output, of a scraper puts together all the points discussed previously in this section. It's based on figuring the production time of a single cycle from loading, through the haul, to dumping, and return for the next load. Then the load and time of one cycle for the scraper is projected to expected time of operation in a clock hour. This gives its productivity.

To summarize the procedure, the key points in the calculations will be reviewed. First, 'the scraper's payload must be figured. Assuming a full load can be obtained with the help of a pusher, or whatever means, the heaped capacity of the scraper's bowl must be known. Then the number of pay yards per cycle is obtained by applying the % swell from Table 5-1 for a swell factor to use in Eq. 5-1. In that equation with V_l as the scraper's heaped capacity, we are solving for the bank measure or pay yardage. Thus, pay yards, $V_b = V_l \div s_w$. The pay load per cycle is found by multiplying these cubic yards times the weight of a cubic yard of material taken from the bank or pay volume. This maximum load should be checked to be sure it does not exceed the load limit of the scraper's axle or other supporting parts.

The variable travel time (VT) for the planned haul route requires separating the route into lengths with common grades and surface conditions. Then, for each such length or part, the total resistance to travel is figured. This resistance includes grade resistance, F_{GR}, and the rolling resistance, F_{RR}, for the type of scraper equipment used. For this total resistance the appropriate gear or gear range is found for the scraper's tractor. The top governed speed of operation for that part of the haul route is taken from the specifications. The time to travel that part of the cycle at top speed is figured from its length. The same procedure is used for other parts of the haul route. Of course, the times that vary with the length do not allow for the extra time to get up to top speed or decelerate at the end. This is covered by fixed-time additions.

There are various fixed times (FT) to include for a scraper's total cycle time. They are practically the same regardless of whether the one-way haul route is 1000 or 5000 feet long. They do not depend on how long it takes to travel along parts of the haul route but only on the number of gear shifts and turns made. The fixed time for acceleration, deceleration, and braking (ADB) accounts for the anticipated shifting of gears. The ADB times were covered previously. The time to allow for turns (TT), as previously discussed, suggests 0.25 minute/180° turned. This might be modified if special consideration is given to the dumping part of the cycle. Generally, the fixed time for dumping the load (DT) is assumed to be half a minute. If the steps to take in the dump area are more precisely known, the time for dumping and turning there may take from 0.4 to 1.25 minutes per cycle. These values were determined by time and motion studies. There is obviously some difference at the above extremes from the time suggested earlier, where the fixed time, assuming a 180° turn in the dump area, would be $0.25 + 0.5 = 0.75$ minute. Actually, the average for this part of the fixed times was 0.6 to 0.75 minute from the time and motion studies.

The most important single item of fixed time for scraper operation is the loading time (LT), also discussed earlier. The total fixed time is the combination, $FT = LT + ADB + TT + DT$. A reasonable minimum total of these for a power-shift, all-wheel type of scraper on a simple loop is 2.5 minutes/cycle. At the other extreme, the fixed time could be more than twice this amount. In any case it should be calculated rather accurately for any important estimate.

The productivity can now be figured, using the total time for a cycle,

$$CT = VT + FT \qquad (5\text{-}6c)$$

This is another variation on Eq. (5-6) given in Sec. 5.2. With the cycle time and pay load per trip known, the scraper's production can be calculated. This is referring to the equipment's output per hour. It means an estimate of the productive, working time per hour. For peak productivity 60 minutes would be used, but this rate of production could not be expected for very long—probably less than a full clock hour and certainly not for a day. A reasonable productivity (q), depending on the basis for the estimate, can be made using the expected working efficiency. Then

$$q = \frac{V_b}{CT} \, (f_w) \, 60, \qquad (5\text{-}13)$$

which is the pay yards per hour, using the definition of symbols and units given previously.

5.6.4 Economic Comparisons for Scrapers

A discussion of the costs of scraper equipment could follow that given for tractor equipment in Sec. 5.4.4. Review that section for the basic costs of using owned equipment. Several unique particulars to realize in dealing with scraper costs will be taken up now.

With the high speeds anticipated for all-wheel scrapers, there is great wear and tear on the tires. The usual experience with reasonable care and attention for the tires is that the equivalent cost of one new tire will be paid for this expense of a scraper each operating year. This is with normal use of approximately 2000 working hours. Variations in use will lead to other wear experience. A review of Sec. 3.2.2 on tires is suggested.

Other parts of scraper equipment with a high rate of wear are the cutting edge, bowl bottom, apron, ejector gate, and cables or hydraulic parts. To extend the life of these parts, higher-strength metals may be used. This will increase the initial cost but reduce the maintenance cost and expense of idle time. Also, the life of the cutting edge can be extended by equipping the bowl with router bits at some minor added cost. These are like small scarifier teeth. They are particularly helpful for scraper work in shot rock or other hard material.

The biggest difference in initial cost of a given size of scraper occurs when it is equipped as an elevating scraper. The elevator mechanism adds 15 to 25% to the original price. However, there is the advantage that the elevating scraper can load itself to full capacity more successfully, and by reversing the loading mechanism, it can unload faster than the conventional scraper. Also, the scraper type of equipment will be more successful with more powerful engines or electric motors. Of course, the larger the power unit, the higher will be the original cost of this equipment. The equipment expense for scrapers is almost directly proportional to the bowl capacity and to the hp rating of its engine. However, with the smallest units commonly used—in the 6-12 cu yds range—the cost is somewhat higher. This is generally because a proportionately larger power unit is used to improve the scraper's self-loading possibilities. At the upper extreme the cost is also proportionately higher because the demand for scrapers over 30-yard capacity is not great enough to reduce their cost.

Scraper-type earthmoving equipment is designed to dig, load, haul, and dump earth material. It is the only construction equipment that can do all these steps of an operation economically by itself for a wide range of haul distances. That range is generally from about 300 feet to more than a mile. The front-end loader equipment, which will be covered next, can do all those steps, but for much more limited haul distances. Of course, the scraper has competition from this and other equipment for some earthmoving operations for which it might be considered.

At the lower limit of its economical one-way haul distance, the scraper competes with a bulldozer. Even though the dozer does not pick up and haul its load, it can push it over the ground surface for several hundred feet economically. A cost comparison will be made next to show this.

Example of Scraper vs. Dozer. This is an example of a cut-and-fill operation with a 300-foot one-way haul distance and a total work area that is generally level. Likely choices of equipment to compare will be crawler-tractor prime movers appropriate to a dozer blade with 3-cy capacity or a scraper combination with a bowl capacity of 10 cubic yards heaped or 12-ton load, whichever is smaller. For the earth material and ground surface we will assume 3000 lbs/pay yard; rolling resistance (RR) = 110 lbs/ton; and coefficient of traction (C_t) = 0.9, for crawler-tractors.

The calculations for the dozer on this operation will be done like the ones in Sec. 5.4.3. For the scraper, in this case with heavy material, the load limit is $(12 \times 2000) \div 3000 = 8.0$ cu yds, governed by the load's weight limit. The speeds of travel for the short distances will be in 2nd forward (direct-drive) gear—the loading gear—moving at 2.0 mph (176 fpm) for 200 feet. The return empty is in top gear at 6 mph (538 fpm). For such a short haul the scraper should be self-loading to be economical. We will assume the load time to get the full 8 cy to be 1.5 minutes and the other fixed times to take 1.1 min. The results of each equipment in the comparison can now be tabulated:

	Scraper	Dozer
Capacity, pay yards	8.0 cu yds	3.0 cu yds
Speed loaded, mph	2(176 fpm)	3(264 fpm)
Speed empty, mph	6(538 fpm)	5(440 fpm)
Cycle times, minutes:		
Loading, LT	1.5 est.	in VTL
Travel loaded, VTL	1.14	1.14
Travel empty, VTE	0.56	0.68
Fixed time, FT	1.1 est.	0.3
Total, CT	4.3 min	2.1 min
Normal productivity:		
$q_n = $ (pay yds $\div$ CT) 50	93 cy/hr	71.5 cy/hr
Costs of production (for reasonable comparison)		
Total with equipment rented	$32/hr.	$28/hr.
Cost per pay yard	34.5¢	39.2¢

Notice that the outcome of this comparison could favor the dozer if there were a greater cost differential per hour. In this example this was estimated to be $4 per hour. Also, note that a higher travel speed could be used by the loaded scraper. But for such a short haul, it may be preferable to have the maximum power at all times. A power-shift, torque-converter-type unit would have this and the maximum speed possible, automatically. That likely would be more expensive equipment.

Example of Crawler vs. Rubber-Tired Scrapers. For a scraper operation with a relatively short haul, in the order of 1000 feet one way, a comparison of types of equipment as shown in Fig. 5-18 should be made. Such a haul distance is not too long for economical use of a crawler tractor for the scraper combination. It is long enough for the higher speed, rubber-tired prime mover to be effective. Therefore, scraper combinations of these varieties should be checked to find the most economical equipment to use. To illustrate the calculations, an example will be worked.

The example compares scraper combinations with a crawler-tractor or a rubber-tired prime mover for a 1000-ft one-way haul cut-and-fill operation at 700-ft altitude.

Assuming: simple, straight, and level haul route with 180° turns in cut-and-fill areas; haul road in natural condition, RR = 110 lbs/ton; and loading dry clayey gravel (40% swell) 3000 lbs/cy in bank.

Alternate #1:   a crawler, direct-drive (dd) tractor weighing 23,000 lbs with two-axle, wheel-mounted scraper of 9-cy heaped capacity, weighing 13,500 lbs empty.

(a) Payload $= 9 \left(\dfrac{1}{1.40} \right) 3000 = 19,400$ lbs, using Eq. (5-1).

Check loadability, i.e., max TE $>$ sum of resistances:

assuming RR = 150 lbs/ton in level cut;

$$\text{total } F_{RR} = 150 \left(\frac{23,000 + 13,500 + 19,400}{2000} \right)$$

$$= 150 \times \frac{55,900}{2,000},$$

$$\text{minus } 110 \left(\frac{23,000}{2,000} \right) \text{ covered by DBPP};$$

net F_{RR} = 2930 lbs to move with a full load.
2nd gear DBPP = 22,000 lbs $<$ traction limit,
minus net F_{RR} = $-2,930$ lbs for loading
$= 19,070$ lbs ≈ 19400 max.

(b) Travel time loaded, with $F_{RR} = 110 \times \dfrac{32,500}{2,000}$

$F_{RR} = 110 \times 16.25 = 1790$ lbs $\ll$ 5th gear DBPP;

$$\text{VTL} = \frac{1000}{5.9 \times 88} = 1.93 \text{ minutes, and}$$

VTE = VTL = 1.93 minutes.

(c) Fixed times, FT = LT + ADB + DT + TT.
Using Eq. (5-12) with a value assumed from Table 5-5,

$$LT = \frac{19{,}400 \times 100}{19{,}070 \times 90} = 1.13 \text{ minutes}, \qquad \approx 1.1$$

accelerating-decelerating-breaking, ADB = 0.5
dumping time, DT = ½ minute   = 0.5
turning, 2 × 180° at ¼ min each   = 0.5
                     total FT = 2.6 min.

(d) Cycle time = 2(1.93) + 2.6 ≈ 6.4 minutes.
So alternate #1 with crawler-tractor as prime mover,

peak productivity, $q_p = \dfrac{9 \times 0.72}{6.4} \times 60 = 60.7$ cy/
hr, assuming $f_w = 50/60$, normal $q_n = 50.6$ pay yards/hr.
Estimate of total equipment cost = \$23/hr.
Cost per pay yard = \$23/50.6 = 45.5¢

Alternate #2: a two-axle, wheel tractor-scraper with 148 hp (dd) and 9-cy heaped capacity, weighing 24,000 lbs empty.

(a) Payload = 9 (0.72) 3000 = 19,400 lbs., which is less than scraper's limit.

(b) Travel time loaded, with RR′ = 130 for rubber tires:

$$F'_{RR} = 130 \left(\frac{24000 + 19400}{2000} \right) = 130 \times 21.7$$

= 2820 lbs. moving in the cut area.

speed, mph $= \dfrac{375 \text{ (hp) eff.}}{\text{rimpull } F'}$ (See Sec. 3.4.1 and assume engine's mechanical efficiency is 65%.)

$$v = \frac{375 \ (148) \ 0.65}{2820} = 12.8 \text{ mph} < 3\text{rd gear speed,}$$

so travel loaded in 2nd (9.5 mph):

$$VTL = \frac{1000}{9.5 \times 88} = 1.20 \text{ minutes, and}$$

return in top gear (4th) at 25 mph:

$$\text{VTE} = \frac{1000}{25 \times 88} = 0.45 \text{ minute, so}$$

total $\text{VT} = 1.65$ minutes.

(c) $\text{FT} = \text{ADB} + \text{others}$ [see (c) in alternate #1]
   $= 1.5 + 2.1 = 3.6$ minutes.

(d) Cycle total, $\text{CT} \approx 5.2$ minutes.
   So alternate #2 with rubber-tired tractor as prime mover, peak productivity, $q_p = \dfrac{9 \times 0.72}{5.2} \times 60 = 74.8$ cy/hr, assuming $f'_w = 45/60$, normal $q_n = 56.0$ pay yards/hr.
   Estimate of total equipment cost $= \$25/\text{hr}$, which is reasonable in comparison to crawler unit.
   Cost per pay yard $= \$25/56.0 = 44.6¢$.

The outcome of this example shows the higher speed, rubber-tired tractor-scraper combination to be more economical. Note that the wheel tractor has speeds at least 50% faster than the crawler tractor. Of course, the faster equipment has greater fixed time for acceleration, deceleration, and braking. Since this is a fixed time that will not change much for different job situations, there will be a break-even operation, i.e., the job situation in which the cost to move a cubic yard by the crawler-tractor or by rubber-tired tractor-scraper combination would be the same. The break-even operation, according to the above example, would have a one-way haul distance less than 1000 feet.

Scrapers with or without Haul Route Maintenance. In Section 5.6.3 the effect of the haul route condition was discussed for a scraper operation. It was indicated that a rutted, muddy travel surface would be particularly bad for an all-rubber-tired scraper combination. Such a condition will reduce the higher speeds possible with that equipment. This is of great significance on jobs with long haul operations. The importance is such that on a moderately large operation, with 5 or more scrapers, the equipment planner should consider the use of a motor grader to maintain the haul route. The added expense of this extra equipment in the fleet for the operation can be economically justified. An example will now be worked to show this point.

This example is an operation where large scrapers are needed to move a million yards of earth material an average of 3000 feet one way. The cost of production will be figured without, and then with, a motor grader to maintain the haul route to reduce the rolling resistance and, so, increase the speed of the scrapers.

Given:

the job to move 1,000,000 cubic yards of wet earth loam (weighing 3200 lbs/pay yard) a distance of 3000 feet one way—1000 ft up a 4% grade and the remainder on level ground, using a two-axle, wheel-type, 420-hp (power-shift) tractor with rear axle power and a single-axle scraper unit (6 × 2 wheel rotation) [This design puts approximately 40% of total weight of loaded scraper on the drive wheels.]; empty weight of total equipment is 77,000 lbs; scraper bowl capacities are 21 cu yds struck and 30 cu yds heaped; scraper's load limit is 72,000 lbs.

Alternate #1:

with a loose, wet haul surface described as "rutted, muddy" earth (see Table 1-2) with no maintenance, estimate $RR = 200$ lbs/ton.

(a) Payload based on a swell factor, $s_w = 1.20$, is limited to $72,000 \times 1.20/3200 = 27.0$ cu yds. Checking traction with a coefficient of 0.5, the maximum tractive effort that can be applied when fully loaded, $TE = 0.5$ $(0.4 \times 149,000) = 30,000$ lbs, which is greater than tractor's delivered power.

(b) Travel times will be figured in four parts: up 4% grade where $GR = 4(20) = 80$ lbs/ton:

$$\text{total } F_1 = 280 \times \frac{149000}{2000} = 20,900 \text{ lbs;}$$

and assuming 65% mechanical efficiency,

$$\text{speed, } v_1 = \frac{375(0.65)420}{20,900} = 4.9 \text{ mph;}$$

then, $$VTL_1 = \frac{1000}{4.9 \times 88} = 2.32 \text{ minutes.}$$

on level for 2000-ft haul, loaded:

$$\text{total } F_2 = 200 \times \frac{149,000}{2000} = 14,900 \text{ lbs;}$$

$$\text{speed, } v_2 = \frac{375(0.65)420}{14,900} = 6.9 \text{ mph;}$$

$$\text{then, VTL}_2 = \frac{2000}{6.9 \times 88} = 3.30 \text{ minutes.}$$

return on level ground, empty:

$$\text{total } F_3 = 200 \times \frac{77,000}{2000} = 7,700 \text{ lbs;}$$

$$\text{speed, } v_3 = \frac{14,900}{7,700} \times v_2 = 13.2 \text{ mph;}$$

$$\text{then, VTE}_1 = 3.30 \times \frac{6.9}{13.2} = 1.72 \text{ minutes.}$$

return down 4% grade where RR = 200 − 80 = 120 lbs/ton:

$$\text{total } F_4 = 120 \times \frac{77000}{2000} = 4,620 \text{ lbs;}$$

$$\text{speed, } v_4 = \frac{375(0.65)420}{4,620} = 22.2 \text{ mph;}$$

$$\text{then, VTE}_2 = \frac{1000}{22.2 \times 88} = 0.51 \text{ minutes.}$$

total VT = 2.32 + 3.30 + 1.72 + 0.51 = 7.85 minutes.

(c) Fixed time, FT = LT + ADB + TT + DT:

to load 72,000 lbs of wet earth loam will require a pusher;

Using a crawler tractor, weighing 65,000 lbs with $C_t = 0.7$ delivers (0.7) 65,000 = 45,500 lbs max. traction plus (0.5) 56,700 = 28,350 lbs from scraper for total TE = 73,850 lbs at the end of loading.

to get 72,000 lbs loaded, assuming digging efficiency is 85% in cut,

$$LT = \frac{72,000 \times 100}{73,850 \times 85} = 1.15 \text{ minutes.}$$

for accelerating-decelerating-braking, fixed time from Table 5-6,

ADB = 0.7 minute

for dumping estimate, DT = 0.5 "

for two 180° turns,

estimate   TT = 0.5 "

total fixed time,   FT = $\overline{2.85}$ minutes.

To summarize times:

travel up 4% grade,	VTL_1 =	2.32 minutes
2000 ft on level,	VTL_2 =	3.30 "
return on level,	VTE_1 =	1.72 "
down 4% grade,	VTE_2 =	0.51 "
total variable time,	VT =	$\overline{7.85}$ "
total fixed time,	FT =	2.85 "
total cycle time,	CT =	$\overline{10.7}$ minutes.

Productivity for 6 × 2 wheel scrapers operating with a pusher to load wet earth loam and traveling on a haul road without maintenance (estimated RR = 200 lbs/ton):

Each scraper produces at best (peak)

$$q_p = \frac{27.0 \times 60}{10.7} = 151 \text{ pay yds/hr;}$$

assuming a working efficiency, $f_w = 45/60$,

$q_n = 0.75 \times 151 = 113$ pay yds/hr.

If the pusher has a cycle time = 2.0 minutes,

it can handle $\dfrac{CT_s}{CT_p} = \dfrac{10.7}{2.5} = 5.3$, say, 5 scrapers.

Estimating the cost of rented equipment:

for each scraper, e_s = $40 per hr with operator;

fleet costs, $5e_s$ = $200 per hr;

with the pusher, e_p = $38 per hr with operator.

Normal production of the earthmoving will cost

$$c = \frac{5e_s + e_p}{5q_n} = \frac{200 + 38}{565}(100) = 42\cancel{c}/\text{pay yd.}$$

Now this productivity and cost will be compared to what could be realized if the haul route is maintained by a motor grader. Assume that the motor grader can reduce the rolling resistance to that for well-maintained earth (see Table 1-2).

Alternate #2: with haul road maintained by a motor grader so that RR = 60 lbs/ton.

(a) Payload as in alternate #1 limited to 27.0 cu yds, or 72,000 lbs.

(b) Travel time figured in four parts:
up 4% grade for 1000′, GR = 4(20) = 80 lbs/ton:

$$\text{total } F'_1 = 140 \times \frac{149000}{2000} = 10,430 \text{ lbs;}$$

$$\text{speed, } v'_1 = \frac{375(420)0.65}{10,430} = 9.8 \text{ mph;}$$

$$\text{then VTL}'_1 = \frac{1000}{9.8 \times 88} = 1.16 \text{ minutes.}$$

on level for 2000-ft haul, loaded:

$$F'_2 = 60 \times \frac{149000}{2000} = 4,470 \text{ lbs;}$$

$$v'_2 = \frac{10,430}{4,470} v'_1 = 22.9 \text{ mph;}$$

$$\text{VTL}'_2 = \frac{2000}{22.9 \times 88} = 0.99 \text{ minutes.}$$

return on level, empty (scraper is 77000 lbs):

$$F'_3 = \frac{77,000}{149,000} F'_2 = 2,310 \text{ lbs;}$$

$$v'_3 = \frac{4470}{2310} v'_2 = 44.3 > 40\text{-mph top speed,}$$

also for returning down 4% grade, so

$$\text{total VTE} = \frac{3000}{40 \times 88} = 0.85 \text{ minutes}$$

(c) Fixed times same as in alternate #1 (*FT* = 2.85 minutes), except time for acceleration-deceleration-braking is slightly higher for higher speeds, ADB = 0.8 minute: estimating total fixed time, FT′ = 3.0 minutes.

Then, total cycle time, CT = sum VTs + FT′: CT = 1.16 + 0.99 + 0.85 + 3.0 = 6.00 minutes.

Productivity for 6 × 2 wheel scrapers operating with a pusher to load wet earth loam and traveling on haul road maintained by a motor grader to give **RR** = 60 lbs/ton:

Each scraper produces at best (peak)

$$q'_p = \frac{27.0 \times 60}{6.00} = 270 \text{ pay yds/hr;}$$

If the pusher has a cycle time = 2.0 minutes, it can handle $\dfrac{CT'_s}{CT_p} = \dfrac{6.0}{2.0} = 3$ scrapers.

Estimating the cost of rented equipment:
for each scraper, e_s = $40 per hr with operator;
for three, $3e_s$ = $120 per hr;
for the pusher, e_p = $38 per hr with operator;
for the motor grade, e_m = $30 per hr with operator.

So normal production of the earthmoving (assuming motor grade works continuously to assure most favorable RR) will cost

$$c' = \frac{3e_s + e_p + e_m}{3q'_n}$$

$$= \left(\frac{120 + 38 + 30}{606} \right) \times 100$$

= 31¢/cu yd ≪ 42¢ in alternate #1.

This shows quite conclusively that when the rolling resistance (RR) of the haul road can be reduced significantly on a large earthmoving job, it pays to add a motor grader to the earthmoving equipment combination.

Other Alternative Comparisons. Various other alternative choices between types or forms of scraper equipment might be made. The Caterpillar Tractor Company has an excellent presentation of important cases in their booklet on earthmoving system selection.[7] In that publication the comparison between crawler and rubber-tired scrapers is shown to depend not only on haul distance but also on total resistance to moving. This is shown by the simple graph of Fig. 5-22.

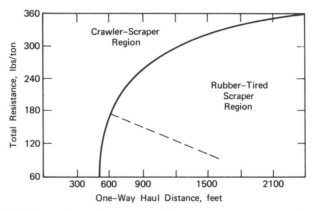

FIGURE 5-22 Crawler vs. rubber-tired scraper, based on travel resistance and distance.

Other scraper comparisons in the Caterpillar booklet are of rubber-tired wheel scrapers. The choice between a two-axle scraper and a three-axle one, in which the prime mover has two axles, is comparing traction and maneuverability against speed. With a haul in the range 2,000 to 3,000 feet one way and various adverse grades, a two-axle scraper proves more economical. On the job with a haul of one mile or more loaded, a three-axle scraper will likely be more economical.

Another scraper equipment choice is between a tandem-engine and a single-engine-powered scraper. The tandem unit has an additional engine powering the rear axle behind the scraper bowl. This provides much more traction, but for about 25% more investment in the scraper. With Caterpillar's comparison the tandem unit is more economical where there are 10 to 15% adverse grades when hauling loaded on relatively short cycles.

They might be considered for the upper third of the rubber-tired scraper region in Fig. 5-22.

The Caterpillar comparisons include various possibilities for pusher help. The statement is made that "pusher selection can be the most important single element of a properly applied scraper spread." An indication of this importance was given in Sec. 5.6.3 on loading time. One obvious comparison is between a crawler tractor and a wheeled tractor for the pusher. The wheeled pusher, generally, can be justified economically only when it can take advantage of its mobility. Tandem pushers can be justified for some job conditions and scraper selections, but they require topnotch operators for a well-coordinated combination.

There are several possibilities for scrapers to be self-loaded. The principle factors were discussed in Sec. 5.6.1. The tandem-powered scraper is generally self-loading. For smaller loads with less power requirement, the specially designed elevating scraper is a self-loader, covering the lower range of shorter hauls for rubber-tired scrapers shown in Fig. 5-22. That is the region between the curve and dashed line. A larger-capacity self-loading scraper is the multibowl type with all wheels serving as drivers powered by electric motors.

The possibility for two scrapers to help each other load by a "push-me, pull-you" method was mentioned earlier in this chapter. The use of twin-bowl scrapers moved by a single tractor is becoming frequent. With all the various methods and combinations of scraper equipment to select from, it becomes a time-consuming chore to make the desired calculations. This is the sort of situation that calls for use of high-speed computers, as mentioned in Chap. 2. The Caterpillar Tractor Company has a computer program for their comparative earthmoving analyses which is called Computerized Vehicle Simulation Program.

5.7 Front-End Loaders for Earthmoving

The front-end loader is a crawler- or wheel-tractor piece of equipment with a bucket container mounted on the front end. An example of this type of construction equipment is shown in Fig. 5-23. The bucket is mounted to dig or load earth or granular material, lift it up, carry it where necessary, and dump the material from some height. The first loaders had rigid, mechanical arms pivoted about the tractor frame with cable controls. This meant that the digging force and any change of horizontal position of the bucket had to be provided by movement of the tractor. The cables could only move the bucket straight up and down. The

FIGURE 5-23 A front-end loader in action (courtesy of Caterpillar Tractor Co.).

bucket had to be pushed by the tractor into a bank of material. When the bucket was full, the tractor backed off, raised the bucket straight up, moved it over a truck or the like, then dumped the load. All these tractor movements were awkward.

5.7.1 Design Features of a Front-End Loader

Now loaders are designed with hydraulic controls and arm extenders as shown in Fig. 5-23. This means that much of the bucket operation is done by the mechanism attached to the tractor.

The buckets range in size from $\frac{1}{4}$ cubic yard to the largest size of more than 25-cu yd heaped capacity. The commonly used and available loaders range up to the 5-cu yd size. With the loader bucket a more permanent attachment to the tractor than a dozer blade, the equipment designer will be sure there is a carefully balanced design between the bucket and tractor size. This takes into account the extreme working condition of a full bucket load supported in a raised position with the arms fully extended in front of the tractor. The safety against pitching forward under that condition is called the static tipping-load capacity. A usual safety factor is 2, meaning that the load which would cause tipping is twice what a bucket loaded to the struck capacity with 3,000 pounds per cu yd material can give. This high safety factor is needed to protect against the more severe load conditions when moving.

To provide this static tipping safety, the tractor's weight, W, is generally 40 to 60% more than the tipping-load capacity. Therefore, the sizes of bucket and tractor that would be designed together for a balanced front-end loader can be roughly calculated. A bucket of nominal rated or heaped capacity, B, in cubic yards would be figured for a load of $3000(B)$ in weight, shown as W_B. With the safety factor, the maximum weight the loader can carry is higher than W_B. The static tipping load, $W_T \approx 2W_B$, and the tractor's weight would be approximately $1.5\ W_T$. Thus, a loader of bucket capacity, B, will be designed for mounting on a tractor weighing approximately $9000(B)$ pounds.

This weight must also be sufficient for the maximum "breakout" force at the front cutting edge of the bucket. The breakout force is produced by pivoting the bucket about its horizontal support pins with the force applied through hydraulic pistons to dig up through tough, sticky material. The maximum angle for pivoting is about 40° from the initially horizontal position of the bucket's bottom face with its cutting edge. Maximum breakout force is designed in some loaders to be as high as 150% of the static tipping-load capacity. The tractor must be heavy enough or have its weight distributed well enough to take this condition.

The loader-bucket mechanism is designed to have a dumping height of between 8 feet and 15 feet above the tractor's footing. Such height is proportional to the loader's size. This makes it possible for the loader to dump into a truck or hauling unit of suitably balanced size. The operation of a loader between its position for loading and that for dumping generally requires a lot of maneuvering. If the space for use between the loading and dumping positions is limited, there may be a problem. A crawler-tractor-mounted loader can pivot slowly on its tracks without much difficulty. But an ordinary, two axle, wheel-mounted tractor needs more space for maneuvering. This has led to the more modern articulated tractor unit.

An articulated frame loader is hinged approximately midway between its two axles. Its turning ability is greatly increased by the more than 30° angle that the front axle can swing either side of the straight-forward position. The photographic series in Fig. 5-24 shows an articulated loader carrying out a common loader operation.

Of course, the bucket will be in many different positions during operation of the loader. With the bucket raised to a dumping height, it has no effect on the lateral interface dimensions at ground level. So the length of the loader is generally given without including the bucket. The turning

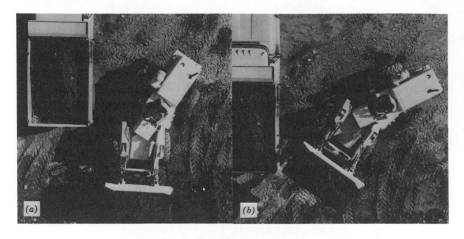

FIGURE 5-24 Sequence of moves in front-end loader operation (courtesy of TEREX, Division of General Motors).

radius measured to the outside rear wheel of an articulated loader is approximately equal to its overall length less the bucket. This means that a well-designed articulated loader can operate in a confined space. It will generally not need more than two machine lengths to dig, maneuver, and dump its load. An articulated loader can swing its front part to load straight into the bank of material. And this same flexibility in maneuvering can help during the dump into a long hauler bed. It allows the loader to spread its load for evenness in the hauler without having to change its entire position. The tipping-load capacity of an articulated loader in the maximum turned position will be about 85% of the limit in a straight-ahead position.

The type of operation a loader does means many direction changes. This is shown in the photographic series in Fig. 5-24. Counting all moves —forward and back, as well as changes of direction for the front and rear parts—there may be 400 to 500 direction changes per hour. This amounts to a directional change about every 6 seconds. Add to these movements the variety of positions for the bucket, and it is not hard to tell that the operator has a difficult machine to control. For this reason, improvements in the design of front-end loaders have concentrated on simplifying the controls for the operator.

One way to make the operator's task simpler was to put the shift and speed control in a single lever. He can shift from 1st or 2nd gear forward to as high as 3rd reverse gear in one move of this lever. Another means for simplifying the operation of a loader is to provide an automatic bucket-control system. This has been done on the most modern front-end loaders to let the operator concentrate on maneuvering his piece of equipment.

The manual operation of the bucket on a loader makes use of two levers within easy reach of the operator's seat. One is to control the vertical position and lifting of the bucket. It will raise, lower, or hold the bucket vertically. The other lever is to control the bucket's position about a horizontal axis at the back edge of the bucket closest to the tractor. It will put the bucket in a tilt-back position for maximum capacity, a full forward position for dumping, or hold in between for digging, scraping, etc. These positions are illustrated on the specification sketches of Fig. 5-25.

Automatic control of the bucket positions relieves the operator of many small points for decision that are tiring to a person. The bucket will be raised in one of three set-tilt positions and automatically stopped at a set height. When the bucket is emptied, it automatically returns to the proper digging tilt angle for the next load. The points of the automatic control can be adjusted as needed or programmed for a given operation.

A special bucket loader has been designed for certain operations with particularly cramped working space. This may be called an "overthrow" loader. The bucket digs its load; then, in lifting, it swings the bucket in a vertical plane completely over the operator for dumping in back of the machine. The tractor is not articulated but is of the single, rigid-frame type and needs to move only back and forth for operating as an overthrow

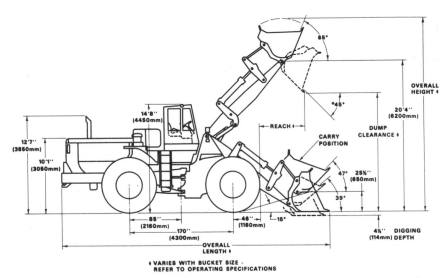

FIGURE 5-25 Loader bucket positions (courtesy of Caterpillar Tractor Co.).

loader. There are certain drawbacks to this equipment design. The biggest problem is the safety and control for the operator, who must look in front and then in the rear for each cycle of his operation.

5.7.2 Operations for Front-End Loaders

The front-end loader is a fairly recent entry on the construction equipment scene. It was apparently introduced as another attachment to make the crawler tractor even more versatile and to take care of the problem of cleanup around a construction operation. A bulldozer could only push the excess material or waste to one side, but a front-end bucket could pick it up for loading into a truck. That original operation was done with the mechanical, cable-controlled buckets on rigid arms of the first loaders following World War II. For the reasons explained in the previous section, the loader did not win much favor until hydraulic control for extendable arm mechanisms was introduced. Now there is wide use and acceptance of the front-end loader.

A most common use of the loader is for loading materials into a hauling unit such as shown in Fig. 5-23. If the area surrounding the material to be loaded is reasonably level, the hauling unit can be located in a convenient position closeby. Then the loader can dig and make the short move to dump its load in the hauler. During the first years of developing the front-end loader, it was used as a substitute for a small power shovel to load trucks. Now, as bigger and larger-capacity loaders—up to 15-cubic yard buckets—are manufactured, this equipment is replacing the power shovel on almost any of its traditional operations.

Another common use for a loader is in excavating a basement or foundation. This would be in an operation where the smaller, horizontal dimension is at least the bucket width, if not several times that dimension. If the shorter dimension of the basement floor is at least twice the loader's length, not counting its bucket, the operation can be arranged for truck loading at the foundation level. A ramp can be dug by the loader so that the truck can drive down to the foundation level along one long side. Then they can be loaded just as in the loader-truck hauling operation discussed previously.

A third important use for a front-end loader is loading blasted material into haulers in the close quarters of a rock excavation, tunnel, or quarry. The loader has an advantage in such a situation over a power shovel with its boom and other protruding parts. Some newer-designed shovels, which have overcome this objection, will be discussed in a following chapter. However, either a crawler-mounted or wheel-type loader with good floatation on rough footing can handle the loose rock excavation.

The front-end loader is also used to dig aggregate or quarry material to be loaded into the grizzly hopper feed for a crusher plant. Generally, the hopper is located at the edge or just inside the raw material pit. The loader will then dig its load and carry it some small distance to the hopper. If the carrying distance is greater than a shovel can reach from its digging position, the loader has a distinct advantage over a power shovel for this operation.

Of course, any construction cleanup operation that involves picking material up and dumping it somewhere else is ideal for a front-end loader. As mentioned earlier, this is the reason the bucket loader was introduced. Some examples include removing stumps, boulders, or other big objects from the work area for grading equipment; backfilling a trench or foundation wall, especially where the material has to be carried some distance; and the variety of cleanup work around an aggregate or concrete materials yard.

5.7.3 Productivity of a Front-End Loader

The productivity of a front-end loader is figured in cubic yards per hour. It can be determined by estimating the actual payload of material and figuring the time it takes to handle each bucketful. In other words, the planner estimates the bucket load and the cycle time for each. Then the productivity can be figured for the average time spent each hour in actual production. This is the way to estimate cubic yards per hour for any piece of construction equipment handling bulk material.

The bucket for a loader is given in stating the size of loader. It will be called perhaps a ½-yd, 5-yd, 15-yd loader, or any of many sizes in between these. That size is according to the Society of Automotive Engineers' SAE-rating, which is the nominal heaped capacity of the bucket. Of course, the heaped material is loose material, and to determine the pay load of bank measure, it will be necessary to apply an estimated swell factor as found in Table 5-1 and used in Eq. (5-1).

The cycle time for a loader to handle each bucketful should be broken down into several key components. This subdividing is to separate the variable times, which depend on the distances the loader moves with each bucketful, from the so-called fixed time. The fixed time, FT, will include those parts of the loader's cycle which are reasonably constant no matter what the operation is. The parts are the times required to load the bucket, to shift gears, to turn, and to dump the load. Regardless of the operation setup and distance to move the load, the value of FT is estimated to be 0.25 to 0.35 minute (15 to 21 seconds) for a reasonably efficient operation. The maneuvering time is naturally the biggest part of this fixed time. A time study made by the U. S. Bureau of Public Roads on crawler-mounted loaders showed a range of 8 to 19 seconds for loading, with an average of 13 seconds. This is a high average because it was observed on early models of loaders. However, it agrees reasonably well with the up-to-date range for FT given above.

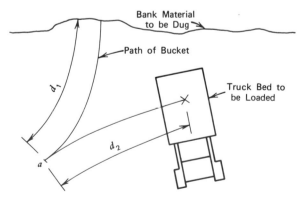

FIGURE 5-26 Plan of ideal loader setup.

The variable travel time is based on the speeds of travel and distances to move between loading and dumping and then returning for the next load. The shifting, turning, and maneuvering, in general, are carried out between the end points of each cycle. An ideal operation setup for high loader productivity is one as illustrated in Fig. 5-24. A sketch of the plan view of this setup is given in Fig. 5-26. The loader's cycle in this ideal setup is to load the bucket with the bank material, back up to a convenient spot (*a*) for turning, move forward to dump into the truck, reverse to point *a*, and move forward to dig the next load. Therefore, the variable time, VT, in a cycle will cover moving each distance d_1 and d_2 in both forward and reverse gears. For some typical values, the VT could be estimated by:

Gear and Speed (fpm)	Times, in minutes	
	for $d_1 = d_2 = 15$ ft.;	20 feet
Forward @ 3 mph (264)	$2 \times \tfrac{15}{264} = .114;$	$\tfrac{40}{264} = .152$
Reverse @ 5 mph (440)	$2 \times \tfrac{15}{440} = .068;$	$\tfrac{40}{440} = .091$
	total VT $= 0.18$	0.24

If the material is easy to load, this would be an ideal loader operation in all respects. Then the total cycle time, CT, could be estimated by CT = FT + VT:

$$CT = 0.25 + 0.18 = 0.43 \text{ minute}$$

for the minimum ones of 15 feet in each direction. This leads to a maximum, peak productivity for the loader operating without any delays found by:

$$q_p = \frac{SAE}{s_w} \times \frac{60}{CT}$$

For a 1-cu yd loader working in dry, clean sand (14% swell) with the ideal operation setup described above,

$$q_p = \frac{1.0 \times 60}{1.14 \times 0.43} = 123 \text{ cu yds/hour.}$$

One cause for major delay in an operation of a front-end loader with hauling units is due to the positioning of a new haul unit for the loader to fill. This can be called "spotting" time, ST, and is estimated to take an average of 0.25 minute per unit. The full amount of ST should not have to be added to the loader's cycle time to figure productivity. If the pieces of equipment are operated in a coordinated way, a haul unit can be positioned or spotted while the loader is getting its next bucketful. Perhaps there is a way to handle this factor and recognize that accuracy to the hundredth of a minute is unrealistic in the estimate for CT. One way would be to assume the value to use for cycle time is that found by adding FT and VT and an amount for ST between 0.1 and 0.2 minute to make the estimate for CT show in round tenths of a minute. For example, the so-called ideal setup described above with some time allowed for spotting haul units could be estimated with a cycle time,

$$CT = 0.43 + 0.17 = 0.6 \text{ minute.}$$

The productivity, as described above for the very short travel distance, is equally applicable to crawler-mounted and wheel-type loaders. If the bucket loads of material have to be moved more than 30 or so feet, it is likely that a wheel-type loader would be more effective. In fact, in an aggregate pit or quarry the loader may move several hundred to a thousand feet to dump its loaded bucket. The higher speeds of 8 to 15 mph for 4-wheel, rubber-tired loaders moving the greater distances will give the desired high productivity.

When the loader operation calls for carrying the filled bucket long distances, productivities will have two major variables—the loader's bucket capacity and the variable time of the cycle. These are given in the equation for peak productivity in pay yards (bank measure) per hour.

$$q_p = \frac{(\text{SAE rating}) \div s_w}{VT + FT + (0.1 \text{ to } 0.2)} \times 60, \qquad (5\text{-}14)$$

where SAE-rating is the bucket size;

$$s_w = \text{swell factor} = \left(1 + \frac{\% \text{ swell}}{100}\right);$$

VT and FT are described above; and
0.1 to 0.2 is added for spotting time, or for extra maneuvering.

If many operations are going to be figured by an estimator with a given loader, it would be helpful to graph major variables in the productivity determination. The VT is a major variable which can be graphed in combination with the FT as shown in Fig. 5-27. The fixed time includes time for loading, shifting, turning, and dumping, as well as travel in reverse gear for 15 feet at each end of the haul route. If higher forward gears are used, the FT should be increased by 0.1 minute or more for shifting and accelerating, deceleration, and braking.

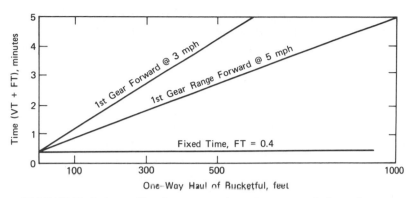

FIGURE 5-27   Cycle time for front-end loader.

5.7.4   Costs for Front-End Loaders

Since a front-end loader is essentially a tractor-mounted piece of equipment, the planner thinks about the same factors as in Sec 5.4.4 for tractor equipment. There are the variations between crawler-mounted and wheel-mounted front-end loaders. Rental rates show the wheel loader costing only 67 to 95% of the rate for an equivalent bucket size crawler-tractor-scraper. The smaller % is for larger units. But the wheel tractor-loader rates do not include the additional charges for tire wear. The cost of wear and tear on tires is so variable it must be handled separately.

For instance, operating a front-end loader continuously, every working shift for a year, can wear out a whole set of four tires in that year's time. For a large unit this would amount to about $25,000, which essentially doubles the hourly cost for the use of this equipment. Obviously, adding that cost makes the wheel-mounted loader more expensive than an equivalent size crawler-tractor front-end loader. High wear and tear on the loader's tires can be reduced by giving special attention to the wearing surface or the operating procedure. The operation can be helped for loading rough shot rock by cleaning and leveling up the work areas as often as possible. Tire wearing surfaces can be protected by using tire chains. The chains are also useful in loose sand or gravel and in soft muck.

A newer development for wheel-type tractor equipment is the Dystred cushion track to cut down wear and tear on tires. The track assemblies look like crawler tracks mounted like chains but cover the whole bearing surface of inflated rubber-tired wheels. A set is shown on the loader pictured in Fig. 5-23. Such tracks combine the ruggedness of crawler tracks with the maneuverability of rubber tires. Another way to reduce the wear and tear of rubber tires as much as 50% is by using deeper treads. Treads up to 4″ deep resist tearing apart by small rocks. Of course, none of these measures can be included without extra cost for the equipment. The good planner seeks the most economical and feasible combination of schemes for his operations.

In selecting the front-end loader, some other cost variables parallel those for other tractor-mounted equipment. The power unit could be a gasoline or a diesel engine. A diesel power unit would probably mean around 10% higher initial cost, but the fuel cost should reduce this difference materially during operation. Either type of power unit might have a direct-drive or torque-converter transmission, and the variations are extended to include a manual shift or power shift. A unit with torque converter and power shift will cost at least 5% more than one with direct drive and manual shift.

The new articulated front-end loader, described in Sec. 5.7.1, has certain distinct advantages in the operation of this type of equipment. A disadvantage of the articulated-frame-and-steering loader over the older, rigid-frame unit has to do with its cost and maintenance. An articulated loader has a more involved mechanism and costs 5 to 25% more than a rigid-frame loader of equivalent bucket size. Of course, in operations with limited working area for maneuvering it may be necessary to use an articulated loader. No further justification is needed for the higher cost. If there is ample room for maneuvering, justification for using the higher-cost articulated loader will depend on the efficiencies of various parts of the operations. The operator of an articulated loader should have some advantage in digging and also in spreading a greater load in the hauler's bowl. With the operator on the back part of the loader, he will not be turning quite so much on an articulated unit as he would on a fixed-frame unit. This should mean less fatigue and higher working efficiency. On the other hand, an inexperienced operator may find the additional movements of an articulated loader more tiring, until he gets used to it. The only way to make determinations on these points is to run some time and motion studies on the two types of loaders.

It is interesting to note the comparison of costs between a front-end loader and a power shovel. Section 5.7.2 on the operations of a loader pointed out that it can do many of the jobs that are done by a power shovel. If the two types of equipment are equally suited for an operation based on their designed capabilities the costs definitely favor a front-end loader. It rents for 45 to 65% of the rate for a power shovel of equal bucket size. Other factors may make the power shovel more economically competitive. This type of equipment will be covered in a following chapter.

5.8   Earthmovers with Belts for Loading[8]

Over the years contractors have been seeking higher-production earth-moving equipment. One scheme that appeared to have potential was for the material to be loaded on high-speed hauling units from a conveyor belt. An early version of this was a moving unit with a scoop to pick up loose or easily excavated material and run it up an inclined conveyor for dumping into haulers traveling alongside the loader. That loader had to be towed by one or two tractors. For less production but also less complexity, an elevating grader was developed. It is attached to a motor grader, or it may be self-propelled, and generally loads loose material from a highway right-of-way.

To reach the higher production desired, the next development occurred around 1960 with machines loading material on stationary belt conveyors. In this arrangement the continuous belt loader is fed by a number of dozers working in a high mound of material to be moved. They push material into a grated trap, and the conveyor dumps it into open-top haulers at a fixed point. This high-production operation was followed by the development of specially designed bucket wheel excavators. The San Luis and Oroville Dams in California, each with more than 70 million yards to move, promoted the design of those expensive excavators. They were special and part of earthmoving systems which included other belt conveyors and railroad hauling units to move 3000 yards of earth each hour.

Most recently the development has turned to continuous-belt loaders that are self-propelled and so, more versatile. These are made for high-production excavation in mining, earthfill dams, and airport and highway construction. In all the variety of belt loader earthmovers it is apparent that the material must be either free flowing or fairly uniform and amounting to a large quantity in one area.

5.8.1   Design and Use of Belt Loaders

The towed loader is moved with the front, ground end of the conveyor outfitted with a scoop like the loading end of a scraper. It may have rotating discs or horizontal augurs to help cut the material and move it onto the conveyor. The conveyor moves up in a line at an angle with the direction of the loader's travel, so that it can discharge into haul units traveling alongside. With the need generally for two tractors to tow this loader, it is a cumbersome train of equipment requiring good coordination. For this reason it is most useful in a large, flat area with a uniform depth of cut. The towed loader with either two detachable tractors or its own crawler-mounted power units will make a "train" of equipment approaching 100 feet long. To be economical, it must be able to load at least a half-dozen hauling units between turns.

Smaller versions of this sort of belt loader have been designed for special uses. These are the pieces of equipment known as elevating graders. An earthmover of this sort is frequently designed to be towed by a motor grader. In that way the grader can serve to cut the material and direct it onto the belt. But with a more specialized unit the belt loader may have its own scoop and rotating cutters to feel the conveyor. This type would be most useful in widening roadways, shaping shoulders, or picking up windrowed material.

The development of stationary belt loaders to attain high-production earthmoving was started by contractor-built setups. Ordinary manufactured conveyor belts were used with shop-made traps and supporting structures. When this was found to be an accepted method, several manufacturers built complete, portable, belt loading units. To get high production, they use 48″ to 72″-wide reinforced belting, which runs up to 400 fpm. The belt is shaped into a deeper-than-usual trough for a conveyor by using five idlers in some units. To load such a greedy earthmover generally takes five to eight dozers, using the advantage of gravity. The material must be free flowing, likely loose granular or rocky, and at least 20 feet deep to justify the equipment setup. It has been suggested that there should be a minimum anywhere from 250,000 to 1 million cubic yards to move for this equipment setup to be economical. The gravity advantage in this system cannot be realized unless the elevation difference between the trap and top of the earth cut is at least 25 feet and the material is not moved more than a few hundred feet. Ideally, it should be like pushing to keep up a perpetual landslide with the trap for the belt at the bottom of the hill.

The new, continuous-wheel excavators are moving pieces of equipment. They have bucket wheels to dig material that will stand in vertical banks. And the buckets dump it onto a continuous belt to move the dug material to the side away from the bank or into hauling units. One of the large units of this sort presently made for general purchase is Barber-Greene's

FIGURE 5-28 Continuous-belt excavator in action (courtesy of Barber-Greene Company).

Continuous Excavator. A picture of one of these pieces in action is shown below in Fig. 5-28. It is similar to that manufacturer's wheel trencher, which is one of a much older type of equipment as described in the next chapter.

The Continuous Excavator is self-propelled by three crawler tracks designed for enough power and stability. It excavates material from a 2- to 13-foot-high bank in a milling fashion by 1/2-yard buckets mounted on a 60″-wide wheel. There are two 13-foot-diameter wheels, and the inner one may work partly to clean up loose material sloughing off the bank. An excavator of this type working in a deep bank does a good job of blending naturally stratified material.

The cut material is conveyed on a line of belts than can be moved at various speeds up to 1000 fpm (99% faster than one for B-G's wheel trencher) to meet job conditons. Several lengths of conveyor, as seen in Fig. 5-28; make it possible to switch the discharge from one hauler to a parallel one without loss of time or material. The excavator-loader moves at speeds up to 2 mph, so hauling units can be loaded in 5 to 10 feet of travel. It is designed for reasonable maneuverability and can turn its nearly 50-foot length and 12-foot width in only a 42-foot radius.

There are a few other mobile excavator-belt loaders now being manufactured with some similar features. Another line of models will be mentioned in the following section.

5.8.2 Productivity and Costs with Belt Loaders

The early towed belt loaders have been able to show productivities between 1000 and 2000 cubic yards per hour under ideal conditions and operating efficiencies. Thus, for jobs with flat terrain and a long cut area to excavate (so turnaround is minimized) this equipment combination can load the earth at that rate for around five cents per cubic yard. It was found some years ago that such equipment had a working efficiency of about 67%. The latest towed loaders can excavate up to 3000 cu yds/hr.

Elevating graders, as the smaller version of a towed loader, can move 600 to 800 cu yds/hr of material having very good loading characteristics. Their production is reduced considerably in more difficult material. The still smaller and more specialized equipment for widening roadways, etc., where they cut a 2- to 4-foot-wide strip up to 15″ deep, does well to excavate 400 yards/hr.

Belt loaders introduced around 1960 to work in a stationary position had higher productivity than the earlier towed loaders. They can turn out 2,000 to 3,600 cu yds/hr, if the belt conveyor is fed enough material by dozers. The belt discharges the material at a 10- to 18-foot height into high-speed trucks or bottom dump wagons. With haulers of 20- to 30-yard capacity, the belt can load one every half-minute. The belt loader with such productivity is relatively inexpensive at an original cost of $30,000 to $75,000. Nevertheless, the cost of loading earth this way will come close to 10¢ per cubic yard for the equipment. One recognizes the five to eight dozers as a major expense in this loading operation. But the higher cost of loading can be offset by the great volume moved in high-speed haulers with a haul distance of one mile or more.

To make a high-production excavator that could have more applications, the manufacturers have designed equipment for somewhat lower productivity. The Continuous Excavator, which is a self-contained loader requiring one operator, can load about 1750 cy yds/hr. The original cost of this specialized equipment is in the vicinity of that for 4- to 6-yard power shovels. Under the right job conditions it should be an economical means for excavating and loading earth material. Another manufacturer of this type of equipment has a line of models that have productivities ranging from 400 to 3500 cu yds/hr.

5.9 Introduction to Compacting Earthwork

The placement of earth or other granular material for an embankment, fill, or base course needs special treatment. As explained in Sec. 5.1.1, material made up of particles in a loose state will have to be compacted to form a well-knit mass in its new location. This is true even if it has a special cementing material added to the bulk for binding the mass together.

Ages ago, the observation of the compacting benefits by animal hoofs tramping on loose fill material led to an innovation. A variation of the ancient solid stone roller used by early Romans was made by using straight lengths of hardwood tree trunks. Forged iron spikes or lugs with the outer end shaped like a sheep's foot were driven radially into the cylindrical trunk. This is the way the original sheepsfoot roller was made. It provided a somewhat different compacting action from that of the smooth cylinder which will be explained in Chap. 12.

Compaction is the effort or mechanical process used to increase the density of the earth or other material, i.e., to pack the particles into a smaller volume, consolidating the mass. A specific density of the material is frequently stated in terms of values ranging from 90 to 100% of a standard maximum density found by laboratory tests with the material. Such standards are set by contract awarding authorities such as the U.S. Army Corps of Engineers. The specifications used involve making field tests that are an attempt to match the lab tests on each part of work as it is done.

FIGURE 5-29 Compacting equipment working on an earth fill (courtesy of Caterpillar Tractor Co.).

Parts of an embankment are the layers or lifts of earth placed successively to build up a solid mass to the designed height of section for the earthdam, levee, or roadway.

The object is to make it strong enough to resist displacement or movement under whatever load or force may be applied on it in the future. A certain amount of compaction is caused by the earthmoving equipment moving on an earth fill or embankment. Crawler tracks supporting the weight of its equipment causes some compaction, and the vibration of their movement has a compactive benefit. Earthmoving scrapers with their wide, low-pressure tires, and other rubber-tired equipment moving on a fill or embankment help to compact the material. It has been found that the natural compaction with earthmoving equipment operating on the loose, dumped material may produce as much as three-quarters of the required compaction, and compactors must take care of the rest.

There are several methods for compacting a deep earth deposit of a granular material. It can be done with sonic compactors which are vibratory probes installed like an open pipe pile and sand drain. Pile-driving equipment that would be used for such an installation is discussed in Chap. 8. Another means for compacting a predominantly granular earth layer up to a dozen feet deep is with a high-frequency vibrator compactor. An earth fill should be compacted in layers of no more than a few-feet-thickness each. A detailed discussion of the more common compaction equipment will be found in Chap. 12 dealing with subgrade and paving equipment.

References

1. "Fundamentals of Earthmoving" booklet by Caterpillar Tractor Company (Peoria, Illinois, revised 1968).
2. "Production and Cost Estimating of Material Movement with Earthmoving Equipment" booklet by TEREX Division, General Motors Corporation (Hudson, Ohio, 1970).
3. *Land Clearing,* Caterpillar Tractor Company (Peoria, Illinois, 1970).
4. "Handbook of Ripping" booklet by Caterpillar Tractor Company (Peoria, Illinois, 1966).
5. "The Revolution—in Earthmoving" booklet by Michigan-Clark Equipment Company (Benton-Harbor, Michigan, circa. 1967).
6. "Optimum Load Time" booklet by Caterpillar Tractor Company (Peoria, Illinois, April, 1968).

7. "Earthmoving System Selection" booklet AE036939 by Caterpillar Tractor Company (Peoria, Illinois, circa. 1970).

8. "New Developments in Earthmoving—Belt Loading," reprint from *Construction Methods and Equipment*, McGraw-Hill Book Company (New York, N.Y., 1965).

9. Carson, A. Brinton, *General Excavation Methods*, McGraw-Hill Book Company (New York, N.Y., 1961).

Chapter 6

■■■■■■■■■■■■■

■■■■■■■■■■■■■■■■■■■■■■■■■■■■■■■■■■

TRENCHING, DREDGING, AND TUNNELING EQUIPMENT

■■■■■■■■■■■■■■■■■■■■■■■■■■■■■■■

Introduction to Operations and Equipment

The variety of construction equipment to be discussed in this chapter is for special excavation. It is equipment to dig, cut, bore, or suck material out of the ground or water for deposit elsewhere. This discussion will limit the dredging to operations of underwater trenching for pipelines. The regular dredging equipment will be as useful for the trenching operation as it is for general dredging. With this limitation on the discussion of this chapter, we can note a common characteristic of the operations for this variety of equipment. It is that the operations are to excavate a long line of restricted cross-sectional area, i.e., the length of opening is much greater than its transverse area, which is purposely kept to limited dimen-

253

sions. This observation on the operations leads to a similar one about the equipment. The common feature of trenchers, dredges, and tunneling equipment is that the length of the excavating part is much greater than its width, compared to the excavating equipment described in Chap. 5. Also, their progress is measured in linear feet rather than cubic yards.

Some specialized pieces of equipment are designed to do nothing but trenching, dredging, or tunneling. Such equipment is the main subject of this chapter. It is also true that some more versatile equipment can be used for these excavation operations. Where that approach to an operation would be advantageous, it will be mentioned. For instance, a crane-mounted backhoe may be most economical to use for trenching, a dragline might be used for underwater trenching, or a front-end loader and haulers may be used with a drill-and-blast approach to tunneling. These pieces of equipment are described more fully in other chapters. But they will be discussed where appropriate with the trenching, dredging, or tunneling operation. Those three types of operations will be discussed separately now in that order.

6.1 Trenching Operation

The trenching operation is excavation to open a slot in the earth's surface for a conduit or other long, thin construction item to be installed. Note that the installation part of this operation is important. It calls attention to the fact that the opened slot or trench will be backfilled on the installed item as soon as it is in place. This is the feature that makes trenching different from ditching. A ditch is also a comparatively long cut, but one that will be left open and may be sloped outward from the bottom to maintain the undisturbed earth's stability. Constructing the sloped slot or ditch with a trapezoidal cross section is frequently done most economically by a dragline. The crane-mounted dragline is discussed in Chap. 7.

A trench can be constructed with steeper, frequently even vertical, sides as long as they will not collapse before the open trench is backfilled. The backfilling part of the trenching operation has a special importance for the equipment used. It means that the trenching equipment must generally provide for stockpiling the excavated material along the trench line. Consequently, the equipment for this operation will either have some form of conveyor to unload the material to the side and away from the edge of the trench, or it will swing each bucketful to the side for stockpiling.

Another important consideration in the selecting of trenching equipment is the loading on the sides of the trench and on the conduit to be installed, if that is a concern. Material through which the trench is cut should be relatively firm, strong, and stable. If the trench sides have a tendency to collapse or fall down under loading close to their top edge, the equipment and operation must be planned to prevent that happening. The trenching equipment used should be mounted on crawlers to distribute the equipment load better. Extra-wide crawlers might be used to further reduce the surface load per unit of area. If collapse of vertical trench sides is very likely, then they could be braced across the trench width, or the sides could be sloped to appear more like a ditch. Either solution has its disadvantages. A trench with vertical sides braced extensively causes interference for equipment and men working in the trench. The efficiency and economy of such an operation will be greatly reduced, but safety of the operation will be greatly improved, and this is all-important.

If the sides are sloped to make the trench seem more like a ditch, loading on an installed conduit will be greater because bridging of the load over the pipe is reduced. For an open pipe conduit this is important. Construction specifications for pipe installed in a trench will generally have provisions to insure that the pipe does not become overloaded. Studies have shown that maximum loading on a pipe conduit occurs at the level of the top of the pipe. Therefore, if the trench width is kept to a minimum to the top of the pipe where installed, loading on the pipe should be within reason. The limitation of loading on the pipe also depends on the care with which the pipe is bedded and backfilled.

6.1.1 Design of Equipment for Trenching

The equipment designed for a trenching operation is of two general types. One is the special attachment known as a backhoe or pull shovel to be used on the versatile crane-type or tractor equipment. The other type is the equipment specially designed for trenching and known simply as a trencher or ditcher.

Design Features of a Backhoe. The crane-mounted variety of construction equipment will be discussed in the next chapter. A backhoe unit mounted on the basic crane equipment is shown in Fig. 6-1.

The backhoe, whether crane-mounted or on a tractor base, has several characteristic design features. The equipment is supported on crawler tracks or wheels with outrigger feet for digging stability and the load distribution mentioned in the previous section. The digging mechanism has a boom, a dipper stick with the bucket mounted on its outer end, and

FIGURE 6-1 A backhoe used for trenching
(courtesy of the Koehring Company).

cables or hydraulic cylinders to control motions. One end of the boom is
attached to the supporting equipment and pivots both vertically and
horizontally. The horizontal swing is by rotation of the turntable on
crane-mounted equipment. On a tractor the boom swings on its base
support. The dipper stick of a backhoe is supported at the outer end of
the boom and pivots about that point in the vertical plane of the boom.
In the same way, the bucket or dipper is attached at the end of the dipper
stick and is pivoted for digging.

With this mechanism the backhoe can reach out horizontally or down
into a trench with the boom, dipper stick, and bucket extended to start
digging. Then the bucket is pulled through the material and back toward
the equipment base to get its load. When it is full, these three parts are
in pivoted positions so the angles between each other are the greatest—
like a man's arm carrying a bundle against his body. To empty the

bucket's load, the boom is raised to clear the sides of the trench and then swung horizontally to dump the bucket away from the edge of the trench. This motion includes extending the basic three-part mechanism which prepares it for the next digging cycle. The described motions of the cycle are repeated from one position of the total equipment until all the trench material readily dug from there is removed.

Because of the strains of digging and the repeated swinging motion, another basic feature of most backhoes, especially tractor-mounted ones, is the outrigger, stabilizing feet supports. This additional mechanism is located at right angles with the trench line. The feet, which may carry the full weight of the working end of the backhoe, are positioned far enough out to avoid overloading the open trench sides. To extend the trench, the backhoe, with the crane-type or tractor base leading the way, moves backward, which necessitates taking the load off the outrigger feet and retracting them for the move. Since this has to be done frequently in a trenching operation, the outrigger motion is controlled hydraulically from the operator's position.

The choice of a hydraulic- or cable-controlled backhoe will depend on the basic equipment to be mounted with the trenching attachment. Crane-type equipment of 1/2-yard to 3-yard bucket sizes may have either cable-driven or hydraulic controls. For the cable controls, which are almost obsolete for backhoes, there must be an A-frame rising above the power source to give needed leverage for the cables to handle the boom and dipper. A tractor-mounted backhoe attachment with 3/8-yard to 1 1/2-yard bucket sizes is handled by hydraulic controls. With any form of equipment, hydraulic controls can do a job with more precision, but they are not as rugged as cable-operated mechanisms. This is because the cable can take the shock of impact with its flexibility and resilience.

The telescoping backhoe first introduced by Gradall has a unique design and is worth special mention as a trenching piece of equipment. It is like a crane-mounted machine on wheels for mobility and versatility. The boom has several parts with one telescoping inside another. It is closed up, like a telescope, to its shortest length for compactness or work in close quarters. For ordinary work and when needing reach, the boom can be extended up to 100 feet. This is not truly a crane unit since it is not designed to lift much load vertically. It is designed to do the sort of work that a backhoe or dragline crane attachment does. This backhoe with its telescoping boom is hydraulically controlled. It has the additional motion, not possible with cable controls, that can turn the bucket's front digging edge to a 45° angle from the horizontal line perpendicular to the backhoe's centerline. In some respects it acts like a hand on an arm. With this ability the trench or ditch sides can be sloped to any angle, if the equipment is positioned right.

To summarize the design features of a backhoe, the most important points are listed now. The design to select for a trenching operation should be based on:

1. dipper bucket width, which can be altered up to the overwidth bucket, which may be 4 feet wide, but in tough digging must not be too big for the power available;

2. digging force available with cable-controlled unit, which depends on engine's power, size of cable, and number of lines to dipper controls; or with hydraulic controls where force can be increased by increasing the diameter of the cylinders, provided the equipment load limits are not exceeded;

3. depth that can be dug from the front of mounting base—crane-mounted units can dig 12 to somewhat more than 30 feet deep, whereas the largest Gradall will dig 25 feet straight down;

4. reach of the bucket, which ranges from 15 to 45 feet with different size units, and is not very important for most trenching operations;

5. swing of the backhoe, which varies from full revolution (360°) with a crane-mounted unit to about 90° with a tractor-mounted unit; and

6. extra operations possible with the equipment, such as a backfilling device on a Gradall or a dozer blade on the opposite end of a tractor-mounted backhoe.

Design Features of Trenchers. The equipment specially designed for trenching operations is made to have continuous digging. This continuous digging and equipment movement is quite different from the action of backhoes. Trenchers get their continuous action by use of an endless chain of scoops or buckets on a wheel and steady forward movement along the trench line while excavating. Such action leads to some advantages but also several disadvantages in a trenching operation. These will be discussed shortly.

The continuous-chain trencher is most nearly like the historical forefather of trenching machines. Originally, crude buckets were pulled by a chain or rope along and through the material to be dug for a trench. Now, as always, lack of rigidity in a chain or rope prevents the buckets from digging into tough material. They tend to twist over or move out of such earth. In loose material the buckets can be expected to get full loads. Trenching close to the earth's surface will more likely involve loose material. This is probably weedy, loamy earth or loose granular material.

Modern continuous-chain trenchers as shown in Fig. 6-2 are designed for fairly shallow, narrow trenches using digging scoops spaced along the chain. The chain is supported on a boom attached to the rear end of the prime moving tractor unit. Or the trench-cutting attachment may be like an electric carving knife in place of the boom. This variety cuts the earth by rapid vibrations (thousands per minute), tearing the trench slot through the surface. Both a chain-type and a knife-type trencher may be attached to the same tractor, as shown in Fig. 6-2.

FIGURE 6-2 A continuous-chain trencher (courtesy of Davis Manufacturing).

In either variety the "boom" can be pivoted in a vertical arc of nearly 100°—about 40° above and 60° below the horizontal position. The diggers for the chain trencher are attached in an alternating pattern on the sides of the chain to cut a trench from 4″ to 24″ wide. Narrow trenches can be dug 6 to 7 feet deep, while the widest ones may be limited to a 2-foot depth. You will note that the cross-sectional area of the trenches with these varying dimensions is almost constant. For instance, a 6-inch-wide by 7-foot-deep trench has a vertical area of 3 1/2 square feet, whereas one 24 inches wide by 2 ft deep has an area of 4 sq ft. The power needed to excavate the trenches is roughly proportional to the cross-sectional area with somewhat more required for the deeper trenches. This is because the chain and diggers have to pull the cut material up steeper slopes. An augur mounted at the prime-mover, power end of the chain takes the loose, dug material off several feet at a right angle to the trench line.

The smaller chain trenchers are mounted on rubber-tired wheels for greater maneuverability. The smallest ones, not much bigger than a power mower, are designed for laying cable or small, assembled pipe in shallow, narrow trenches and are operated by a man walking alongside. Most of the other chain trenchers are designed for the operator to ride on the trencher. They may be mounted on rubber tires or crawler tracks. Wheel-mounted ones can turn curves up to 20° and toe their wheels at a similar angle for varying trench alignment. The trencher is pivoted on its crawler supports to cut the trench vertically when the machine is on an incline.

The other main type of trencher is the one described as a wheel-type, continuous-bucket trencher. An example is shown in Fig. 6-3. The wheel

FIGURE 6-3 A wheel trencher in action (courtesy of the Cleveland Trencher Company).

is made of two ring plates supported 8 to 18 inches apart by braces, with the buckets spaced every 20° to 30° around the wheel's circumference. The buckets can be 8 to 22 inches wide with cutting teeth that extend outward as much as 4 inches to cut a trench 30 inches wide. The maximum width dug by a wheel trencher is 6 feet wide.

This type of trencher is mounted on crawler tracks for stability and better load distribution. Stability is an obvious concern when it is realized that the several main parts of the wheel trencher are suspended out beyond the ground supports for the equipment. The wheel is suspended from a boom or cables at the required height straight back from the prime-moving tractor base. The height is adjusted to get the desired digging to the maximum of 7 to 11 feet deep. And the buckets on the wheel cannot be resting on the bottom of the trench without binding or causing excessive power demand. Some trenchers will use a crumbing shoe attached to the boom's end in a nearly vertical position just beyond the outer circumference of the wheel to pick up and carry part of this suspended load on the trench bottom. The "crumber" also serves to catch loose material in the trench and feed it back into the wheel buckets. In this way a clean trench is more certain. Another suspended load in the rear with the wheel is the conveyor to carry material from the trench off to the side for storage. The conveyor may be projected out either side 3 or 4 feet at right angles to the trench line, producing an additional unbalanced load. To counterbalance these suspended loads at the rear of a wheel trencher, the prime-mover engine is cantilevered out in front of the equipment's base supports. It is because of such unusual loading that continuous-wheel trenchers are mounted on crawler tracks for the best load distribution possible. With

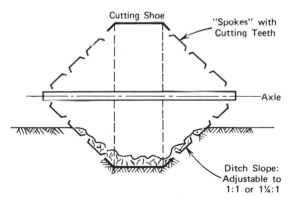

FIGURE 6-4 Cross section of ditcher's wheel.

the desired working speed of much less than 2 mph, a crawler-track mounting is certainly satisfactory.

At least one manufacturer of trenching machines, Barber-Greene, also makes a ditcher which operates in a manner similar to a wheel trencher. The wheel can be thought of as three-dimensional with the axle expanded outward as shown in Fig. 6-4. The "spokes" of the wheel have teeth to cut the side slope, while the cutting shoes or buckets cut the ditch bottom. With a level ground surface to operate on and a set of good-grade stakes with taut stringline, the ditcher can automatically do its work at a governed speed of about 30 feet per minute.

A ladder-type trenching machine is still another variety. It combines the advantages of a continuous-chain trencher and a wheel trencher. A ladder-like or point-down triangular projecting frame serves as the boom that can be raised or lowered into the trench being excavated. This depth can be adjusted while moving, so the ladder-type trencher can operate on a more irregular surface than the wheel type. It has continuous chains on both sides of the projecting frame with buckets like those on the wheel trencher attached to and between the two chains. So it can get the high production, along with lateral rigidity and power, of the wheel trencher. However, the ladder-type can reach a greater depth, up to 25 feet, by adding extensions to the ladder boom and chains. The excavated material is cast to the side of the trench by a conveyor belt as with a wheel trencher.

6.1.2 Use of Equipment for Trenching

The laying of cable is one of the easiest trenching operations. This conduit is generally not placed very deep in the ground. One reason is that the cable is not likely to have structural damage. Also, its repair is done by digging it back up so the closer to the surface, the better. The top few feet of earth is relatively loose material and easy digging. Furthermore, the cable does not need special bed preparation, so no handwork is done in the trench and it is not kept open any longer than the time to put the cable in the trench. This all adds up to the use of a continuous-chain or oscillating-knife trencher. A trench for cable may follow an irregular path along curbed roadways or around corners. The highly mobile and maneuverable rubber-tired mounting of a chain or knife-type trencher is well suited for laying such cable.

In the operation to install pipe conduit much more care and difficulty is encountered than with installing cable. The open pipe must have help to carry the loads transmitted to it through the ground. It must have proper alignment and grade according to the design for the pipe's use well below the frost depth. Generally, such trench bottoms must be at least several feet below the ground surface and wide enough to accommodate a working man. This means a trench width about one or two feet wider than the pipe. To provide added safety, a trench box as shown in Fig. 6-1 should be used. Of the equipment described in Sec. 6.1.1, the wheel-type or ladder-type trencher and backhoes could be used to handle the pipe-laying operation.

To help decide between the two prime sorts of equipment for a piping operation, both the economics and physical conditions must be studied. If the surface is practically level or at a constant grade, even with the pipe's, and the bottom is to be within ten feet of the surface, a wheel-type trencher may be ideal for the job. This is particularly true with fairly long, straight runs of trench, generally up to 30 inches wide. For a deeper or wider trench the ladder-type trencher may be best. An operation that includes many obstructions to the continuous travel of these trenching machines on the ground surface will tend to rule out such equipment.

For a trench line that is to have a modest depth down to 6 or 7 feet maximum but has considerable interference on the surface, the hydraulic-controlled, tractor-mounted backhoe may be best. This trenching equipment needs a minimum of headroom and can readily maneuver into various positions for digging. It will not dig as clean a trench bottom as the wheel- or ladder-type trenchers, so the labor to shape the pipe's bed will be greater. But the versatility of a hydraulic-controlled backhoe can be an advantage in placing the pipe, etc. If the digging becomes real tough, this type of backhoe may not be as effective as desired. In that case a crane-mounted backhoe is more desirable.

A backhoe for trenching can work on a variable terrain without much difficulty. This is a key advantage over the continuous-wheel-type trencher. But all backhoes require more labor effort in the trench. If the sides of the trench will not stand vertically and must be sloped, the choice of equipment is limited to either a backhoe, a Gradall unit or possibly a clamshell excavator (discussed in Chap. 7). The Gradall will handle a trench as much as ten feet deep. For trench depths greater than 10 feet a crane-mounted backhoe or clamshell of suitable reach and bucket size will be chosen. This type of equipment would also be chosen in shallower trenches where rock is to be removed.

In summary, we can note certain key factors in the selection of trenching equipment. These are:

1. type of conduit—whether it can be dropped in like cable, can be assembled at the ground surface, or need to be bedded and connected in place;

2. width of trench—for conduit only, or working space needed and with extra amount at joints (will govern trencher width);

3. depth of trench—generally less than three feet for cables, less than five feet for welded pipe carrying petroleum products under pressure, and variable depths where certain grades or penetrations are to be reached;

4. conditions of material to be excavated—at the extremes are tough or rocky material to be dug and material so soft and unstable that it has to be braced;

5. shape of ground surface—level, uneven, or quite variable terrain; and

6. length and alignment of continuous trench—how long and straight is each stretch of trench to be dug?

Of course, it must be recognized that the final choice of equipment or method will be the one that is most economical, all things considered. This is the general basis of decision for any construction operation. The application of this general rule to trenching is discussed further in Sec. 6.1.4.

6.1.3 Productivity in Trenching Operations

The rate of production in a trenching operation is stated in terms of linear feet per unit of time. Trenching productivity is most frequently stated in feet per hour. A careful distinction must be made between the rate of excavating the trench material and the production rate for the whole trenching operation. The latter includes the excavation, installation of the conduit, and backfilling of the trench. If the trench must be braced to work in, it may be left open longer than necessary because there is no rush to backfill before a cave-in. This time element would be shortened by a need to reuse the bracing material. In any case the rate of production of the total trenching operation is quite variable, except in the simple cable-laying operation.

For the remainder of this section, trenching productivity will be limited to excavating the trench. When applied to the trenching equipment discussed in this chapter, the productivity is the rate the excavator can move along the trench line. The trench is generally planned for a minimum width—just enough for the conduit and working space where needed in the trench. Thus, the rate of production in terms of volume excavated, such as cubic yards per hour, is not important. The versatile equipment, such as a backhoe, can handle a greater volume than it does when excavating for a trench. It is the linear feet of trench excavated per minute or hour that counts.

The rate of excavation for a continuous trenching machine is a controlled productivity. It will vary with the depth and width of trench, the toughness of material to excavate, and the power available or selected by the operator. When the trench has been initiated for the specified conditions, the trencher's production in linear feet/minute can be maintained until something interrupts it. A general representation of this productivity, q, is shown in the curves of Fig. 6-5.

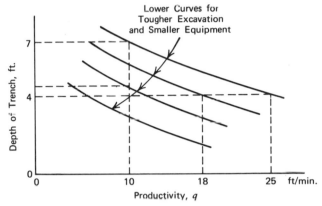

FIGURE 6-5. Production rates of trench excavation.

The family of curves shows the sort of variation to be expected in rates of production for different trenching machines under varying job conditions. The highest line is for the largest wheel trencher in easy digging material; the lowest is for the smallest, continuous-chain trenching machine. Where a larger trencher can dig a 7-foot-deep trench at 10 fpm, a smaller one could dig a trench only 4 1/2 feet deep at that rate. For all the trenchers, a deeper trench will result in a slower rate of advance, fpm, along the trench line.

Such a family of curves can also be used to account for harder digging. If the largest trencher runs into tougher material to excavate, its q rate would be found by moving to a lower curve. This is illustrated by referring to the longer dashed lines in Fig. 6-5. Assuming that the trencher digging four feet deep in easy material can move at 25 fpm, it may be able to dig that depth of trench at a production rate of only 18 fpm in tougher material.

TABLE 6-1 Trenching Equipment Data

Type of Equipment	Trench Section		Total Equipment			Travel Speed, mph	Digging Speed, max. fpm
	Depth, feet	Width, inches	Plan Area, ft, est.	Weight, 1000-lb	Engine Power, hp		
Cable layer	2–5	3–6	5 × 12	4–16	30–260	10	20–35
Walking-chain tr.	1½–6½	3–12	3 × 7	0.3–1 +	4–7	towed	6–20
Driven-chain tr.	2½–11	4–30	5 × 14	1.2–18	12–60	3–20	12–36
Wheel trencher	4–8½	9–66	6 × 26	10–50	42–110	5	10–110
Ladder-type tr.	5–15	6–54	6 × 20	7–40	47–90	3.4	10–22
Tractor backhoe	–20	18±	7 × 25	15–20	–100	–22	var.
Telescoping backhoe	–10	24±	8 × 25			40	var.
Crane backhoe	12–27	24–48	11 × 20	40–98	70–280	1–40	var.

The rate of production in feet per minute of trench advance for a back-hoe is not as high as for a continuous trenching machine. This is to be expected because the backhoe does not have continuous, automated action. Practically each move of this excavating operation is governed by the equipment operator. The advantage of a backhoe is realized where there is variability of digging conditions, i.e., toughness of material to be excavated, uneven ground surface, or frequent changes in trench dimensions and alignment.

To summarize the trenching abilities of the variety of equipment discussed in this section a tabulation can be made, as in Table 6-1.

6.1.4 Costs for Trenching Equipment

The cost of the primary equipment for a trenching operation will vary from only a cent per foot to considerable expense. One would guess this to be true with the variety of equipment. The difference in equipment expense from the smallest chain trencher operated by walking along beside it to the largest crane-mounted backhoe may be 20-fold. Of course, it must be recognized that there will not be much difference in the manpower cost for operating the smallest to the largest trenching equipment. To minimize the comparatively large labor expense for the walking-type of chain or knife-like trencher, that equipment is designed to automatically pull itself by winch along a stake line set for the trench. This frees the operator to set other lines or do other things while the trenching machine is working.

The cost of trenching equipment varies more in proportion to the trench width than to its depth. As shown in Table 6-1, a small chain trencher can excavate six feet deep, whereas the largest wheel trencher costing ten times as much will go only a few feet deeper. Regarding the other dimension, to get a wider trench will necessitate using a larger, more expensive trenching machine. With this in mind, a rough analysis will be made to show a comparison of the bare equipment expense to the operating manpower cost. The data used for this is given in Table 6-2.

The curves in Fig. 6-6 show dramatically the relative effect of the operator's expense in the cost of this equipment. Even with reduced operator attention to an automatically running continuous-chain trencher, the cost of the operator is unreasonably high. The horizontal measure from the zero cost line (vertically through 0 ¢/min) to the first curve is the bare equipment cost. The horizontal distance between the two curves represents the cost of the operator.

As the trench width gets larger and bigger machines are needed, the operator expense is less significant. For the smallest trenches this signifi-

TABLE 6-2 Cost Data on Trenching Equipment, Including Operator (based on 1969 information)

Type of Equipment	Trench Dimension			Equipment Cost		Digging Speed (v) fpm	Cost, ¢/linear foot, at		Excavating Cost, @ avg. v ¢/cu yd
	Width, inches	Depth, feet	Section, sq ft	Bare, approx. per hr	With Operator, approx. /min		max. v	avg. v	
Walking-operated chain trenchers	3–6	1½–3	3/4±	$ 1	7*¢	1–8	0.8	1.6	58+
	8–12	4–6½	4±	$ 2	9*¢	3–10	0.9	1.5	10+
Operator on chain or ladder trencher	8–12	3½–6½	3±	$ 2.75	12¢	1–12	1.0	2.0	18+
	12–18	3–6	4±	$ 4	14¢	1–16	0.9	1.7	12+
	22–24	–7	8±	$ 6	18¢	1–36	0.5	1.0	3+
	26–30	–11	15±	$ 8	21¢	—	—	—	—
Wheel trenchers	9–24	–6	6±	$ 5.75	17¢	1–100	0.2	0.4	2±
	30–36	–7	9±	$13	30¢	3–60	0.5	1.0	3+
	–66	–8½	20±	$27	53¢	1–35	1.5	3.0	4+

* Machine operated "automatically" on winch, so assume only 75% operator attention.

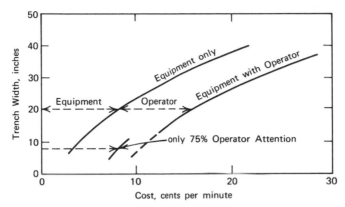

FIGURE 6-6 Comparing trenchers with operator expense.

cance suggests a planning point. When a trenching operation is for a relatively narrow trench, do not use the smallest machine possible. Such a machine may have great difficulty in the hardest parts of the excavation, whereas the next larger machine might do those parts with relative ease. Selecting the larger machine will result in an economy due to higher overall productivity. It must be mentioned that the smallest trenching machines do serve well for the owner-operator or the business that has many short cable or flexible-pipe installations with much moving between jobs.

Next, the cost of trenching equipment will be related to its productive work. The planner selecting the right machine wants to know how much it will cost per linear foot of excavated trench. He might also like to know how this cost of excavation compares with other excavating operations. These costs, based on the values shown in Table 6-2, are graphically shown in Fig. 6-7. It is important to note that these values include only the bare equipment and its operator. These two parts are the most significant elements of operating cost but account only for the actual operating time (peak productivity).

The curves in Fig. 6-8 show that trenching machines are designed and priced so the cost per linear foot dug is essentially constant at a few cents per foot. However, the excavation cost per cubic yard removed increases considerably as the trench width is reduced below 20 inches. These curves seem to also show that a two-foot-wide (24 in.) trench is the most economical with present trenching machines.

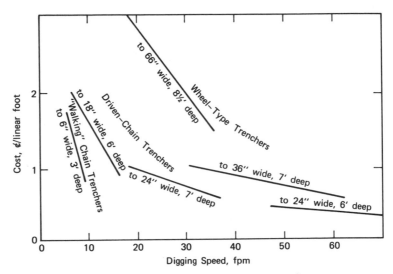

FIGURE 6-7 Trencher cost related to digging speed (referring to Table 6-2).

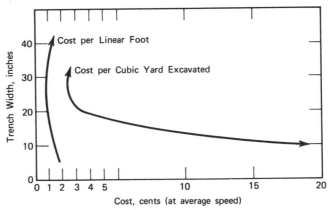

FIGURE 6-8 Cost for excavating by trenching machines.

6.2 Dredging Equipment

A dredging operation is a means for excavating soils from under water. Frequently, the material will be suspended in water to transport it to the place for deposit. Such an operation may be for the general excavation under the water of a bay, harbor, river or lake. In that case, if the area to be dug has considerable width as well as length, the dredge is set to undertake a sweeping action on the water surface about a pivot point. At other times the dredge will be excavating a channel or trench and needs to move along the line as the excavation proceeds. The same equipment can be used for either type of operation. In this chapter the emphasis will be on the more lengthy type of operation.

6.2.1 Design of Dredging Equipment

The basic dredge designs all consist of equipment mounted on a barge for floating and moving on the water surface over the soil to be excavated. Another common design feature is the projecting frame at the forward end to support the excavating part of the dredge. This part may take different forms according to the particular dredge design, but the frame or boom is a fundamental part. Also, there are generally several spuds, or retractable anchor posts, mainly at the stern end of the barge, to hold it in a stable position and for pivoting. These are moved vertically up and down by a gantry hoist at the same end.

The basic designs are known as the suction dredge, the cutter-head pipeline dredge, the bucket-ladder dredge, and the dipper dredge. The design features of these will be discussed in the next sections. A typical arrangement for an excavating dredge is shown in Fig. 6-9.

The Suction, or Pipeline, Dredge. The suction, or pipeline, dredges have a closed pipe running down the boom to the soil being excavated. They are called by the diameter of this pipe, such as a 20-inch dredge. Their operation is based on a suction created by a centrifugal pump on the barge which pulls the soil solids up the pipe suspended in water, as shown in Fig. 6-10. With the excavated soil being transported this way, it is a simple matter to keep it moving in a pipeline extended to the chosen point of deposit, which will generally be some distance from the barge. Consequently, the pipeline must be supported over the water by floating it on pontoons. The pipeline must be longer than the straight-line distance to allow the dredge freedom to sweep from side to side or to move back and forth in its excavating operation. In addition, the pipeline may extend

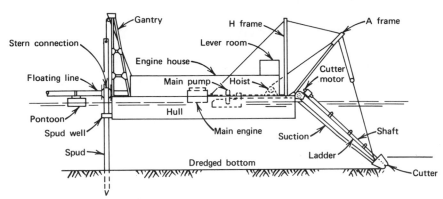

FIGURE 6-9 Dredge components.[1]

some distance from the shore overland to the discharge point. That latter stretch of pipe can be more permanently located for the operation.

The size of pump for a suction dredge will depend on the:

1. proportion of solids to be transported in the water—usually 10–20%;

2. suction lift or velocity of fluid required at the intake to get the solids; and

3. total dynamic head, i.e., the total effective hydraulic force needed to get the water with solids from the intake to the discharge point.

The dynamic head will include not only an actual elevation difference but also frictional resistance to the fluid moving in the pipe and some velocity needed at the discharge end of the pipe. If the frictional resistance is too great, a booster pump is installed along the line. A review of pump equipment in Chap. 4 would be desirable.

In selecting a dredging pump, the equipment planner may speak of the velocity of flow. This is partly because he recognizes that the pump must force water by the intake at considerable speed to produce the needed suction. How this works is seen in Fig. 6-10. The usual way of expressing velocity of the fluid in a pump is in a rate such as gallons per minute, gpm. The chosen gpm rate may be the minimum to move the material and results in a low percent of solids carried in the fluid. Or it may be based on a desired rate for excavating the material. If the desired amount of solids in the water is 20%, then one could estimate a specific gravity of 1.2 and that it will be necessary to pump five times as much volume of water as the soil to be excavated. This high percent of solids will require a high-velocity flow. The rate of flow and friction head will depend on the size

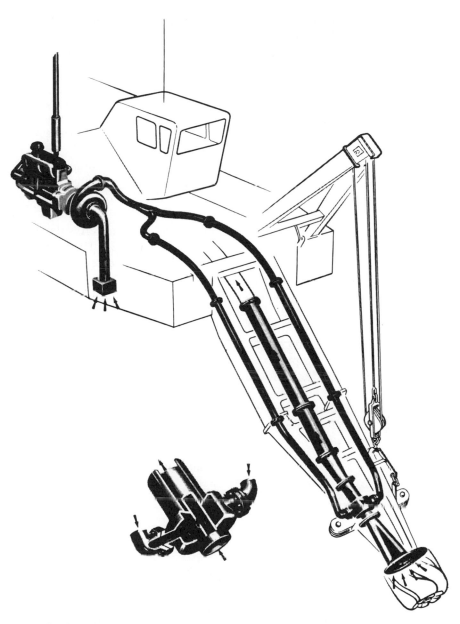

FIGURE 6-10 The design of a suction dredge (courtesy of PACECO, a Division of Fruehauf Corporation).

of pipe used. The pipe diameter will vary from 6″ on the smallest dredges to 36″ on the largest. For a certain desired gpm flow, a larger pipe will result in less frictional head loss, but also a lower velocity of flow through the pipe.

The smallest suction dredge with a 6″ pipe can be built on a barge covering only 14 feet of beam and 30 feet end-to-end on the water surface. Furthermore, it may need only two feet of draft, or depth of water, in which to operate. The largest pipeline dredge may need a barge four times as wide and seven times as long, and may require a 14-foot draft. It can excavate material that is located in a water depth of more than 50 feet. The pipeline dredges are designed so that the ladder frame and suction-pipe lengths can be extended with inserted sections to excavate at greater depth.

A normal hydraulic (suction or pipeline) dredging operation is designed for a fluid flow rate of 12 to 50 cubic feet per second (fps). Using the approximate conversions suggested above, these flows should result in 300 to 1300 cubic yards of solids moved each hour with a 20% suspension in the transporting fluid. At 15% solids in the flow, which certainly should be expected, the excavated material moved would be between 240 and 1000 cu yd/hr. This range of fluid flows, converted to gallons per minute, ranges from 5400 to 22,500 gpm.

The specification of a pump for dredging equipment is generally done in terms of the brake horsepower (bhp) to meet the job conditions and quantities desired. The bhp will be related to the quantities discussed above and the pump's efficiency (eff). In dredging work the expected efficiency is between 50 and 60%. The desired size of pump is found by

$$\text{pump's bhp} = \frac{\text{gpm (sp. grav.) head}}{3960 \text{ (eff.)}}. \tag{6-1}$$

For the range of pumping rates given above, the bhp should be between

$$\text{bhp}_1 = \frac{22,500(1.2)h_t}{3960(0.5)} = 13.6 \text{ per ft of total head}(h_t) \text{ and}$$

$$\text{bhp}_2 = \frac{5400(1.1)h_t}{3960(0.6)} = 2.5 \text{ per ft of total head } (h_t),$$

To be able to transport the solids through the pipeline for a mile or more, suction dredges will have pumps driven by power units of 1000- to 8000-hp capacity.

Most modern suction dredges can improve on their effectiveness with a special digging or cutting mechanism at the intake of the pipeline. The cutter-head pipeline dredge has this advantage over the simple suction type shown in Fig. 6-10. This type has a revolving cutter that looks like a cross between a marine propeller and a partly open orange-peel bucket with the peels held rigid for cutting or breaking up tough clay or rocky material. It rotates about an axis through that of the suction pipe. The drive shaft for the revolving cutter-head runs from power on the barge and is supported on the ladder frame which holds the dredge's intake pipe. The ladder frame with the intake pipe and cutter-head can be raised or lowered according to the required depth of excavation. This is done by winch lines running over the tower frame at the forward end of the dredge barge. The power unit for these winches also operates the spud winches and the anchor and swing cables for holding or moving the dredge.

Other Forms of Dredges.   The forms of dredging equipment, other than the pipeline variety described in the previous section, do not move the material great distances in one continuous operation. A bucket-ladder dredge is designed to operate very much like a ladder-type trencher. Its ladder frame projecting from the forward end of the barge supports a continuous chain of buckets that excavate like the trencher. The buckets dump their loads, generally, into a hopper-like container on the barge. When the hopper is full, the barge will have to stop digging and move to deposit its load. A variation to avoid the interruption of unloading a hopper on the dredge's barge can be arranged. This would involve separate hopper barges lashed alongside the dredge and loaded by a conveyor from the buckets' dumping spot.

The dipper dredge has the advantage over the previously discussed varieties in its digging flexibility. It is essentially a barge-mounted power shovel. The common power shovel will be discussed completely in the next chapter. A dipper dredge can also be compared to the backhoe discussed earlier in this chapter. They both have a boom, a dipper stick with a bucket on the outer end, and cable controls to operate these parts. However, the dipper dredge bucket digs when it is moving away from the power unit on the barge. This is like the power shovel and opposite to the backhoe or pull shovel action.

When the dipper bucket is full, it is swung laterally to unload. If the excavated material is to be moved some distance, the hopper is mounted on the dredge's barge or one along side. This easy variation in arrangement plus the flexibility that the dipper can dig at any spot in the semi-circle ahead of the dredge makes this variety a frequent first choice for dredging. The arrangement of spuds to position and hold the dipper dredge is somewhat different from the other dredges. Generally, there are two spuds at the forward end with the dipper mechanism. These support the dredge for more digging forcefulness. The third spud is at the aft end and movable to an angle of 20° or 30° with the vertical to give the dipper extra crowding advantage. This movable spud can also help the dredge "walk" along the bottom for repositioning.

Still another variety of dredging equipment uses a clamshell bucket as the digging tool. This is designed for working in loose sand and gravel or silt where not much digging force is needed. The dredging bucket is controlled from either a swinging boom or an overhead traveling gantry. In the latter case, the bucket may dig through an opening in the deck of the dredge for the confined quarters of a boat slip or between piers. The bucket is generally emptied into a hopper on the dredge's barge. Then the material is either processed right on the barge or moved by conveyors into the sand-and-gravel plant or disposal barge nearby.

6.2.2 Uses and Costs of Dredging Equipment

The many dredging operations to excavate earth material from under a few feet to more than a hundred feet of water could have any of a number of objectives, such as :

1. canal and waterway construction;
2. trenching for pipeline for transport tube;
3. river dredging for navigability;
4. harbor improvement;
5. marina construction;
6. land reclamation by hydraulic filling;
7. reservoir maintenance;
8. sand and gravel recovery operation; and
9. other processing with earth materials.

The discussion of dredging equipment in the previous sections suggested the possible choices for these objectives. In making a choice, careful study must be made of the material to be excavated. It may be anything from soft muck or silt, peat, or sticky clay, to granular sand and gravel of different sizes, heavy minerals, or loose coral. The material may have tree roots, swamp grass, or other growth materials or trash in it. Added to this complex set of factors in the dredging choice is the extreme variation in job conditions. These factors are indicated by recognizing the variations of calm or rough wavy water with varying water depth and tides on the sea coast. Also, there are the concerns for getting the equipment to, and maneuvering it in, the area for dredging.

These points, which suggest the complexity in choosing appropriate dredging equipment, are made to indicate why it is unwise to generalize about their selection. It can be observed that land filling or depositing the excavated material not too far from relatively calm water suggests a pipeline dredge should be the best choice. Or if the digging is tough and variable but not too deep, a dipper dredge is probably the choice to make, whereas an operation to recover loose sand and gravel to be processed close-by will undoubtedly find a clamshell dredge the best selection of equipment.

In making an economical selection of dredging equipment, it should be noted that frequently the best solution is custom-made equipment. Or at least part of the equipment is built especially for the given job. The total expense of such equipment, including the initial cost, generally, must be covered by income from the one job or series of operations in one locale. Such work often takes a long time—months or even years. So maintenance costs are of great significance. With a pipeline dredge, maintenance of the pipe presents a problem. Because of the solids being transported in the water, it is essential to use an abrasion-resistant material for the piping. Also, since sections of pipe that can be handled and changed readily are each approximately 30 feet long, the joints between sections must be carefully designed for flexibility and maintainability. There is some variety of materials and designs that can be selected for the pipeline. This aspect of a pipeline dredge must be well planned for an economical operation.

There is now available a sectionalized pipeline dredge, such as those designed by Ellicott, that can be transported from one job to another. The sections are the hull with control housing; the pump module with pipeline; the digging depth modules with hoisting frames, spuds, and ladder; and the excavating modules with cutterhead, winches, and driving mechanism. These can be disassembled for moving or changing sizes of the sections. Thus, this type of dredge is versatile and can be used on a variety of jobs. In that case its initial cost can be prorated over a number of jobs and years. Other dredges should be equally flexible for economical equipment use.

6.3 Tunneling Operations and Equipment

The third variety of equipment to make long, narrow cuts in the earth is for tunneling. Unlike the trenching equipment described first in this chapter, tunneling equipment bores a hole completely surrounded by earth materials, including the overlying material. So it cannot lift its excavated material up through the top open side to dispose of it, except where there is a vertical access shaft. It must pass the material back through the tunnel already bored like the animal, known as a mole, does his boring under the ground surface. From this similarity came the name for the modern tunneling machine called a "mole."

In ancient times a tunnel was bored through earth materials with the crudest hand methods. Originally, they amounted to picking at the rock with antlers and other sharpened hand tools and wedging, then wetting, wooden pieces in the natural cracks to split out pieces of the rock. The excavation changed somewhat when it was discovered that the rocks break up better if they are dried out and subjected to great temperature changes. The method tried with this discovery was one of firing, i.e., setting fires in the tunnel beside the rock to be excavated. This scheme was used until the 19th century, when the drilling-and-blasting method took over. At first the drilling was done by hand-hammering steel chisels into the rock. However, by the turn of the present century, drilling was being done by hand-guided tools driven by mechanical power. The compressed-air-powered drill hammers were the first form of equipment used in tunneling.

The drill-and-blast method of tunneling, which has been the prevailing scheme in hard ground for the past century, makes extensive use of manpower. Men handle the drilling tools, pack the explosives, install bracing and tunnel lining pieces, and load the excavated material in many operations. The use of manpower for so many of the steps in tunneling has been dictated by the variability of ground conditions and the efficiency and economy of labor. The man-handling of air drills in large tunnels has been helped by movable frames with two or more working platform levels. These are known as jumbos and may be mounted on steel wheels to run on rails or on rubber tires.

The use of other power machinery in a tunnel had been avoided because it might create a dangerous atmosphere in an already limited source of air. Certainly, internal combustion power units that exhaust carbon monoxide should be avoided. So as long as labor has not been too expensive and a high rate of tunnel advance was not expected, the drill-and-blast method has prevailed. It resulted in tunnel excavation advancing up to 20-25 feet per week at the beginning of the 20th century. The drill-and-blast method was further encouraged with the introduction of tungsten carbide bits which doubled the drilling life, speeds, and efficiency of the tools.

6.3.1 First Equipment for Tunneling

More equipment was introduced into the tunneling operation in the early 1900's. Mechanical loaders were designed for the mucking step to dispose of the excavated material. This equipment was developed for the extensive mining use, but is equally useful for tunnel construction purposes. The loaders were designed to use with small-gauge rail cars to haul out the loose material. A picture of such a modern muck loader is shown in Fig. 6-11.

FIGURE 6-11 A transloader used for mucking (courtesy of Joy Manufacturing Company).

Muck haulers have been needed with the ability to side dump their load when run completely out of the tunnel or to a vertical shaft, where a cable hoist lifts the material box up for disposal. The track gauge for the muck cars and their prime mover is generally between 18 and 36 inches. The cars are almost twice as wide as this distance between rails. The variety of sizes ranges from $\frac{1}{2}$ to 50 cubic yards, depending on the track gauge which can be used. It is necessary to design such a mucking system with room for passing cars on parallel tracks within the tunnel width.

Therefore, a tunnel 8 feet wide would possibly use an 18-inch gauge rail system, whereas one 14 feet wide could use 36-inch gauge.

If the tunnel is 18 feet wide or more, diesel-powered, tire-mounted trucks can be used to haul the muck. Two such trucks should be able to pass within the tunnel width for an efficient operation at the tunnel face more than several hundred feet from the portal. They are loaded by a diesel- or electric-powered muck loader. This may be a power excavator specially designed for work in the cramped space of a tunnel. Also, an overthrow-type front-end loader has been used with success in tunneling. A popular loader for tunnel excavation is called the transloader (see Fig. 6-11). It has a scoop bucket to load its own cargo dump unit, both parts being in front of the operator's cab for more efficient operation.

Diesel engines used to power these pieces of equipment in the tunnel will have to be given special treatment. There is not the deadly carbon monoxide exhaust from a diesel engine as there would be from a gasoline engine used inside a tunnel. However, a diesel power unit exhausts nitrous oxides, which can produce nitric acid in a man's lungs, so an exhaust conditioner must be used on the engine. Even with this precaution, the U.S. Bureau of Mines standard of 75 cu ft of free fresh air per minute per horsepower must be provided. That gives one of several important demands on a ventilating equipment system for tunneling.

Returning to the equipment for mucking, we need to know some more particulars of units designed especially for this operation. When rail cars are used to move the excavated material away from the tunnel face, it is most economical to use the largest muck cars that will fit in the tunnel cross section. Of course, the car size will depend somewhat on the prime mover. This could be a diesel- or diesel-electric-powered locomotive, but often a battery-operated locomotive is used to avoid the added ventilation requirement. In either case the locomotive will weigh as little as a few tons or up to 50 tons. The size depends on the power requirements as a prime mover.

The size of a narrow-gauge locomotive for tunneling work depends on the weight of the loaded muck cars, the maximum number of cars to make up a train, and any adverse grade to go up with the loaded train. The required power or tractive effort, TE, can be determined by the formula,

$$\text{TE} = (L + nC)(F \pm 20G) \tag{6-2}$$

where L = weight of locomotive, tons;
C = weight of a loaded muck car, tons;
n = number of muck cars/train;

$F =$ rolling resistance or friction of wheels on rails, 10 to 30 lbs/ton; and

$G =$ percent of grade ($+$ if up).

The maximum tractive effort to be expected from a locomotive with all wheels driving in a tunnel is 20% (based on the coefficient of traction) of the unit's total weight. If that fact is used to pick the prime mover, the unknown locomotive's weight appears on both sides of the equation. To find the smallest locomotive that could be used for a job would be an involved calculation.

Equation (6-2) does not account for power to start the train moving nor to overcome the extra resistances of rough track and curves. It is well to have 50% or more reserve in the prime mover for a tunnel train. Therefore, the recommended way to use Eq. (6-2) is by trying an available size locomotive and checking its adequacy. This is based on a chosen size of fully loaded muck cars and planned maximum length of train for the tunnel job layout and conditions.

For example, 5-cu yd muck cars might weigh 4000 lbs empty. If they are loaded with material averaging 3000 lbs/cu yd, a loaded car will weigh 19,000 lbs. Planning for a train of 8 cars operating up a 1% grade with maximum rolling friction may require a 12-ton locomotive. This can be checked by Eq. (6-2) thus:

$$\text{train's weight} = 12 + 8 \left(\frac{19000}{2000} \right) = 88 \text{ tons;}$$

$$\text{required TE} = 88 \, (30 + 20) = 4400 \text{ lbs.}$$

The locomotive should deliver 0.2 (24000) = 4800 lbs, so the 12-ton diesel locomotive's 4800 $>$ 4400 is enough.

Of course, the number of cars or length of train may have been arbitrarily chosen. A shorter one would allow the use of a smaller locomotive. With a given size prime mover the number of cars can be determined by solving for n Eq. (6-2). It should also be noted that the larger locomotives are diesel powered and are in competition with diesel loaders and independent haul units.

A rail system is at more of an advantage for smaller tunnels with the lighter battery-operated locomotives. There should be enough battery power to reach the locomotive's maximum tractive effort and produce this limit as required during a working shift. Then the batteries can be recharged for the next shift.

The extensive use of such a mechanized mucking operation with air-powered drilling led to better efficiencies in the first half of the 20th century. The tunnel advance rates during this period increased to 100-200 feet per week.

6.3.2 Geology for Tunneling Machines

An understanding of the geology in the site for a tunnel is extremely important. In fact, the planner of equipment and method for excavating and constructing the tunnel almost has to have a geologist with whom he consults. Even so, a personal knowledge of the geological conditions influencing his selections will be very helpful.

The Earth's mineral material may be classified simply as rock or soil. The category known as soil is the uppermost material near or on the surface that has been weathered, decomposed, transported, and deposited. It comes from the rocky material that is generally under the surface, though it may be at the surface because of an earthquake or volcanic upheaval followed by centuries of the weathering and reforming process.

A tunnel driven under the Earth's surface will generally go through rock of varying consistency. It may be dripping or saturated with water due to a high water table and the rock's cracks or variations. The rock may be subdivided into soft, medium, or hard rock. Soft rock includes beach sand, consolidated clay, chalky bedrock, and loose or fractured rock. At the other end of the scale, hard rock includes granite, gneiss, schist, taconite, and hard diorite. The medium rocks are sedimentary rock, sandstone, limestone, chert, and shales.

In order to make a more definite determination for the tunnel equipment planner to use, it is desirable to have a positive scale of the rock hardness. This measurement of the rock is actually an indication of its compressive strength. That strength tells what load the rock can take pushing on it just before the material breaks apart. A tunneling machine has to top the rock's compressive strength to bore through it.

The recognized rock hardness scale is one geologists use known as the Mohs scale, named after the German mineralogist Friedrich Mohs (1773-1839). The Mohs scale (Table 6-3) arranges ten minerals in the order of increasing hardness.

TABLE 6-3 Mohs Scale of Mineral Hardnesses

1.	Talc	6.	Feldspar
2.	Gypsum	7.	Quartz
3.	Calcite	8.	Topaz
4.	Flourite	9.	Corundum
5.	Apatite	10.	Diamond

Then the common rocks, through which tunnels may be bored, can be given a Mohs' rating, as in Table 6-4.

TABLE 6-4 Approximate Hardness Ratings of Rocks*

Shale, usually less than 3 (2,000± psi)
Limestone and marble, 3 to 6 (10,000–15,000 psi)
Slate, 4 to 5 (15,000± psi)
Sandstone, between 3 and 7, depending on cementing agent
Granite, 6 to 7 (more than 20,000 psi)
Gneiss, 6 to 7
Schist, 6 to 7
Quartzite, 7

---

* Based on Mohs' scale, with estimated compressive strength.

6.3.3 Tunneling Machines

The scheme of using a tunneling machine to advance a hole through the ground is not entirely new. In fact, a machine was used in the 1880's to bore two 7-foot-diameter pilot tunnels under the English Channel. Each of those tunnels was driven more than 6,000 feet long. They were discontinued when the British feared invasion from the continent.

Around that time, the drill-and-blast method was becoming more efficient and economical so that elaborate machinery to do the whole tunneling operation was not justifiable. However, some equipment was being developed to help with the tunneling job. These were designed for mining operations or converted from such machinery. They included mechanical drills on movable carriers, muckers, and mechanical miners.

The first complete rock tunneling machine designed for a construction operation was built for Oahe Dam in South Dakota during the mid-1950's. The machine was basically a cylinder on its side with its front end rotating and pushing against the shale to bore through it. This original tunnel machine, known as a mole, had counter-rotating, concentric heads to balance the twisting force on the machine. However, a second version for another phase of Oahe Dam eliminated the counterbalancing head and found no problem of twisting.

Design of Tunneling Machines. The equipment for doing a complete operation of driving a tunnel and removing the waste (muck) material is now well recognized. It is not designed to do a cyclic operation like the drill, blast, and muck method. The tunneling machine is designed as a continuously operating piece of equipment like the wheel trencher. However, it may be stopped by its supporting, material-handling equipment or by unpredictable ground conditions.

The first full-cross-section tunneling equipment used extensively in this country was the shield introduced in the 1920's. A shield was not a self-propelled piece of machinery. It was used as a protection while digging through soft ground to temporarily hold up the surrounding earth material until the tunnel supports could be placed. The shield protected against the ground caving in or ground water flooding the tunnel. To do this, greater-than-atmospheric air pressure is needed to counteract the flow of material into the working space at the heading of the tunnel. This space is sealed off in the full-face shield by a bulkhead between the tunnel face and the installed rib supports or tunnel lining. To get workers into or out of the space under higher-than-atmospheric air pressure, there is an air compression or decompression chamber in the bulkheaded section. Workers can exert themselves only for limited periods in the higher-than-atmospheric air pressure. Otherwise, they will get the bends from too much oxygen in their bloodstream.

A modern, complete tunneling machine can work in soft ground. It serves as its own shield, and its mechanical cutting head can serve as the bulkhead. The design has to have just enough openings to allow the cut or dug material to pass through while the machine moves forward. This balance is difficult to achieve. If the material is too soft or saturated with water, the tunneling machine will be flooded inside. If it is too stiff or hard for the material to move through limited openings, the machine cannot be advanced. So a flexible design for a mole-type machine is necessary to operate in soft ground. This will have large enough openings to work in stiff material and use air pressure 4 to 14 psi (about one atmosphere) above the normal atmosphere to control runny soft material between the heading and a special bulkhead in the machine.

To be more certain of tunneling through earth material of good consistency, several methods have been used to stabilize loose, wet, runny material. One method is to inject a cementing material such as grouting. This material often is a solution of sodium silicate and calcium chloride to form a solidifying silica gel. The jobsite conditions should allow the injection and solidification to take place before the tunneling machine reaches the treated ground. Otherwise, it will probably be uneconomical to use the method because of holding up or slowing down the expensive tunnel equipment.

A tunnel boring machine, known as a TBM or a mole, is designed to operate best in uniformly stiff clay dense enough to keep ground water

7'-6" TUNNEL BORING MACHINE.   FRONT VIEW
EQUIPPED WITH STANDARD DIGGING TEETH.
NOTE: BUCKET SYSTEM CONVEYOR-HOPPER
        ARRANGEMENT.
TM-1   Photo No. 007-09

CALWELD, Division of Smith Industries International, Inc.
9200 Sorensen, Santa Fe Springs, California   90670

FIGURE 6-12   An open face type of tunneling machine
(courtesy of Calweld, Division of Smith Industries
International, Inc.)

from seeping in at the heading. In that case a cutting wheel, such as the unshielded one pictured in Fig. 6-12, can be used.

Tunneling machines of the open-face and the full-face (Fig. 6-13) types are made in a wide range of sizes. The rotating head generally is between 7 and 30 feet in diameter. Some are adjustable for a diameter variation of plus or minus 2 feet. The cutter wheel's rotating speed varies between 3 and 10 rpm, or approximately 80 rpm divided by the diameter in feet. Electric-powered motors are the primary source of power for all tunneling machines. This form of power unit is coupled with hydraulic rams needed for thrust, torque, and movement of the machine.

The cutters to break out rock ahead of the machine are designed and arranged in a variety of ways as equipment manufacturers try to improve their effectiveness. Early designs of the cutters relied heavily on the cutter wheel's thrust and dragging action of the cutters to break out rock on the working face. This resulted in rapid wear and too frequent a need to replace the cutters. Newer designs require the cutters to give a splitting, a chipping, or a pulverizing action such as a planetary-type cutter will produce.

FIGURE 6-13 A rock-cutting TBM (courtesy of the Robbins Company).

The design of the cutters in the tunneling machine's head is vital to the success of this equipment. There are several basic cutter designs— roller or disc-type or planetary cutters. Roller cutters are conical in cross section with the cutting from a crushing or pulverizing action. These are designed for compressive strengths of rock up to 35,000 psi. In the case of the hardest rock the tunneling machine has to exert high pressure by its thrust. The roller cutters working in very hard rock will wear down fast and have a high maintenance cost. They work well in combination with disc cutters. The rollers cut kerf rings in the face. Then the disc-type cutters are designed to split the ridges of rock between these kerfs. These produce excavated rock with about 40% of the largest chunks being hand size. A cutter head with this design is good for use in rock of compressive strengths up to 20,000 psi. Tunneling machines with the disc-type cutters require less energy, less horsepower, and less thrust than one of all roller cutters. Also, with larger pieces of excavated material less swelling is possible if the muck is soaked with water. However, the mucking operation may be more difficult with the large pieces, especially on narrow belts, and it would be impossible if a slurry method is to be used.

The planetary type of cutter works to pulverize the rock. These cutters rotate in different directions on their separate axes so they are not so dependent on the leading face of the TBM. In fact, they can operate on a movable arm at the machine's front end. Such mobility makes this type of cutter useful to bore a tunnel face that is not circular.

Hydraulic power controls are used to hold a tunneling machine in position or move it along the tunnel path. This power gives an axial thrust of 500,000 to 1,500,000 pounds, or at least 50,000 lbs per foot of diameter, and a rotational torque up to one million inch-pounds. The thrust works to crowd the cutter head into the rock. The machine's reaction is provided through ribs and radial jacking feet operated against the tunnel sides at right angles to the machine's direction. These retractable feet also help to move the machine along its path.

The power units for a tunneling machine vary with the difficulty of digging expected, the way the TBM or mole will cut through the material, and its auxiliary parts. For instance, the cutting head of a rim-drive machine for soft material may need only a 75-hp motor for rotating its wheel. At the extreme to rotate the cutter head with conical roller cutters and a 1,500,000-lb thrust against hard rock may require a motor arrangement totaling 2000 hp. In rock tunneling the horsepower for rotating the cutting wheel must be about 50 times the diameter in feet. For rock of medium hardness the required horsepower could be related to the expected rate of tunnel advance. The relationship would be about 100 hp per foot of theoretical, maximum advance per hour. Other power requirements for a tunneling machine include those to operate the rib jacks and other mechanisms for moving or steering the mole. These will necessitate 100 to 150 hp. And the Teredo mole's slurry pump is driven by a 30-hp electric motor.

The torque ability of a tunneling machine gives the cutter head the power to excavate the rock. The jacking feet prevent the rotating cutter wheel from twisting the machine out of its intended alignment. In spite of this design measure to resist it, the TBM's head still tends to wander from the intended line and grade. A special machine designed to overcome this wandering tendency was originally called the Alkirk hardrock tunneler. It used a principle developed in mining for driving vertical or sloping shafts between vertically separated tunnels. This is called "raise" drilling in which a rig drills a small pilot hole of 8- to 12-inch diameter to connect the tunnels. This hole serves as the center guide for the larger-diameter shaft. In the Alkirk tunneler the pilot anchor is a small tunnel cutter operated ahead of the main machine. When it has cut ahead and anchored into the rock tunnel's face, the pilot anchor serves to hold the tunneling machine on course against the face. This anchor also provides some of the thrust capacity of the machine.

Recent schemes have been developing to be more assured that the common tunneling machine stays on course. One such method uses a laser guidance system. A pinpoint concentrated beam of light the full length of the excavated straight tunnel can be a constant guide for the machine's advance. The jacking feet can adjust the machine's direction if it gets off course.

The arrangement for handling the excavated material in a tunneling machine is tied directly to the cutter-head design. With a big, open wheel design, useful in stable soil and relatively soft rock, the cut material falls into buckets around the wheel. The bucket-type arrangement is all around the circumference of the cutter wheel. Buckets at the bottom catch the material, and when they rotate to the top, they dump the material onto a continuously moving belt conveyor at least 30″ wide. This conveyor carries the muck material back through the tunneling machine anywhere from 50 to 300 feet or more. The same arrangement is developed for tunnelers with a more closed cutter head. They would still have openings for the cut material to pass through. On the side of such openings away from the tunnel face are buckets to catch the excavated material. A Japanese model made primarily for soft ground has knife-edged radial slots on the rotating wheel and buckets behind these slots.

Another method for handling the excavated material is by a water slurry. This is possible for loose, small particles or the powdery material resulting from pulverizing the rock. Specifically, it could be used with the planetary- or roller-type cutters operating with considerable thrust behind them. Or it might be used with soft ground. The Teredo tunneling machine uses a slurry method for handling the muck. The cutters are moved by a continuous chain (like a chain trencher) on an arm that pivots around the central axis of the machine. This tunneler has a watertight bulkhead just behind the rotary cutter-head arm. The space between the tunnel face and this bulkhead is kept about half filled with water—either ground water or other water is pumped in to form the slurry with the excavated cuttings. This slurry of the muck material is pumped out through a 6-inch opening near the bottom of the bulkhead and connected to the same size pipe for discharge back of the tunneling machine. The machinery and operator of this tunneler are protected from flooding by the same bulkhead. It should also be noted that the arm of cutters rotate in this slurry to keep the material from settling out and also to keep sticky material from clinging to the cutters.

Three other key features of a tunneling mole's design need to be mentioned to give a complete picture. They are dust control mechanisms, roof opening for tunnel supports, and hoods for soft or loose ground around the tunneling machine. The dust control is designed to gather or hold down the great amount of dust caused by the churning, pulverizing action of the machine's cutter head. A mechanism to create a suction like the common vacuum cleaner has been used on some moles. Other tunneling machines have made use of water spray nozzles in the cutter head and along the conveyor belt to hold down the dust. The proper amount of water is important—too little may make the fine material sticky, whereas too much will cause sloppy working conditions.

Openings are designed in the tunneling machine back of its cutter head. These give working room for rock bolts or rib plates to be installed to hold the earth's material surrounding the tunnel until the permanent lining is placed. For loose, soft, or running ground, the tunneling machines have detachable hoods. One such hood piece can jut out ahead of the machine, like the visor of a baseball cap, and serves to define the shape of the tunnel cut in soft ground. Hood pieces will be used throughout the length of the machine to protect against loose material that could cave into its working space. A trailing hood at the back end of the machine can be just the lining thickness away from the cut tunnel dimension. This will allow such a trailing hood to serve as the form for a concrete lining to be poured as the machine moves away from it. It is then a slip form.

Productivity in Tunneling Construction. The rate of carrying out the construction of a tunnel can be stated in several ways. It might be expressed in terms of the length of tunnel centerline excavated, supported, and finished with lining in a given time span. This would be helpful to the owner or final beneficiaries because it would tell how quickly the job might be done. However, productivity expressed in that manner might be misleading. The reason for this is that there can be a great delay between placing of supports for an excavated tunnel and finishing it off with a permanent lining. There may be delays caused by treating the exposed natural rock, by handling the ground water, or by installing pumping and ventilating equipment. The timing for any step of the tunnel work will depend on the urgency to finish the construction and the problems encountered in reaching that goal.

Tunnel excavation generally governs the speed of completing tunnel construction. That is, the tunnel cannot move ahead any faster than the opening can be bored through the ground. Supporting the walls of the opening will be done as needed when the excavation moves forward. Installing any equipment and the lining can be carried out with as much dispatch as good planning, equipment, and the location will allow.

Excavation for a tunnel can be measured in cubic yard quantities like most other forms of earth excavation. However, it is more frequently stated in terms of length of tunnel excavated in a given period of time. To be most meaningful, such a "rate of advance" should recognize the diameter of tunnel as well. This mention of diameter calls attention to the volume of material handled and size of equipment to do the job.

Historically, the first tunneling done hundreds of years ago with only slave labor advanced a few inches per week. With the introduction of explosive black powder by the middle of the 19th century the rate of tunnel advance was up to about 7 feet/week. Mechanically driven drills increased the rate to more than 20 feet/week by the start of the 20th century. Then, with the use of greatly improved drill tools and materials, mucking equipment, etc., the drill-and-blast method of tunneling was able to make advances of 200 feet/week by the 1930's.

The complete tunneling machine, christened the "mole," was introduced at Oahe Dam, South Dakota, in 1954. Four different tunnelers working in soft shale during a period of five years at Oahe reached average advance rates of 30 to 63 feet/day. These were for 24-hour operation and would amount to 210 to 440 feet/week if worked steadily for seven consecutive days. Such early machine records are even more impressive when it is realized they were made in 25- to 30-ft-diameter tunnels. During the mid-1960's even faster rates were made on several tunnels in New Mexico boring through sandstone and shale. Maximum advance rates of 10 feet/hour for a 20-ft-diameter up to 17 feet/hour for a 10-ft-diameter tunnel were possible.

The history of increasing the tunneling excavation rate makes a very impressive show of growth in one segment of the construction industry. The growth has been a sort of geometric progression. Note that the plot of rate values in Fig. 6-14 does not account for changes in the size or diameters of the tunnels.

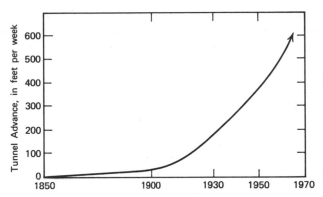

FIGURE 6-14 Increasing tunnel-advance rates vs. years.

Unfortunately, the average rates of advance with the high-production tunneling machines used in New Mexico varied from one-fourth to one-half of their maximum potential. This seems to be due in part to frequent need for adjustments and maintenance. However, the full production time of these tunnelers is perhaps more dependent on the backup equipment, such as that for the mucking operation to dispose of excavated material. The best that a tunneling machine planner should expect at this stage of development is 50-60% of the maximum productivity. A statement has been made that if this utilization results in an average advance rate of less than 2 feet/hour, it will probably be uneconomical to use a mole compared to the drill-and-blast method.

Costs of Tunneling Equipment. It is interesting to observe that the cost of tunnel construction was fairly constant from the mid-1950's through the 1960's. That was the period of the mole's early development. During the same period the ENR Construction Cost Index showed a 67% increase.

Tunneling machines were not the only innovation to hold the cost of tunneling down. Perhaps the experience on the Hetch Hetchy tunneling construction in San Francisco is the best indication of other saving factors. That 14-foot-diameter tunneling was done by the drill-and-blast method. The project was able to save 50% of the usual blasting expense with new explosives. Loading the explosives pneumatically and other benefits re-resulted in a 15-20% saving of labor. And the use of large, 42-inch-gauge mucking equipment saved on trips to be made. The combined benefits resulted in an average advance rate of 480 feet/week.

The initial cost of modern tunneling machines ranges upward from a quarter of $1 million. That is a sizeable outlay, even for construction equipment, so a basis for comparison with the drill-and-blast method is needed. Since the mole essentially eliminates the drilling equipment and labor in the older method, an analysis of those costs will be useful.

A comparative cost analysis was made by Hill.[4] It was based on the 13'-3"-diameter Azotea Tunnel in New Mexico moving at an advance rate of 8 feet per shift by the drill-and-blast method. The labor cost per shift was figured to be about $1218, which reduces to $152 per linear foot. For the tunneling machine the cost of manpower was $761/shift, a savings of 38% over the older method. Furthermore, the mole's rate of advance would generally be greater. When it could bore through the tunnel at a rate of 24 feet/shift, the labor cost was about $32 per linear foot—one fifth as much as for the drill-and-blast method.

The big expense of the mole tunneling machine is its initial cost, which may have to be written off on the one job. For a 14-foot-diameter tunneler the original cost may be $500,000. It has been said that a mole would not be economical for a tunnel less than 5,000 feet long. One of that length would be charged $100 per linear foot to write off the above machine. Combining this machine and labor cost amounts to $132 per foot. A higher rate of advance for the machine would reduce the labor-cost part of the expense. However, that would take a more powerful machine which costs more initially. The curves of Fig. 6-15 show the general variation in original costs of tunneling machines.

The cost of the tunneling machine is recognized as most important

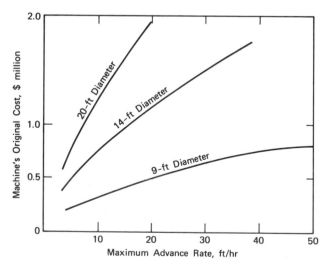

FIGURE 6-15 Cost of tunneling machines.

when it is realized that all the moving equipment for a drill-and-blast operation may only cost $250,000. This total includes the air compressor, rock drills, jumbo, loader, muck cars, and ventilating equipment. Even if their total initial cost had to be written off entirely on the 12.8 miles of Azotea tunnel, the charge would only be $3.75 per linear foot. In terms of the excavation they help do, their cost is only 73¢ per cubic yard.

A tunneling operation using a mole will have additional major expense if a lengthy conveyor system is used to back up the tunneler. This may run $25 to $50 per foot, and for a system to extend for half this tunnel length would cost $1 million. A more economical scheme might be a combination of conveyor and muck train. The conveyor is designed as part of the tunneling machine and moves the excavated material through it to the trailing end. Then it could be discharged into muck cars for hauling out of the tunnel. The cost of a locomotive with muck car system will be in the order of two-thirds of a conveyor system to handle tunneling at 5 ft/hr.

The advance rate for a tunneling machine is a most significant factor. We noted previously the effect of a tunneler's ability on its initial cost. Now let us relate it to the combination of this major cost plus the labor to work with a mole. A comparison will be made, using the Azotea Tunnel again. Taking the labor cost figured by Hill as $761 per shift for excavation, this segment of the cost can be figured for different rates of tunnel advance. For instance, at 2.5 feet/hour the labor cost is $38.05 per linear foot, and at 10 ft/hr it is $9.52/linear foot. Now to add the major equipment cost, we might base that on charging off the original cost of a 14-ft-diameter mole plus conveyor and muck train. Referring to Fig. 6-15, we will estimate the total cost of equipment which can move at 10

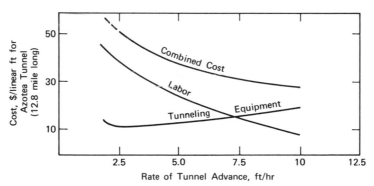

FIGURE 6-16 Cost vs. tunnel-advance rate.

ft/hr to be $1,200,000. For the 12.8-mile tunnel this is $17.75 per linear foot. To advance at the slower rate of 2.5 ft/hr will take equipment estimated to cost $800,000. This amounts to $11.80 per linear foot. Now we can combine these major expenses in tunneling with a machine. The curves in Fig. 6-16 show the labor expense and the initial equipment cost separately, and then their combined cost.

This comparison of major tunnel costs clearly shows that as the rate of advance falls below five feet per hour, the labor costs go up rapidly. At the rate set in Hill's analysis of the drill-and-blast method (8 feet/shift), the labor cost working with a tunneling machine would approach $100/ linear foot. For a shorter tunnel the equipment-cost curve will make a shift vertically upward. It is not difficult to recognize that for the slower rates of advance with a much shorter tunnel the drill-and-blast method would be more economical than the use of a tunneling machine.

The other equipment costs for tunneling are relatively minor. For the almost 13-mile-long, 13-ft-diameter Azotea tunnel, the mucking equipment cost 30¢/ cu yd, or a few dollars per linear foot. The other reusable equipment (jumbo, drills, and air compressors) ran the total of this expense to less than $5/linear foot.

With a tunneling machine operation, one significant cost is related to keeping the cutters in their effective, designed shape. If the rock material is of varying consistency or too hard for the cutters, they will wear down fast and have to be resharpened or replaced. This can cost from as little as 20¢/cu yd with soft rock where very little repair is needed to almost $40/cu yd in granite of 30,000 psi compressive strenth. The general variation is shown in Fig. 6-17.

On the Azotea Tunnel the cutter expense of about 20¢/cu yd amounted to slightly over $1/lin ft. In harder rock with compressive strength of

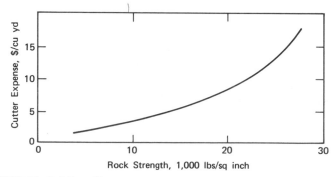

FIGURE 6-17 Cutter cost with tunneling machine.

15,000 psi, the cutter expense in a 20-ft-diameter tunnel might be as high as $60/lin ft. An even higher cost will be encountered if the electric power for operating the mole and its auxiliary equipment must be generated and distributed on the construction site. If permanently installed power lines are available in the vicinity of the site, it will be cheaper to establish a substation with appropriate transformer, lines, etc., to run the electrical power to the equipment.

Advantages in Using a Mole.   As previously mentioned, there are certain limitations on length of tunnel and composition of rock that restrict the economical use of tunneling machines. On the other side of the coin are a number of advantages in using a mole. They have been listed by Armstrong:[2]

1. faster tunnel advance;
2. round, smooth, unshattered bores;
3. little overbreak, averaging about 5% compared to 20+% with drill-and-blast method;
4. savings in concrete for lining;
5. less support required—fewer rockfalls;
6. adaptable to continuous-type system of operation, for continued improvement;
7. less hazardous with reduced exposure of men to excavation operations and unsupported rock; no explosives required; and
8. smooth operation with little or no disturbance to surface or nearby facilities.

References

1. Huston, John. "Dredging Fundamentals," *Journal of the Waterways and Harbors Division*, Vol. 93, No. WW3, American Society of Civil Engineers (New York, N.Y., August 1967), pp. 45-69.

2. Armstrong, Ellis. "Development of Tunneling Methods and Controls," *Journal of the Construction Division*, Vol. 96, No. CO2, American Society of Civil Engineers (New York, N.Y., October 1970), pp. 99-118.

3. Juergens, Ralph editor "New Developments in Tunneling Machines," reprint from *Construction Methods and Equipment*, McGraw-Hill Book Company (New York, N.Y., March and April 1966).

4. Hill, George. "What's Ahead for Tunneling Machines," *Journal of the Construction Division*, Vol. 94, No. CO2, American Society of Civil Engineers (New York, N.Y., October 1968), pp. 211-231.

5. "Materials Handling for Tunneling," Report No. FRA-RT-71-57, prepared for the Office of High Speed Ground Transportation, U.S. Department of Transportation (Washington, D.C., September 1970).

6. Parker, Albert D. *Planning and Estimating Underground Construction*, McGraw-Hill Book Company (New York, N.Y., 1970).

7. Williamson, T. N. "Tunneling Machines of Today and Tomorrow," HRR No. 339 Symposium: Rapid Excavation, Highway Research Board (Washington, D.C., 1970), pp. 19-25.

Chapter 7

██████████████

██

POWER EXCAVATORS
AND CRANES

██

Variety of Equipment

The power excavators and cranes grouped together in this chapter show some variety. This equipment includes primarily power shovels, mobile draglines and clamshells, and the basic mobile lifting crane. These various pieces of construction equipment are related by their common mountings and revolving superstructure. Because of these similarities most manufacturers of this variety of equipment are joined together in the Power Crane and Shovel Association (PCSA). Their equipment is standardized to provide assurances to the users and operators. The variety of equipment is most obvious in the front-end operating parts attached to the revolving superstructure.

The mobile-base mountings for this equipment ride either on crawler tracks or rubber-tired wheels. Mountings of these types were discussed in Sec. 3.2. Bases of this sort support the revolving superstructure directly. This is a platform on which are mounted the engine and the mechanism for transmitting the power to hoist, swing, and move the piece of equipment. The mechanism is primarily several drums for cable or cylinders for hydraulic power lines. The drums or cylinders are linked to the engine. To support these operating parts, a machinery frame is attached to the base of the revolving superstructure. Also, there may be an A-frame (so-called because of its shape) which allows a boom hoisting cable to have a good angle with the boom for leverage from the height at the top of the "A". Other cables prevent the boom from rebounding over the cab. To one side of the power plant of this type of excavator or crane is a control center at the operator's station. The controls are either levers for mechanical or hydraulic transmission or push buttons for electrical power transfer. All of these key parts mounted on the revolving superstructure are surrounded by a housing or cab as shown in Fig. 7-1.

The equipment to be covered in this chapter will start with the forefather of them all. In 1836 William S. Otis was granted a patent for his mechanically powered, land-type excavator—the first power shovel. It was steam powered and soon adapted for use on the expanding railroad lines. The early railroad-mounted shovels could swing through 180 to 270 degrees horizontally. They were a real boon to the original work of digging the Panama Canal. It has been reported that more than 100 railroad-operated power shovels were used on that gargantuan excavation work from 1907 through its completion in 1914. Power excavators were all steam powered until the gasoline engine was first introduced in 1912. About that time in history, fully revolving shovels of the crawler-mounted type were being developed.

Other power excavators will also be covered in this chapter. These include the pull shovel or backhoe, sometimes simply called the hoe, which is very similar to the power shovel but with reverse motions of the excavating bucket. Two other digging or loading varieties of this basic equipment that use the lifting boom, front-end attachment are the dragline and clamshell excavators. In order to understand their capabilities more fully, it is necessary to be familiar with the power crane. So this chapter will cover the mobile crane as a basic lifting piece of equipment. This will serve as an introduction to the crane's use for pile driving and material handling taken up in following chapters.

7.1 Power Shovels and Hoes

The power shovel and hoe excavators have very similar front-end working parts. Both have short, sturdy booms attached directly to the front of the revolving superstructure. The boom then supports a dipper stick with a dipper or digging bucket at its end. A basic difference between the shovel and the hoe is the direction of moving the buckets to get their loads. The shovel moves its bucket up and away from the power source and operator's controls to dig material from a bank above the equipment's standing level. The hoe moves its bucket down and toward the operator to dig material from below the equipment's mounting. To be effective in these motions, the dipper sticks have different points for connection to the booms. A new design known as the Skooper, introduced by Koehring, combines the upward motion for digging by the shovel bucket with a front-end mechanism like a hoe.

Shovels support the dipper stick at about the mid-length as shown in Fig. 7-1. The dipper stick is moved back and forth through a slot in the boom, as well as rotated about this pivot point on the boom. A forward motion of the dipper stick allows it to "crowd" the bucket into the bank of material. It is retracted by the dipper stick moving back on the boom. In contrast, backhoes support the dipper stick at the end of their booms as shown in Fig. 6-1. The dipper stick is only rotated or pivoted about this point of the boom. The bucket is swept through the material to be dug by this pivoting action. However, the digging arc can be readily changed by changing the boom's angle. Operating details for this type of equipment can be found in specifications or suitable manuals.[7]

7.1.1 Design Features of a Power Shovel

The power shovel has six basic movements. Most of these are common to the other related excavators, or they have counterparts in their operation. These counterpart actions will be pointed out when discussing the dragline, etc.

The six basic shovel movements are tied into parts of the power mechanism in the superstructure. The shovel's movements are illustrated in Fig. 7-1 and can be described as follows:

1. the main hoist power mechanism or drum cable lifts the dipper bucket up through the material being dug;

2. the secondary hoist operates the dipper stick for crowding the bucket into the bank; this action can be made together with the main hoist for positive, powerful digging;

3. retracting the shovel's dipper stick is done, generally, by a separate cable from that rigged for the crowding action but operated on the same secondary hoist drum;

4. the boom hoist is used to pivot the boom about its base point on the platform of the revolving superstructure; it may be raised to a 65° maximum angle with the horizontal or lowered to a minimum 35°

FIGURE 7-1 Power shovel movements (courtesy of Northwest Engineering Company).

angle, with lowering under power as a safety measure against accidental dropping of the boom;

5. swinging the superstructure with the power shovel calls on separate cable, mechanical, or hydraulic mechanism with clutches and a brake to save on clutch maintenance and to prevent drifting on the turntable; and

6. the traction movement of the whole power excavator on a crawler mounting may be tied in with the swinging mechanism, since these motions would not be done together; this movement for travel can also take care of steering by moving only one crawler track while the other is braked.

As suggested in the description of these controlling motions of the power shovel, there can be variations. They may be due to the type of power mechanism used in the equipment. Obviously, if it is wheel-mounted, there will be a difference in the traction and steering for travel.

A majority of the power shovels are crawler mounted. As such they can travel only 2 mph, at best. This is generally satisfactory because once such a power shovel is on its job, there is very little reason for moving far. Mobility is sacrificed to gain great stability. The PCSA standards provide equipment that is designed to climb a maximum slope of 40% with the boom pointing upgrade and at its lowest angle. The crawler-mounted power shovel is moved from one job to the next on a low-bed trailer or dismantled to be moved on railroad cars.

The sizing of parts in any power shovel has been practically standardized by the PCSA. For instance, the dipper bucket sizes for most commercially available shovels are 3/8, 1/2, 3/4, 1, 1$\frac{1}{4}$, 1$\frac{1}{2}$, 1$\frac{3}{4}$, 2, and 2$\frac{1}{2}$ cu. yd. These sizes are "struck" capacity, which approximates the payload volume, eliminating the voids in a heaped bucketful. An exception must be made to this statement when handling only large pieces of rock. Custom-made power shovels in use extend the maximum size up to a 50-cu yd bucket. Such larger sizes are built specifically for strip mining.

The PCSA standards show working dimensions as given in the sketch of Fig. 7-2. The maximum values generally vary directly with the size of power shovel and its boom angle. Representative values of the maximum

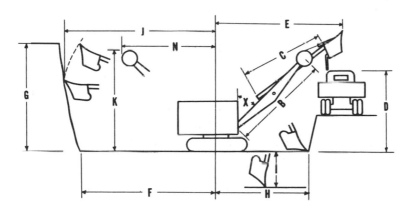

X Boom Angle	J Digging Radius (Max.)
B Boom Length	H Floor Level Radius
C Dipper Stick Length	I Maximum Digging Depth
E Dumping Radius at Maximum Height	Below Ground Level
F Dumping Radius (Max.)	K Clearance Height of Boom Point Sheave
D Dumping Height (Max.)	N Clearance Radius of Boom
G Height of Cut (Max.)	Point Sheave

FIGURE 7-2 Shovel clearance diagram.[2]

digging radius and height will be helpful in later selection of a shovel for given work. These dimensions for the standard sizes with a 45° boom angle on the shovel are shown in the graph of Fig. 7-3. The maximum digging depths for the standard shovel sizes range from five feet to less

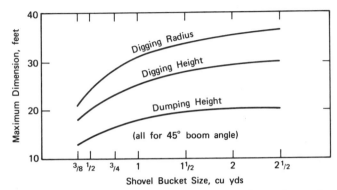

FIGURE 7-3 Working ranges for shovels.

than ten feet for the 2 1/2-yard shovel. The power shovel is not designed specially to dig below the level on which it stands. But some uses require it to dig below that level, so this limit is worth knowing.

The parts of a power shovel are designed for a good balance of sizes from the bucket to the supporting ground. Starting with the loaded bucket of a particular size, each element of the shovel's front-end attachment is designed for the anticipated load. The operating mechanisms and cables are sized to have the necessary strength and also flexibility to handle their loads. A counterweight will be added to the rear of the superstructure to supplement the power parts for balancing the weights of the front-end parts. A shovel's front-end attachment weighs about one-third as much as the superstructure with its power parts and cab. Of course, the base mountings must be designed to take all these weights and transmit them to the supporting surface or ground satisfactorily. This is based on causing between 5 and 12 pounds per square inch (psi) pressure between the supporting tracks and bearing surface. Other standards of design lead to a maximum of tractive effort that is 60% to 70% of the machine's weight. The balance of parts and stability of the whole piece of equipment will be discussed with the design features for a mobile crane.

7.1.2 Design Features of a Pull Shovel or Hoe

The design features of a pull shovel, backhoe, or simply hoe, are very similar to those described for the regular power shovel in the previous section. The features of the hoe were discussed in Sec. 6.1.1 in connection with trenching operations. One significant difference between the power shovel and the hoe, in addition to their direction for digging, should be noted.

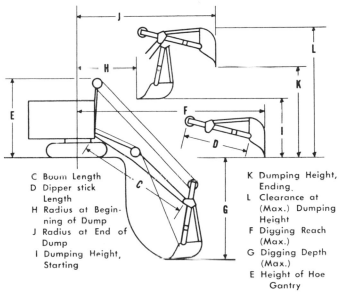

C Boom Length
D Dipper stick Length
H Radius at Beginning of Dump
J Radius at End of Dump
I Dumping Height, Starting

K Dumping Height, Ending.
L Clearance at (Max.) Dumping Height
F Digging Reach (Max.)
G Digging Depth (Max.)
E Height of Hoe Gantry

FIGURE 7-4 Hoe clearance diagram. (Note: Dimension K varies with dumping radius chosen. G varies somewhat depending on character of the material being dug. J and K vary with the dumping height.[2])

The basic crowd and retract movements of the shovel do not apply to the hoe (Fig. 7-4). The hoe's dipper stick is supported almost at its end connected to the outer end of the boom. It is attached by a hinge connection so it can be pivoted about this point. The hoe's dipper is moved forward or back by the combined movement of this pivoting and changing of the boom's angle in each digging cycle. Obviously, the hoe's boom is not kept within the narrow range of angles as those for a shovel. The hoe is designed to dig much deeper than a shovel. The range is from 12 feet for the small 3/8-yard backhoe to almost 30 feet deep for the largest standard hoe. These depths are two to three times the limits for comparable size shovels. Obviously, the shovel is not used for trenching except for shallow depth when it can be done along with other shovel excavation work. A backhoe is used extensively for trenching, as discussed in Chap. 6.

7.1.3   Uses for Power Shovels

When steam-powered shovels were first used extensively, they were the main type of excavating equipment. They were used for a variety of operations—quarrying, excavating embankments, the earthwork to clear a horizontal plane, digging for side slopes, opening shallow ditches and trenches, etc. Now there is such a variety of earthwork or excavating equipment available that the power shovel is used in a more specialized way.

The power shovel is most useful where there is considerable hard digging of rock or consolidated material from a bank that will stand with a fairly vertical face. This situation exists in earth cuts of rocky or clayey material or in quarry work. The embankment should be able to stand up at least one-fourth as high as the shovel's maximum digging height shown in Fig. 7-3. Under those conditions the shovel will be more productive than the front-end loader discussed in Chap. 5. This is certainly true if the embankment is extensive and does not force the shovel to move much to do a lot of excavation. Also, if the earth being dug will stand as high as half, or more, of the shovel's maximum digging height, the shovel can be more productive than a loader of equal bucket size. This is partly because the front-end loader will tend to undercut the bank and be inefficient in its digging. Also, the shovel can more easily load its material into trucks alongside the excavator than can a loader.

A crawler-mounted shovel is well suited for rock quarry work because of its ruggedness, power, and stability. Quarrying means much excavation. Also, the surface available for the excavator to move on may be rocky and rough. That is hard on rubber-tired equipment, building up the high cost of maintenance to be expected with a front-end loader used for this work.

Another good use for a power shovel is the initial excavation of a side-hill cut for a roadway running with the contour. In that case the material dug from the uphill side can be swung around and dumped or cast on the downhill side of the cut. While the cut is narrow enough, the dumped earth will roll down away from the cut surface. When the cut is too wide for the shovel's cast to dump material over the side, a dozer can be used to push it over the edge of the cut level.

Shovels with special equipment are used in large-diameter tunnel excavation. It is best to use an electrically powered shovel to avoid noxious fumes inside the tunnel. The shovel is equipped with a short boom and dipper stick for the work in such limited space. It generally takes from

2 to 4 hours with an auxiliary crane to change the shovel's front mechanism with all its rigging on the superstructure.

7.1.4 Productivity of Shovels

In this section we are interested in knowing how to determine the amount of material handled by this equipment in a certain length of time. The usual determination is to find the number of cubic yards excavated per hour. This will depend on (a) the size and designed features of the shovel used, (b) the variety and condition of material being dug, and (c) the job setup along with the operating abilities for the given excavation operation. For instance, on a given job a small shovel might work, operating close to its limits of power for the tough material and close to its maximum digging and dumping heights as indicated in Fig. 7-3. A larger shovel would have more power, operate well within its efficient working ranges of dimensions, and have a higher productivity. Of course, the larger one would cost more per hour. Which one to use will be based on determinations of the productivity and cost for each possibility.

A shovel's productivity will depend greatly on the designed features of that equipment. The amount of material it can handle in a given time period depends on the size of bucket, the speed of moving the dipper in a vertical direction, and the rotational speed of the superstructure on its horizontal turntable. These are all designed features. For a cable-controlled shovel the speeds are governed by the line speeds of the cables. The hoisting and other front attachment line speeds may vary between 75 and 180 feet per minute (fpm). Another designed speed is that for swinging the superstructure on its base mounting. For cable-controlled shovels that speed may be three to four revolutions per minute (rpm).

The cable line speeds and the swing speed are the key factors that determine the best cycle time an operator can get with a shovel. The shovel's cycle can be described in relation to the basic movements of the equipment discussed in Sec. 7.1.1. The secondary hoist crowds the dipper into its load while the main hoist lifts the bucket up through the embankment. Then the shovel's superstructure swings around to dump the load where desired. The dipper-bucket is then swung back and lowered to get into position for the next cycle.

For a 10-foot depth of cut, the least time to hoist the dipper up through the material may be five seconds. This is with a hoist-line speed of 120 fpm. Then, the minimum time to swing through an angle of 90° to dump and back again at 4 rpm is about 8 seconds. These times do not account for time getting up to the governed speeds nor back down again, i.e., acceleration or deceleration, in each movement of the cycle. Also, the time needed to dump the load with some accuracy is not included. These elements may add three to five seconds of necessary time. Therefore, the best cycle time an operator could expect to get with the shovel in this digging situation would be 16 to 18 seconds. That time is governed mainly by the design features of the shovel.

Job conditions and setup also greatly affect a shovel's productivity. Most efficient production for varying height of cut is obtained with the shovel working at its "optimum depth" of cut. This is the height of embankment that a shovel works in where the dipper stick lifts the bucket up through a depth of material from which it just manages to get a heaped bucketful. Digging in this height of cut requires no re-digging for a full bucket nor spillage of material over the bucket sides during the cutting movement. A loose, flowing material will fill a digging bucket in a somewhat shorter sweep through the embankment than will a sticky, chunky material that does not easily fill spaces in the bucket load.

The optimum depth of cut will vary between 4 and 11 feet for loose, granular, or earthy material. For hard, sticky material it ranges from 6 to 14 feet. These values vary directly with the size of the shovel. Comparing these heights (depths of embankment) with the curve shown in Fig. 7-3, it is found that the optimum depth of cut is from 25 to 50% of the maximum digging height.

The specific variation in the optimum depth of cut for different size shovels and materials dug is shown in Fig. 7-5.

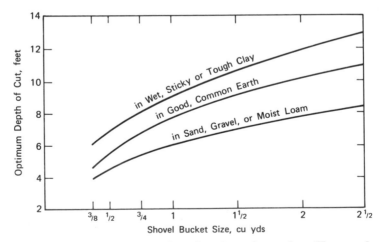

FIGURE 7-5 Best cut depths for shovels. (From data published by the Power Crane and Shovel Association.)

It is frequently observed that a shovel operator will make more than one pass at the embankment to load his bucket. Such a tendency occurs more often where the height of embankment is less than the optimum depth of cut for the shovel. It happens once in five or six cycles with cut depths less than five feet. The frequency and depth limit is greater with an embankment containing oversize material that does not readily pass through the bucket. Making extra passes to get a loaded bucket is a questionable practice by operators, but it may be difficult to prevent. Assume the loading part of a cycle takes six seconds in a total cycle time (CT) of 18 seconds for the most efficient operation. A second pass will add at least another six seconds. This makes CT = 24 seconds, for a time increase of more than 30%. The justification would have to be based on increasing the bucket's load at least that much. Also, it can be noted that the greater portion that load time amounts to in the total cycle time, the less desirable it will be for the shovel to make extra passes to increase its load.

The amount of load in a bucket compared to its capacity is a vital factor in determining a shovel's production. The ratio of the bucket's load to its capacity is called the bucket efficiency (E). For the materials dug by shovel, E will vary from 110% to 50% as the best to be expected. Materials that are easy to dig and could be described as flowing, such as sand, gravel, or loose earth, should easily fill the bucket to capacity. At the other extreme are the hard, rocky materials that will, at best, only half fill the shovel's bucket. This sort of variation is shown in Fig. 7-6.

In sec. 7.1.1 it was noted that the capacity of a shovel is the struck volume of the bucket. Though the material frequently is heaped in the bucket, the actual payload of material, deducting for the voids, is closer to the capacity. Figure 7-6 shows that in easy digging with a somewhat sticky material the heaped material can be compact enough that the payload exceeds the bucket's capacity slightly.

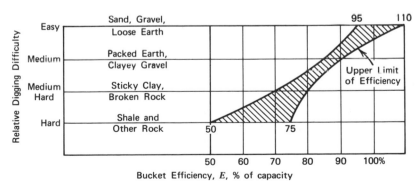

FIGURE 7-6   Shovel loading effectiveness.

If the material dug by a shovel has a significant amount of oversize chunks or is exceptionally sticky and does not pass through the bucket readily, the average bucket efficiency will be reduced significantly. This reduction is practically the same as the proportion of oversize parts in the total material. So, if there is 20% oversize in the total, the bucket efficiency will be 20% less than it would be without the oversize material present. Obviously, if there is very much oversize material for a shovel to handle, it would be well to consider using another shovel with a larger bucket, or to break down the size of the largest material pieces, or to push this material aside for handling by other means.

The job condition for dumping a shovel's load is a productivity factor that management can control closely. The points to control are (a) the angle of swing for the shovel to dump its load and (b) the balance of sizes between the shovel's bucket and a hauling unit's container loaded by the shovel. For a good balance of sizes the haul container should have a capacity that is nearly equal to 4, 5, or 6 times the shovel's bucket capacity. Planning for this will eliminate the waste in dealing with a partially loaded bucket and will use the haul unit's capacity to full advantage. Also, it is sound planning to pay attention to the top opening of the haul unit's container. The dimensions of that open target for the shovel bucket should be at least several times the bucket width in each direction. At the other extreme, the dimensions should not be so big that the shovel or haul unit would have to move during the loading time to fill the hauling container.

The necessary angle of swing for the shovel's excavation cycle can be controlled by good planning. In a regular cycle made up of digging, swinging the load, dumping it, and return swinging for the next load, the loads may be dumped to one side of the excavation or into haul units to be taken away. For a side hill cut the shovel may be casting the dug material down the slope opposite the embankment. In that case the swing angle is fairly fixed between 130° and 180°. An operation with that much swing will have a relatively long cycle time and, so, low productivity. When the dumping is into haul units that can be located in any position on the ground surface that supports the shovel, good planning and control will help get high production.

In studies made by the U.S. Bureau of Public Roads in the decade after World War II, more than 80,000 shovel cycles were observed. Most frequent swing angles were from 45 to 90 degrees. Of all the shovel cycles where the swing angle was measured, the average swung through an angle of 79° in one direction. Two distinct areas over which the swing is made are recognized. These are (1) the digging area and (2) the haul-unit spotting area outside of the digging area. We might assume that there will be an average of 45° in the digging area. This would leave a swing of 34° for the haul-unit spotting area in the average cycle. It should be possible to reduce this latter part of the swing to realize more efficiency in the shovel operation. Good coordination between the operators and planners can do it.

The effect of swing angle on a shovel's productivity is shown in Table 7-1. This shows, as would be expected, that for any given depth of cut, a smaller angle of swing will result in better output. As explained earlier and shown in Fig. 7-5, there is an optimum depth of cut for each size shovel digging in a particular material. A shovel digging in an embankment of depth other than the optimum will be less effective. This is true whether the depth is greater or less than the optimum depth to cut. The use of Table 7-1 is with the actual cut divided by the optimum expressed as a percent on the left for each row.

TABLE 7-1 Effect of Depth of Cut (D) and Angle of Swing (A) on Shovel Production

Percent of Optimum Depth of Cut	Angle of Swing in Degrees						
	45°	60°	75°	90°	120°	150°	180°
40%	.93	.89	.85	.80	.72	.65	.59
60%	1.10	1.03	.96	.91	.81	.73	.66
80%	1.22	1.12	1.04	.98	.86	.77	.69
100%	1.26	1.16	1.07	1.00	.88	.79	.71
120%	1.20	1.11	1.03	.97	.86	.77	.70
140%	1.12	1.04	.97	.91	.81	.73	.66
160%	1.03	.96	.90	.85	.75	.67	.62

Data by the Power Crane and Shovel Association.[2]

The information given in Table 7-1 shows a combined effect of 1.00 for the shovel operating with a 90° swing and at the optimum depth of cut (i.e., 100% D). This means that the shovel working under those conditions should produce a normal or standard output. Other amounts of swing or depths of cut lead to different productivities found by corrections applied to the standard production.

The cycle time for the standard conditions is found, as discussed earlier in this section, using line speeds, optimum depths of cut, swing speed, and allowance for acceleration-deceleration and dumping.

TABLE 7-2 Shovel Cycle Times in Seconds for
Optimum Depth of Cut and 90° Swing

Relative Digging Difficulty	Shovel Size by Bucket, cu yds						
	$\frac{3}{8}$	$\frac{1}{2}$	$\frac{3}{4}$	1	$1\frac{1}{2}$	2	$2\frac{1}{2}$
Easy	16	16	17	18	19	21	22
Medium	19	19	20	21	23	25	26
Med. Hard	22	22	23	24	26	28	29
Hard	24	24	25	26	28	30	31

Data by the Power Crane and Shovel Association.

These cycle times are based on no interruption to the designed motions. Actual productivity has to allow for any variation, such as a delay moving to a better digging position. One motion and time study observed that this type of necessary move took an average of 0.6 minute, or 36 seconds. Also, it was made after about 20 dipper loads. Since this is a necessary sort of move for efficient shovel operation, the effect might be added to each dipperfull cycle time. That would amount to adding approximately two seconds to each CT shown in Table 7-2. Other delays in shovel productivity will be discussed later in this section.

Now that we know the factors that govern a shovel's productivity, a formula can be written for it:

$$q_s = \frac{3600 \, B_c \, (E)(A{:}D)}{CT_s}, \qquad (7\text{-}1)$$

where q_s = maximum production, cu yds/hr;
  B_c = bucket "capacity," cubic yards;
  E = bucket efficiency (see Fig. 7-6);
  $A{:}D$ = combined factor for the shovel's angle of swing and depth of cut; and
  CT_s = cycle time, seconds, for "standard" shovel operation of 90° swing and optimum cut.

The application of Eq. (7-1) is fairly simple with the determining factors as explained earlier in this section. An example will help us to understand its use.

Given: a 1-yard shovel is digging packed earth ("medium" difficulty) from an embankment 6 feet deep and swinging the material 75° to dump.

Determine: (from Fig. 7-6) $E = 87\% = .87$;

optimum depth of cut (from Fig. 7-5) $D_o = 7.8'$;

$$\frac{D = 6.0}{D_o = 7.8} \times 100 = 77\%; \quad \text{(estimating from Table 7-1)}$$

$A{:}D = 1.02$;

cycle time (from Table 7-2), $CT_s = 21$ seconds;

production, $q_s = \dfrac{3600\ (1\text{-yd})\ (.87)\ (1.02)}{21} = 152$ cu yds/

hr, based on working 60 minutes an hour.

The productivity found as above, by using Eq. (7-1), will be the best possible output rate and should not be expected for even a full hour. A normal hour's work should be possible working the shovel 50 minutes out of the 60 available. This would mean a working efficiency, $f_w = 50/60 = 0.83$ and normal production $q_n = (f_w)\ q_s$. For the 1-yard shovel of the previous example, $q_n = 0.83 \times 152 = 127$ cu yds/hr.

Actual shovel production, as with most construction equipment, is less than should be expected. Several studies made under sponsorship of the U. S. Highway Research Board (HRB) showed the actual productive time for shovels on a number of highway excavation operations.[3] The actual time for excavating was 50 to 75% of the available working time. This is equivalent to working 30 to 45 minutes of each available 60-minute hour. The balance of time was lost due to such delays as short moves for better digging position, special handling of oversized material, cleanup of loading area, hauling unit exchange, lack of a hauling unit at the shovel, coffee breaks, etc. Such delays cause the actual shovel efficiency, $f = 0.50$ to 0.75, and if they applied to the previous example where the best $q_s = 152$ cu yds/hr, the actual production might be as low as $q_a = 0.50 \times 152 = 76$ cu yds/hr.

For a better understanding of such variation, the findings of HRB studies will be mentioned. These compare available and actual productive time in hours or days. If the total available contracted work days are noted as T, then the net available working time might be shown as N equal to some percent of T. A major difference between T and N is caused by weather and other equipment to work with the excavator. The so-called major delays are ones that take more than 15 minutes each due to weather, equipment repairs, and a few lesser causes. They have been observed to total 4 to 80% of T. The average of observed shovel operations was between 35 and 45%, leading to $N \approx (0.6)T$ for the average net available working time. The next adjustment or correction is to find the actual productive time in T or N. This will account for short "minor" delays during working time. The minor delays, generally of less than 15 minutes each, are due to the interruptions caused by the hauling-unit operations, special shovel maneuvers, and short repairs. Their total effect can vary between 10 and 50% of the net available working time, N. The HRB studies found that the minor delays took up, on the average, between 20 and 26% of T. This means that the productive time, $P = (100\text{-}40\text{-}23)\%$ of $T = 0.37T$, or $P = (37/60)N = 0.62N$ on the average.

These observations and averages lead one to a basis for estimating an overall, average efficiency factor, $f_a = P/T$. If the efficiency is to be based on use of the net available time, then the efficiency, $f = P/N$, as applied in the previous example.

7.1.5 Costs of a Shovel or Hoe

The cost of using a power shovel for a construction operation will have to cover the ownership and the operating costs of the equipment. Such costs were listed and discussed in Chap. 2. A review of that discussion should prove helpful to the reader before going further into this section. If the shovel is rented, the rate charged will cover both ownership and operating costs, except the fuel and the wages of the operators. The rental rate, of course, will provide for a reasonable profit to the owner of this equipment. Further discussion in this section will concentrate on the concerns of the owner-user of a power shovel.

Since a major portion of the chargeable cost is dependent on the original investment in the piece of equipment, special attention must be given to that outlay. The initial cost of a shovel is directly proportional to its size and weight. In fact, in the 3/4-yard or 1-yard sizes the initial cost is now about $1 per pound, with the smaller ones somewhat higher and the larger ones less costly per pound of weight. With each manufacturer including some special features on his shovels, the initial cost of a given size unit, such as a 1-yard shovel, will vary slightly. A decision for one over another is likely based on personal preference, special design features, dealership differences, and the like. It must be recognized that for a given size-capacity, a truck-mounted unit is 10 to 25% heavier and, so, will cost proportionately more than that size of crawler unit. This applies only for smaller units which are made with either mounting. These variations affecting the initial investment of a shovel-type piece of equipment are shown in the graph of Fig. 7-7.[4]

The manufacture and sales of construction equipment is highly competitive. Consequently, variations in weights and costs of the units of a given size are relatively insignificant. A variation of 5 to 10% in the initial

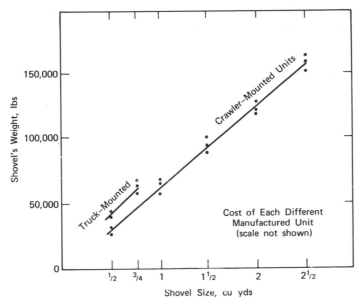

FIGURE 7-7 Shovel size/cost comparisons.

cost will cause a much smaller difference in the total ownership and operating cost to produce each unit of output. This reasoning can certainly be applied to the freight cost, which amounts to only a few percent of the initial cost from the manufacturer's factory. These points suggest that the purchaser of a new unit look for one that will satisfy his preferences. It must be able to do the jobs he plans for it. And he should get reasonable service on the unit from the manufacturer and his agents.

A much more significant point for variation in the cost of owning and operating a power shovel is its use per year and in its lifetime. The normal working time with 8-hour, one-shift days adds up to around 2,000 hours a year. A shovel owner should hope to work his equipment at least that much each year. The U. S. Department of Internal Revenue (IRS) has a schedule for the expected rate of write-off or depreciation expressed in a given number of years. This is based on 2,000 hours of use each year. The useful life for tax write-off purposes may be 5 to 20 years, depending on the shovel size and use. The larger ones are expected to last longer.

If the shovel is worked less than 2000 hours in a year, it might be charged more depreciation per hour to keep on schedule with the IRS. Or each hour of use could be charged the same depreciation based on a greater number of years of useful life. Following the latter practice would tend to defy the obsolescence or outdating of construction equipment. If the practice of taking the IRS-allowed depreciation for the year is followed, and the shovel is used for only 1,000 hours in a year, the depreciation chargeable would be twice the expected amount. With depreciation amounting to perhaps 25% of the total shovel use rate, a significant increase might be competitively disastrous. A low amount of use for a shovel in any year suggests it would be more economical to rent such expensive, specialized equipment.

If the shovel is worked more than 2000 hours in a year by using it say, two or three shifts a day, other concerns must be faced. With a maximum depreciation allowed per year, the extra hours of use should be helpful. That would allow charging less depreciation per hour. However, suppose the IRS schedule calls for a minimum of five years write-off; that is suggesting the shovel is useful for 10,000 working hours. To work the equipment steadily for two shifts a day, or 4000 hours in a year, will change the useful life in years. The IRS would not allow a reduction to $2\frac{1}{2}$ years write-off from its scheduled 5 years. Yet the shovel may last only three years at this rate of use. There will be two years of book value for the owner to write off, if he has not made provisions to extend the shovel's life. Extension of the life of equipment is the purpose of a maintenance program.

To carry out regular maintenance of a shovel, the operator has help from an oiler. For a large machine there may be an oiler required full-time, so the operating expenses include total wages, etc., for the two men. With small machines the oiler may be able to divide his time for servicing two or more machines. This will help to minimize the proportionately higher labor cost in operating smaller machines.

Maintenance Costs for a Shovel.   The cost of maintenance for a piece of equipment is both an ownership and operating expense. The time spent, frequently less than 15 minutes each, to lubricate parts or change a filter or the like, is an operating expense. On the other hand, there is need for periodic, preventive maintenance such as an engine overhaul that is an ownership-type cost. An overhaul may be needed for every 5,000 hours of use. It will probably include new cylinder sleeves, piston rings, connecting rod bearings, main bearings, and valve repair. The hours of shovel use before these repairs are needed may be less than 5000 with poor attention to operating techniques and maintenance, or they may be more with good operation. In any case a good maintenance program is necessary to insure a shovel's usefulness beyond 10,000 to 12,000 hours of use.

Some other special costs are included in shovel maintenance. Wire

ropes are damaged by overloading or being cut by wear. Each rope on a
cable-controlled shovel may last from three months down to only a week
or two of strenuous service. The dipper teeth of a shovel are subject to a
lot of wear and tear. They need frequent repointing, hardfacing, or replacement with new teeth. Hard, poorly blasted rock material will run this
cost up. Crawler shoes, traction chains, clutches, and brake linings can be
made to last much of the shovel's lifetime, if they are properly maintained.

The costs of maintenance for shovels with similar treatment are
directly proportional to the shovel size. The relative importance of these
costs is shown in Fig. 7-8. The total maintenance expense for a shovel, as
well as a hoe, amounts to $2 to $5 per operating hour for the sizes shown.

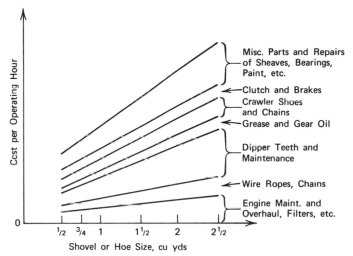

FIGURE 7-8 Relative maintenance cost.[4]

The time required to make shovel repairs is not in proportion to the
cost of the parts involved. For instance, in one Highway Research Board
study 40% of the maintenance and repair delay time was spent with the
power transmission system of clutches, brakes, gears, and drums. Another
26% was spent to repair the boom and dipper assemblies. Changing cable
and other parts took less than 20% of the maintenance time. The average
power-transmission-system repair or boom-and-dipper-assembly repair took
about two hours each. Maintenance crews spent less time on other types of
repair.

Combined Costs for Shovel Use.   In the previous sections key costs for
owning and operating a shovel have been discussed. The use rate or cost

per hour will vary with extent of use and skill in operation, method and time for depreciation, salvage value, and the maintenance program followed. When a use rate is translated to cost per unit produced, there are other variables to take into account. These include the factors governing productivity which were discussed in Sec. 7.1.4. The cost per unit also is affected considerably by the number of hours the shovel is used in a year. This is particularly true as the hours decrease much below 2000 for the year.

The graph of Fig. 7-9 offers some basis for the relative importance of the key costs in a shovel's use rate. A shovel's costs are grouped into fixed costs, variable machine costs, and labor costs. The fixed costs include depreciation, which here is figured for a 5-year, 10,000-hour estimated useful life.

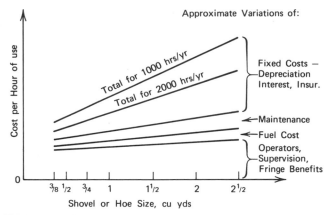

FIGURE 7-9 Relative hourly cost for shovel or hoe.[4]

Note that all the costs included in Fig. 7-9 vary with the shovel size, except the operator's wage and fringe benefits, which cost is assumed to be the same for all sizes. The oiler is included full-time for the 2½-yd shovel but only half-time for the ½-yd size. To be able to divide his time evenly in proportion to the sizes between these extremes is unlikely, so that part of the lines should be stepped instead of straight. The same reasoning should be applied to the allowance for the time and expense of supervision.

7.2 Mobile Cranes

The mobile crane is a basic and versatile type of equipment in the family with the power shovel and other excavators. Crawler-mounted units have the same base part and mechanism. The superstructure platform, power mechanism, and A-frame are very similar whether for crane, shovel, or other excavator. These fundamental parts are supported and revolve on the base mounting unit. The several forms of this family become distinctively different in their front-end attachments.[5]

The crane type of unit has a boom supported at the front of the superstructure platform. The boom may be a latticed, tower-type metal framework with cable-controlled moving lines. Or it may be a telescoping boom with hydraulic controls. The power unit as a prime mover to operate the crane controls, just as with a shovel, may be a diesel engine, gasoline engine, or electric motor. For the hydraulic controls there may be a separate motor from that for the crane's prime moving power.

A mobile crane is a basic piece of equipment in the family for which the Power Crane and Shovel Association of manufacturers was organized. The crane is built primarily to lift, swing, and lower loads held by movable lines. It serves as the common base for excavators, such as the dragline, clamshell, and orange-peel bucket units. These will be mentioned in following sections. The crane is also the basic unit for pile-driving or caisson-drilling equipment, which will be discussed in a following chapter. In combination with the crane's fundamental capacity to lift and handle loads, it must have the necessary stability at its footing and throughout the equipment to keep from overturning. To judge the crane's ability to serve its function, the key factors of its design will be discussed next.

7.2.1 Design of the Mobile Crane

The distinct part of the mobile crane is its boom, which may be an open-lattice framework (as shown in Fig. 7-10) or a closed telescoping structure (as shown in Fig. 7-11). In either case the boom is hinged at its base support point and can be pivoted in its vertical plane. To move a load laterally, the boom is swung with the entire superstructure in a horizontal plane.

A cable-controlled unit has two drums in the power package on the superstructure. These are the same as the main hoist and secondary hoist drums described for the shovel in Sec. 7.1.1. In the case of the crane they provide, respectively, the power line for the main hoisting cable and a jib line if one is used. A third drum is provided for the boom-hoist controls. For each crane size there is a base length of boom. To get greater reach—horizontally, vertically, or probably both—longer booms or additional sections of boom can be purchased, or a jib can be added to pivot in a vertical plane about the top end of the boom. The jib is a smaller section of boom that is generally rigged to be more horizontal than the main

FIGURE 7-10 Mobile Cranes with laticed booms (courtesy of Insley Manufacturing Company).

boom. Any means of lengthening the boom's reach will reduce the lifting-weight capacity of a given crane.

To increase the reach a little and maintain lifting capacity, extra counterweight might be added to the back of the superstructure. This way of increasing the crane's ability is limited by the structural load capacity of the boom and stability of the total equipment. A crane is carefully designed for a close balance between its parts. So if extra reach or load capacity is required for more than an unusual occasion, the planner should go to a larger crane.

A hydraulically controlled telescoping boom can change its length readily with the basic equipment. The boom is made up of several concentric rectangular, triangular, or round pipe sections that telescope up and out with the biggest section at the base support point (Fig. 7-1). The telescoping action is governed by hydraulic cylinders and rams. There is also an articulated boom for a hydraulically controlled crane. This gives the same sort of horizontal reach benefit that the jib on a latticed boom provides. As in the case of latticed booms, a longer extension of the telescoping boom at a given angle will mean lower load-carrying capacity.

FIGURE 7-11 Hydraulic crane with telescoping boom (courtesy of Grove Manufacturing Company).

Basically mobile cranes are equipped with a single line cable for lifting loads. This leads from the main hoist drum over the top end of the boom to the hook, sling, tongs, magnetic grip, bucket, or other loading device. A given drum is designed to be used with a certain-diameter cable. This is part of the total balance of the crane. If the crane is going to be used mainly with the boom at a high angle and for heavier loads than a single line of the cable can take safely, it can be rigged differently. A two-part line of cable for lifting a load can be made by running the cable through a sheave block above the hook and anchoring it back at the top end of the boom. This gives essentially twice the lifting capacity but sacrifices the speed of lifting by a similar proportion. A multi-part line can also be arranged with sheave blocks at the boom point and the loading hook, but this is not a recommended practice for ordinary mobile cranes.

The mobile crane may be mounted on several different types of base unit. It may be truck-mounted, where the superstructure and boom arrangement are carried on the back bed of a special truck chassis. The power unit for using the crane lifting mechanism is on the revolving superstructure. There is the second engine in the truck for moving the whole equipment from one location to another. To gain the necessary stability when the crane is handling a load of any size, outriggers are used. Generally, two sets extend outward from the truck chassis under the turntable for the boom. Feet at the end of each outrigger are planted firmly and jacked against to take load off the truck's rear wheels. A basic discussion of mountings is found in Sec. 3.2.

Some mobile cranes are mounted on single-engine, self-propelled wheel mountings. These do not have the separate truck engine for moving the equipment. Their top highway speed is between 10 to 20 mph, only about half that of the truck-mounted crane. They are designed without springs for the wheel axles to ensure better stability when the crane is used without outrigger supports. Outriggers can be used for better stability, but they take time to set.

Mobile cranes for heavy lifting without much moving between loads should be mounted on crawler tracks. This mounting gives the best stability on natural ground because of the greater bearing surface than with tires or outrigger feet. If broader footing is needed, either wider tracks can be used or, on some modern cranes, the crawler tracks can be extended out from the center of the crane's rotation. Extending the crawler tracks outward is done from the operator's cab position at the crane controls in the modern design. When a crawler-mounted crane is to be moved very far, it will be loaded on a low-bed trailer for hauling on roadways.

7.2.2　Lifting Capacity and Stability of a Crane

A crane's lifting capacity is carefully regulated for the safety of its operation. Standards for the limits have been set by the Power Crane and Shovel Association (PCSA) and adopted for U.S. Government usage.[5] The crane is said to reach its load capacity when it is at a specified percentage of the tipping load. The percentage is from 65 to 85%, depending on the use, and amounts to a factor of safety against the crane tipping over in the direction of least stability. That stability can be improved by extending the crawler tracks outward. A crane-type of machine is considered to be at the point of tipping when a balance is reached between the overturning moment of the load and the stabilizing moment of the equipment. This must be with the machine firmly supported on a level surface.

The question of safety against settlement is important. If the total loaded weight of a crane and the area of its footing on the surface are known, the safety of its total bearing can be checked. To do this, the allowable bearing capacities of the earth surface must be known or estimated. A current building code should have the bearing-capacity information. The following are representative values:

Surface Material	Capacity, tons/sq. ft
Hard, sound rock	Up to 100
Soft rock	12
Hardpan over rock	10
Compact sandy gravel	6
Firm sandy gravel	5
Compact clay/sand/gravel	5
Loose sandy gravel	4
Firm coarse or medium sand	4
Compact fine sand, or sandy clay, or stiff clay	3
Firm fine sand or medium clay	2
Loose fine sand or firm inorganic silt	1½
Loose sand-clay or soft clay	1

The equipment planner should be aware that the crane's loaded weight must be fairly evenly distributed on the total bearing area. Otherwise, the concentration of the load over part of the area may cause unequal settlement in the area with concentrated weight. This sort of failure will tend to compound the trouble if it is not corrected immediately.

To guard the crane against tipping over backward when it is not handling a load, the counterweight must be limited. This is with the shortest recommended boom, since the boom acts opposite to the counterweight in the balance of moments. The counterweight for a crawler crane cannot put the center of gravity for the equipment more than 70% of the distance back from the center of rotation to the tipping point of the tracks. In other words, referring to Fig. 7-12, the horizontal distance s between the center of gravity for the equipment and the axis of rotation cannot be more than 0.7 of the distance from the center of rotation to the backward tipping fulcrum; i.e., $s \leq 0.7f$. The measure for stability against "backward" tipping of a truck or wheel-mounted crane is based on any wheel taking at least 15% of the equipment's weight regardless of the boom direction. This condition is more critical with the boom crossways to the direction of the wheels or tracks. When outriggers are used, the wheels or crawler tracks within the smallest circle containing the outriggers shall be relieved of all weight by the outrigger jacks or blocking. These are bases of the PCSA standards.

The stability and carrying capacity of a crane are best described with a sketch of the equipment and load positions (Fig. 7-12).

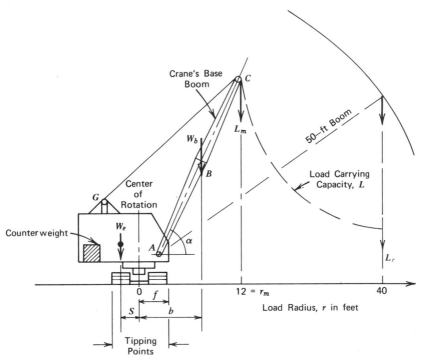

FIGURE 7-12 Crane capacity and stability measurement standards by PCSA. (Notes: W_b = boom's weight, W_e = total equipment weight, including W_b, and s = distance from center of rotation (r = o) to equipment's center of gravity.)

Points to understand about safety against the crane equipment alone tipping over backwards with too much counterweight were explained. Stability for a loaded crane depends on the balance between moment due to load and that due to the equipment parts. The total weight of the equipment causes a moment that counteracts the load's moment. However, it must be noted that the boom's part of equipment weight acts in the same direction about the center of rotation as the load's moment.

With a usual safety factor applied, the maximum load capacity of a crawler crane can be found by

$$L_m (r_m - f) = 0.75\ W_e (s + f);$$

$$L_m = \frac{0.75\ W_e (s + f)}{12 - f} = \frac{3\ W_e (s + f)}{48 - 4f} \tag{7-1}$$

where terms are as shown with Fig. 7-12. The PCSA recommended safety factors are:

75% tipping load for crawler-mounted machines;
85% tipping load for rubber-tired, mounted machines; and
85% tipping load for machines on outriggers.

As the load radius is increased by lowering the boom angle, α, the load capacity is not reduced in direct proportion to the radius, r. This is because the boom's effect in W_e is reducing the equipment's moment. Therefore, the crane's carrying capacity is reducing faster than the load radius is increasing. The load (L) curve in Fig. 7-12 shows this relative variation.

The PCSA standards rate a conventional mobile crane by its maximum load capacaity at the minimum radius with the base boom length and its load capacity with a 50-foot boom and a load radius of 40 feet. Thus, if the crane sketched in Fig. 7-12 has values of $I_m = 40$ tons and $L_r = 9.8$ tons $= 19,600$ lbs, it would be described as:

a 40-ton crane (class 12-196),

where the 12 is for minimum radius, r_m, and 196 is the 100-weight(cwt) at $r = 40$ ft.

Just as the discussion above suggests, the Power Crane and Shovel Association has standardized the key features of their manufactured cranes and shovels. This insures good balance and safety in their design and operation. The balance includes standardization of the rope cables, sheaves, and drum diameters for each given size of equipment.

A crane is used for a variety of purposes and with the full range of safe boom angles. Therefore, it will be helpful in any further reference to the load capacity of crane equipment to have a set of capacity curves (Fig. 7-13).

For any particular mobile crane equipment, the limits should be determined using the unit's specifications, including cable and bucket sizes. Hydraulically controlled cranes with telescoping booms have similar lifting limit curves. This type is not available in as great variety as cable-controlled cranes. Available sizes range from a small 5-ton unit to 40-ton capacity.

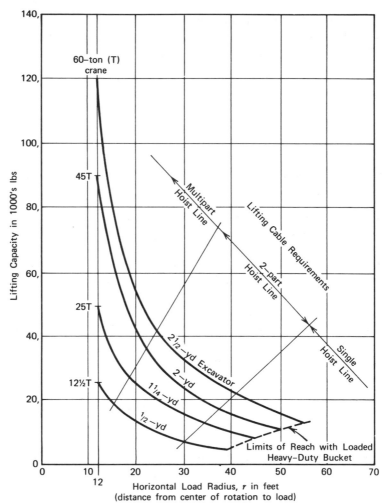

FIGURE 7-13 Load limit curves for crane equipment. (Comparison of representative crane capacities to the basic equipment used for clamshell or dragline excavators.[4])

The lifting capacities set by PCSA are for specific positions and ideal situations of the crane, as previously described. Actual use will involve a wide range of variations on the "standard" settings. An operator lifting a load with his boom in the direction of the crawler tracks may think he can arbitrarily increase his capacity by 20% or more. Or he may think that with a multipart line he can load up his crane to the limit of a single line of cable times the number of parts. Thinking in this way will tend to put excessive load on the boom. As suggested before, the various parts of a crane are designed to be balanced together. Changing the use of any part will affect the intended balance. One common change is with the length or make-up of boom from the original crane's base boom.

Lengthening the boom by inserting standard sections (at point B in Fig. 7-12) between each end section is an intended part of the design by the manufacturer. However, this increases the boom's column action, which is accented by any lack of alignment. The longer boom can take less load, and the misalignment reduces the capacity still more. It is essential that the user of any crane which has been modified check the manufacturer's specifications for load limits under his conditions of use. A modern addition to this equipment is a crane load indicator to help avoid overloading the crane.

In some disastrous cases of the use of cranes extended for high lifts, notably in New York City, there have been bad boom collapses. This has led to city requirements that the manufacturers prove their cranes can satisfy a column formula good for permanent structural design at all times. Such restrictions could be prohibitive. The manufacturers claim their machines meet the safety requirements of the widely accepted USAS B30.5 Code for cranes and derricks. But the city is inclined to accept only strain-gage tested cranes and then only for the booms used as satisfactorily tested. This is very limiting to the manufacturers in PCSA who have set reasonable standards. A solution that holds the possible answer is an automatic control system to refuse loads that would be unsafe for the crane in its given setup. In the meantime, the solution seems to be in a well-accepted system of indoctrinating qualified crane operators.

7.2.3 Uses and Productivity of Cranes

The mobile crane is a most versatile piece of equipment. It is primarily used for lifting loads or loaded containers. The devices or containers for doing this lifting are pictured in the sketches of Fig. 7-14. The crane must be able to lift the total weight of the load and the lifting device, bucket, or platform. In other words, its lifting capacity is based on all the weight truly suspended from the hoisting cable.

A crane is used for high lifting with its boom extended by the insertion of extra boom sections or a jib boom. Usually, the A-frame or gantry (shown at point G in Figure 7-12) is extended upward with the boom extensions to give better stability to the loaded boom. These modifications

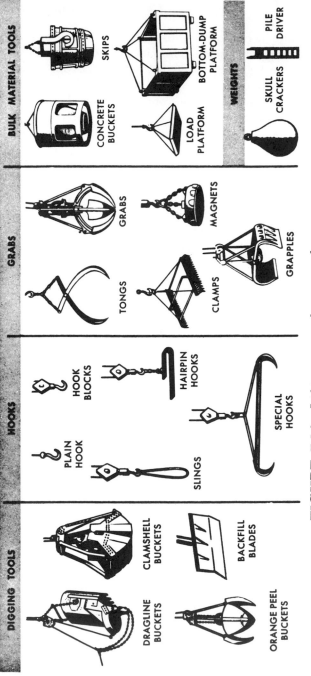

FIGURE 7-14 Lifting attachments for crane use (printed by permission of the Power Crane and Shovel Association).

are made so that the crane can be used for setting structural steel or handling concrete buckets on high-rise buildings. Another similar use for a high reach may be in the case of the wrecking operations for a tall structure. A jib boom is generally used to extend the horizontal reach of the crane over the edge of a high structure. Of course, the load limits of these uses are much less than the maximum capacity of the basic crane equipment.

The productivity of the mobile cranes in the above-mentioned uses is meaningless in the usual terms. That is, the equipment's production in tons per hour or cubic yards per hour is not worthwhile. The reason is that the crane does not govern this sort of productivity. Another construction force, such as the erection crew, likely will govern the tons of steel lifted per hour, and the pouring crew will govern the cubic yards of concrete placed per hour. The crane will control only a part of the total cycle times in these cases.

The speed of lifting a load is controlled by the crane. It is based on a single-hoisting-line speed governed between 100 and 400 feet per minute. For a two-part or multipart rigging to the lifting point, the maximum speed is cut about in proportion to the number of parts.

Swinging a load is another part of the work cycle that is governed by the crane's design. The maximum is generally at a rate of about 4 rpm. This is a significant part of the crane's cycle time when using a long boom, because of the starting and stopping inertia. The load effect from inertia causes many accidents with long crane booms.

The mobile cranes described in the previous sections are used as the basic unit for pile-driving and caisson-drilling equipment, mobile tower cranes for material lifts, and other material handling operations. Some of these common crane uses are discussed in following chapters. The basic crane unit is also used for other power excavators, such as the dragline and clamshell, discussed later in this chapter.

7.2.4 Costs of a Mobile Crane

The ownership and operating costs for a mobile crane could be discussed as those for a power shovel were in Sec. 7.1.5. Pieces of equipment that are basically similar, as the crane and shovel are, will have very similar costs. The shovel with its special boom and dipper design will generally cost 5 to 15% more initially than the same base size of crane. The larger sizes are not quite so comparable. Some large shovels are more expensive because they are designed with special equipment or for special use. The large cranes that are more expensive have special rigging for safety or flexibility on large, well-built truck mountings.

The maintenance cost of a mobile crane is not as great as a shovel, which has considerable expense in maintaining its dipper teeth and mechanism. This was shown in Fig. 7-8. The crane's maintenance expense should run about 60% as high as that of a shovel of the same base size. However, as shown in Fig. 7-9, the maintenance is about one-quarter of the total cost, excluding wages, for the owner-user of this type of equipment. So the lower maintenance cost reduces the total cost of using a crane to perhaps 90% of that for a similar power shovel.

Other variations of cost for mobile cranes depend on the make-up of the equipment. The different designs in significant parts, such as gasoline vs. diesel engine power, crawler- vs. truck-mounted, and cable-controlled vs. hydraulic, telescoping boom, all affect the cost of the crane. For instance, a gasoline-engine-powered crane will generally cost 85–95% as much as a diesel engine unit initially. This saving is realized mainly on smaller units. But it is generally used up in operating expenses because the diesel fuel is cheaper and is more efficient with long periods of continuous use. However, the gasoline-powered crane is advisable when there are to be frequent changes of work location.

Another variation of cranes to compare is the type of base or mounting. For a given size of crane, a crawler-mounted one will cost only 75 to 95% as much as a truck-mounted crane. This difference is even more for the biggest units of equal capacity, say, over 100 tons.

In a similar way the cable-controlled, mechanical crane can be compared with the hydraulic crane. The latter, with its telescoping boom and other flexibility for changing job conditions, is more expensive. Cable-controlled cranes cost initially only 75 to 85% as much as a hydraulic crane of the same capacity. This is partly due to the relative newness and shorter supply of the hydraulic units. Of course, to equal the maximum reach of a fully extended telescoping boom will require the addition of a latticed boom section in a basic mechanical crane. The additional boom section will cost extra and take time and labor to install, bringing the two types closer to the same ownership expense.

For the selection of the crane to buy and/or use, frequently the initial cost is not the key factor. Work conditions and requirements will probably determine what mounting is better and whether a mechanical or hydraulic crane should be selected.

7.3 Dragline and Related Excavators

The Power Crane and Shovel Association's variety of equipment includes several excavators in addition to the shovel. One is the dragline, which is primarily a versatile excavator with greater reach but less digging power than a shovel. Another type of equipment in this group is the clamshell digger and loader. A variation on that type is made by just changing the bucket. This gives the orange-peel digger and loader, which is useful for getting material from underwater.

All three of these types of diggers operate primarily from the hoist line of mobile cranes. They all can pick up load from below the crane's base footing and can get it from under water. The clamshell is better than the dragline for underwater digging because it has a four-sided container, as shown in Fig. 7-14. It is also a more expensive bucket than the dragline. The most specialized and expensive of these three digger types is the orange-peel bucket. However, it is more effective than the clamshell in getting soft, mucky, runny material from a river bottom or digging out a large-diameter pipe pile or caisson. This is because of its design like the segmented skin of an orange. When it is closed on a load, there is not much opening for the material to wash out and yet the water can drain through the orange peel gaps.

A clamshell or orange-peel excavator gets its load by the dynamic force and weight of the open bucket dropping vertically into the material. When the bucket has surrounded enough material, the operator uses his closing line to complete the loading. At that point the crane-type excavator lifts the loaded bucket vertically using the closing line, with the lifting line following, and swings it to a suitable dumping point. This is a typical crane-type operation without special design problems. The buckets have their unusual design features which should be studied in the particular case for efficient use. In the case of a dragline excavator, special design features must be understood.

7.3.1 Design Features of a Dragline Excavator

A dragline excavator is made simply by hanging a drag bucket on the hoist line of a crane and providing a drag cable from the bucket to a second operating drum. The conversion from a basic crane to a dragline can be the easiest conversion between power excavator units discussed in this chapter. Sometimes a third drum is added to give all the pulling action on the dragline bucket. The rigging involved to add this drum will take time. The front-end attachment on the superstructure of the basic equipment is the crane's boom with a dragline bucket. This is to be compared with the power shovel's special boom, dipper stick, and dipper bucket. The shovel's front-end mechanism is designed for maximum digging power and, so, has a minimum of flexibility. The dragline's front attachment is made of the parts shown in Fig. 7-15.

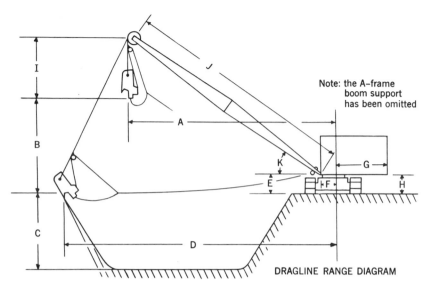

Note: the A-frame
boom support
has been omitted

DRAGLINE RANGE DIAGRAM

A Dumping Radius
B Dumping Height
C Maximum Digging Depth
D Digging Reach (Depends on Conditions
   and Operator's Skill)

E Distance from Ground to Boom Foot Pin
F Distance from Center of Rotation to Boom
   Foot Pin
G Rear End Radius of Counterweight

H Ground Clearance
I Length of Bucket (Depends on Size and
   Make)
J Boom Length
K Boom Angle

FIGURE 7-15 Dragline with limiting dimensions (from PCSA Technical Bulletin No. 4, printed by permission of the Power Crane and Shovel Association).

Effective operation of a dragline excavator requires great coordination in the movements of key parts in Fig. 7-15. Loading the bucket is done by pulling it along the top layer of material toward the machine with the drag cable. When it is filled, the operator takes in on the hoist line to lift the bucket up while letting out on the drag cable. A holding effect by the drag cable and chain keeps the bucket from dumping until desired. The dumping may be several feet beyond or inward from the point of the boom. The empty bucket is swung out from under the point of the boom by slacking off from the hoist line and dropping the bucket at a suitable position for the next loading effort.

The dragline is a versatile machine that can reach a wide digging area with considerable height. This excavator will usually be operated with its boom at a 40° angle (i.e., angle K in Fig. 7-15). At that angle, typical dimensions of dragline excavators are given in Table 7-3.

TABLE 7-3 Typical Dragline Excavator Dimensions (as shown in Fig. 7-15)

	Bucket Size, cu yds				
	¾	1	1¼	1¾	2
Dumping radius (A)	30	35	36	45	53 ft
Dumping height (B)	17	17	17	25	28 ft
Max. digging depth (C)	12	16	19	24	30 ft
Digging reach (D)	40	45	46	57	68 ft
Boom length (J)	35	40	40	50	60 ft
Bucket length (L)	11'6"	14'8"	11'10"	13'1"	14'0"

It is worthwhile to compare these maximum working distances of the dragline with those of the power shovel. The shovel's dimensions were shown in Fig. 7-2 and 7-3. Comparing the digging reach and dumping height for the same sizes of these two key excavators shows a consistent difference. These dimensions for the dragline excavators are about 50% greater than those for the power shovel. Furthermore, the dragline's digging range could be increased by using an extended boom. The greater digging range and flexibility of a dragline is shown very clearly by such a comparison. With this greater flexibility is the disadvantage in loss of digging force. The dragline is designed to work in loose material and clays. However, its digging force can be improved in firm material by attaching the drag chain hitch to a higher point on the bucket. Thus, the bucket teeth will dig in deeper and can apply more digging force. A lower hitch will produce a shallower but longer cut in loose material.

Since the dragline bucket is swung around freely and digging may cover quite an area from one position, the operator has to be concerned about the stability of his equipment. It is for this reason that the safe load capacity for a dragline is taken as two-thirds, or 65%, of the tipping load. This gives the largest safety factor for any crane use. To give added assurance of stability, a dragline may be operated with 36"-wide tracks instead of the more standard 30" crawlers. If the dragline's base is so equipped, the tracks can be extended outward in a direction at 90° from the track length to further increase the excavator's stability.

One specific reason for needing greater stability is the direction of applying the digging force. This direction is more nearly horizontal with the tension in the drag cable. An equal resulting force must be applied by the dragline excavator through its crawler tracks to the ground. This will tend to concentrate the total reaction from the equipment over a smaller area of the ground. The sketch of Fig. 7-16 shows these forces.

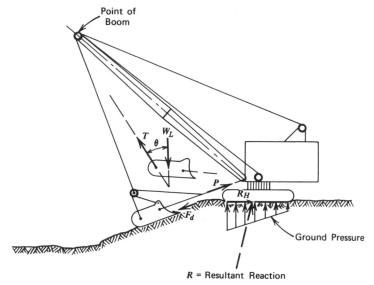

FIGURE 7-16 Dragline forces of importance.

During the peak of digging in the dragline's cycle, the maximum digging force is F_d. This is resisted by horizontal reaction, R_H, and some more vertical reaction to add to the equipment's total weight distribution. The resultant total reaction, R, is therefore, off-center and angled from the vertical, causing extra ground pressure under the front of the crawler tracks. More stability is needed for this sort of weight distribution on the ground.

Another load concern in dragline operation is the maximum pull required by the hoist line. This situation is shown in the second dragline bucket position (with dotted lines for the cable supports) sketched in Fig. 7-16. The loaded bucket has a total weight of W_L. A balance of vertical forces on the bucket is made by $W_L = T \cos \theta$. Therefore, the tension in the hoist line is

$$T = \frac{W_L}{\cos \theta}. \tag{7-2}$$

For example, with certain angles:

θ	$\cos \theta$	$W_L/\cos \theta$
15°	0.966	$1.04 W_L$
30°	0.866	$1.16 W_L$
45°	0.707	$1.42 W_L$

Note that when the loaded bucket is picked up so that the hoist line makes an angle of 45° with the vertical, it will have a tension 42% greater than the loaded bucket's weight. The boom must carry this greater load. The dragline's capacity will be dependent on this increased load in the hoist line.

The operating load capacity for a dragline depends a great deal on the material dug, the bucket used, and the boom's angle. The dragline's boom is usually set for an angle between 25° and 40° with the horizontal. As in the case on any crane-type equipment, the dragline's load capacity is increased by using a higher angle. When the boom angle is set, the maximum load the dragline can handle is determined by the bucket used and the unit weight of the loaded material.

In addition to several basic designs of dragline buckets, each manufacturer has his own variation to make his more efficient. Such variety is based on the fact that the bucket is the key to the dragline's digging ability. The power shovel's bucket depends on the crowding action by the dipper stick to help get a full load. A dragline bucket swings freely, so the operator counts on the bucket design to help him get full loads. Three types of dragline buckets classified by weight are available and identified as light, medium, and heavy buckets.

Lightweight dragline buckets are designed for digging loose, dry soil or granular material, which is easily dug. This material will heap some and have some void space in its dug condition. It would be practical to assume the bucket is loaded to its nominal capacity (i.e., a 1-yard bucket has 27 cubic feet of material) at the bank-measure weight of the material being dug. If the loaded granular material is wet, it will not heap much in a dragline bucket, but most of its void spaces will be filled with water. Using the nominal bucket capacity in that case is also satisfactory because the unit weight of the wet material is its bank measure.

Heavy-duty dragline buckets have reinforced metal plates to permit the handling of broken rock or other abrasive materials. Such materials will likely have much void space in the loaded bucket. The result is less net weight of material but in a heavier bucket. A medium-weight, general-purpose bucket is used for clays, compacted sands and gravels, or any small-grained material, in which the bucket has more difficulty loading. This is denser material that will heap. Consequently, the medium-weight bucket will likely have the highest load for a given size on the dragline's hoist line.

The extreme condition of any dragline operation should be checked for load capacity. This involves the limits of the type and weight of bucket, maximum weight of loaded material, lowest boom angle anticipated, maximum angle the lifted load will make from the boom point, and the tipping load for the dragline under these conditions. The condition causing an extreme loading is to be checked against the dragline's resistance to tipping. If the ratio gives a safety factor of two-thirds, or about 65%, or better (lower %), then the operation should be all right. This sort of check can usually be made with a dragline's specification sheets. Since the dragline, clamshell, and orange-peel diggers are crane applications, their load capacity can be checked, using curves such as those in Fig. 7-13. A suitable curve should be drawn and used for operations of a given piece of equipment.

7.3.2 Uses for Draglines and Related Excavators

A dragline excavator is a versatile piece of equipment used mainly for what has been called loose, bulk material digging. This excavation is contrasted with rock excavation or unclassified excavation, involving a variety of materials. The loose, bulk materials include dry sands and gravels, loose and wet clays and silts, and soil completely saturated or beneath a water surface. These materials all have a low angle of repose. So, to be effective, the excavator must have a good reach. A dragline with the ability to swing its bucket out, and especially using an extended boom, is appropriate for such loose, bulk excavation, whereas a power shovel would not be useful.

Draglines have been used successfully for sand and gravel pit production, strip mining, dredging, irrigation and drainage canals, and open sewer ditches with sloping sides. For these the excavator operates on firm, undisturbed ground and generally digs below its level, moving away from the edge of the cut. It can dig material from many feet below to a few feet above the machine's level of support. It will dump its load anywhere in the range from several feet beyond to an equal distance toward the cab from the boom's end point. This is in the process of casting the dragline's load. The operations resulting in a levee, stock pile, or spoil bank are frequently produced by a casting dragline.

A dragline bucket hanging free and operating with a large horizontal swing cannot dump its load with great precision. If the dragline is used for loading haul units, its production will be slower with much spillage.

For such an operation the target area of the hauling container should be larger than required for any shovel-type unit. Another choice would be for the dragline to dump its spoil in a mound and have a front-end loader fill the haul units and clean up the spillage.

The matter of loading a haul container is not quite such a problem with a clamshell, or similar, type of excavator. The reason is that this digger operates more in a vertical direction. The clamshell or orange-peel digs its load by a vertical drop and lift. Horizontal movement to dump is in an arc swung around the equipment's pivot point. Therefore, any container to be loaded must be on this arc. And the operation's efficiency can be improved if the container has its longer top dimension in the direction of a tangent to the arc.

The clamshell with its vertical digging is useful for excavating foundations, footings, pier holes, cellars, and similar operations. It also is used extensively for moving loose, bulk materials from stockpiles to storage bins, and for loading hoppers or conveyors. Clamshell buckets are available in a wide variety of sizes generally classified by weight, as in the case of dragline buckets. Heavy-duty clamshells are used for digging, the light buckets are for rehandling loose material, and the medium-weight buckets are for general use. The clamshell is usually operated with a higher boom angle than the dragline. This is to take advantage of the better lifting capacity with its vertical actions.

7.3.3 Productivity of a Dragline Excavator

The production rate of a dragline can be figured very much like the productivity for the power shovel discussed in Sec. 7.1.4. The same sort of factors apply:

1. nature of the material dug;
2. depth of the excavation;
3. angle that excavator swings load;
4. loading cycle time of excavator;
5. clearance and movements required during operation;
6. volume of excavation in one place; and
7. balancing haul units where used.

However, for a casting operation the dragline will probably swing through a greater angle.

The loose material handled by a dragline is generally fine-grained or granular, though it may be common earth. Swell of the excavated material may or may not be significant. If it is fine-grained soil taken from under water, the particles may be suspended in the water being moved. In that case swell means nothing and the amount of material moved in each bucketful is the percent of solids in the bucket's volume. If the loose material is sand or gravel that is very wet, it will not heap much, and its weight will be that of saturated soil. This will be 5 to 20% heavier than the dry material (see Table 3-3), so the load capacity of the dragline with a full bucket operating with a long boom and low angle must be checked.

For dryer material that will swell when dug out of an excavation the volume of load in a bucketful can be figured, using the explanation in Sec. 5.1.1. Another way to account for swell is the method used for shovels in Sec. 7.1.4 with Fig. 7-6. That method is not so good for a dragline handling loose material without the crowding action of a shovel. Using the swell factor method, the dragline's load is found by Eq. (5-1) rearranged as

$$V_b = \frac{V_l}{s_w}, \tag{7-3}$$

where V_b = volume of bank-measured material in each bucketful;
V_l = the nominal capacity of the dragline bucket; and
s_w = swell factor = $(1 + \%$ swell$/100)$ with % swell from Table 5-1.

The optimum depth of cut for a dragline, similar to that for a shovel, is the minimum distance the bucket must be moved through the excavated material to just get a full bucket. Since a dragline bucket is loaded while moving more horizontally than a shovel's, that distance is not truly a vertical depth. However, this factor is based on a "depth" to have the dragline's optimum comparable to a shovel's. With draglines the optimum depths of cut are higher than for the same sizes of shovel. At the other end of the size scale, a $2\frac{1}{2}$-yard dragline generally is more effective in a shorter cut than the same size shovel. This variation is shown in Fig. 7-17. In practical terms an optimum depth of cut for a dragline is not as significant as for a shovel, because the dragline is more flexible in its operation. The operator will frequently vary his loading direction and length of cut to shape it from any one position for digging.

The angle of swing for a dragline is generally greater than that for a power shovel. Working with loose material on flatter slopes causes the dragline to reach farther for its loading and dumping. If it is dumping into a haul unit, that equipment generally must be farther away than for

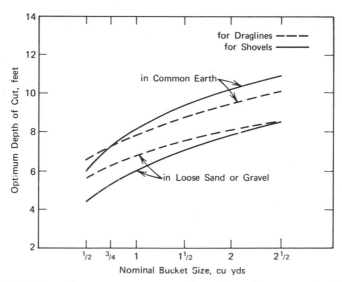

FIGURE 7-17 Comparing cuts by dragline and shovel.

a shovel because of the slopes. When a dragline is casting its loads, it more than likely will have a swing of between 90° and 180°.

A factor combining the effect of depth of cut and angle of swing is recommended by the PCSA to find a dragline's productivity. This is similar to the values given in Table 7-1 for a power shovel; values of this factor for a dragline are shown in Table 7-4.

TABLE 7-4 Effect of Depth of Cut (D) and Angle of Swing (A) on Dragline Production

Percent of Optimum Depth of Cut	Angle of Swing in Degrees						
	45°	60°	75°	90°	120°	150°	180°
20%	.99	.94	.90	.87	.81	.75	.70
60%	1.13	1.06	1.01	.97	.88	.80	.74
100%	1.19	1.11	1.05	1.00	.91	.83	.77
140%	1.14	1.06	1.00	.96	.88	.81	.75
180%	1.05	.98	.94	.90	.82	.76	.71

Data by the Power Crane and Shovel Association.

The table is entered with a value for the percent of optimum cut. For example, if a 1½-yard dragline is digging an average cut of 5 feet in loose, sandy soil, the percent is 5/7.5 (see Fig. 7-17) times 100, or 67%. This will make it necessary to interpolate between the 60% and 100% rows. With an average swing of 120°, the factor $(A:D)$ to use is estimated to be 0.89. If the 1-½yard dragline with the same swing is cutting a 12½-foot average depth, then it has a $D = 167\%$ and $(A:D) = 0.84$.

The cycle time for an excavating dragline is figured in a way similar to that for a shovel. The variables included for the shovel's time were discussed in Sec. 7.1.4 with Table 7-2. One designed feature that generally differs is the governed speed of swing on the superstructure's turntable. The dragline has a slightly lower top speed, say, 3.5 rpm as compared to 4 rpm for a power shovel. With that difference and the flexibility of a free-swinging bucket, the dragline's cycle times are somewhat longer than a shovel's working under the same job conditions. Suitable values for a dragline excavating at the optimum depth of cut and with a 90° angle of swing are given in Table 7-5.

TABLE 7-5 Dragline Cycle Times in Seconds for Optimum Depth of Cut and 90° Swing

Relative Digging Difficulty	Dragline Bucket Size, cu yds						
	⅜	½	¾	1	1½	2	2½
Easy digging	19	19	20	22	25	27	29
Sand or gravel	20	20	22	24	27	29	31
Common earth	24	24	26	28	30	32	34
Hard clay	—	—	30	32	34	37	39

Data by the Power Crane and Shovel Association.

These cycle times are based on no interruption to the designed motions. Notice, in comparing these times to those for shovels in Table 7-2, that the dragline takes 10 to 30% longer for each cycle. Any changes in the dragline's position would add time between regular loading cycles.

Now it is helpful to put these factors together in an equation to figure a dragline's productivity. This gives

$$q_d = \frac{3600 \, V_b \, (A:D)}{\mathrm{CT}_d}, \tag{7-4}$$

where q_d = maximum production, cu yds/hr;
$\quad V_b$ = bank-measure volume in bucket, cu yds;
$\quad A{:}D$ = combined factor for dragline's angle of swing and depth of cut; and
$\quad CT_d$ = cycle time in seconds for dragline operation at 90° swing and optimum cut.

The application of this equation makes use of Eq. (7-3) for V_b and values from Fig. 7-17 and Tables 7-4 and 7-5. An example of its use would be very similar to the one following Eq. (7-1). It is suggested that the reader review that example, including the discussion that follows it on working times and efficiencies.

7.3.4 Costs of Draglines and Related Excavators

The important points about the costs of owning and operating dragline-type excavators are similar to those for power shovels. The shovel cost considerations were discussed in Sec. 7.1.5. No significant nor consistent difference enters into the original investment cost of draglines and shovels. The small, ½-yard size, fully-equipped draglines are likely to be cheaper than that size of shovel. At the other extreme of ordinary PCSA sizes, with the usual diesel power, a shovel may be cheaper. The variation in original cost seems to be governed more by the availability or number of units manufactured than any other factor. The convertible part of the crane-type excavators—the dragline, clamshell, and orange-peel buckets and rigging—amounts to a cost between 2 and 5% of the original investment.

Some equipment owners have noted differences in the allowable depreciation for the various power excavators. They say that a dragline or other crane-type excavator cannot be depreciated as fast as a power shovel. The difference in time is probably based on less wear and tear on the crane-type excavator. However, the equipment owner finds that the government's IRS will approve any reasonable schedule of equipment life that can be justified by previous practice or industry conditions as long as there is consistency for the lifetime. So a case may be made that could result in the same depreciation life for both a dragline and a shovel. Where there is a difference, it would likely be with the larger 2- or 2½-yard excavators. In that largest PCSA size a dragline's lifetime may be taken as long as 12 years, possibly 50% more than the same size shovel.

The hourly cost for a dragline or other crane-type excavator will generally be lower than for a power shovel, provided they have the same number of hours' work estimated for a year. The working-time factor is subject to much variation, the effects of which were discussed in Sec. 7.1.5.

A lower hourly cost will be due in part to a longer lifetime and so less depreciation per year. Even more predictable is the estimated maintenance, repairs, and supplies for the total ownership-operating expense. Comparing the maintenance costs with those discussed in Sec. 7.1.5, it is noted that a dragline's are about 90% and a clamshell's about 80% as much as those for a power shovel. The big difference is most likely in the bucket maintenance compared to maintenance of the shovel's dipper and its mechanism.

In making cost comparisons, the equipment planner must take every precaution to avoid being fooled. For instance, there may be a choice between using a shovel or a dragline for a job that has conditions between the ideal for either excavator. Basing the decision on cost per hour would probably favor the dragline. That basis would be realistic only if the work were on a contract to be paid by the hour for an indefinite period of time. It is much more likely that the work would be paid for on the basis of yardage excavated. In that case, the shovel with a higher productivity may be the more economical choice. These cost comparisons will be illustrated with an example.

Given: 2-yard excavators digging in medium-packed common earth with an average 8-foot-deep cut; operation is opening a sidehill cut where the shovel would dig from the bottom of the cut, whereas the dragline would dig from above and to the side of the cut; assume the same relative efficiencies for each unit; excavated material is to be dumped with an average of 90° swing for either excavator.

for a 2-yd shovel:

$$\text{cut is } \left(\frac{8}{10.2}\right) 100 = 78\% \text{ optimum (see Fig. 7-5);}$$

∴ angle: depth factor, $(A{:}D) = 0.98$ (from Table 7-1);
dipper load, $(E)B_c = (.87)2 = 1.75$ cy (see fig. 7-6);
cycle time, $\text{CT}_s = 25$ seconds (from Table 7-2);
then, using Eq. (7-1),

$$q_s = \frac{3600\,(0.98)1.75}{25} = 247 \text{ cu yds/hr.}$$

for a 2-yd dragline:

$$\text{cut is } \left(\frac{8}{9.5}\right) 100 = 84\% \text{ optimum (see Fig. 7-17);}$$

∴ angle: depth factor, $(A{:}D) = 0.99$ (from Table 7-4);
    swell factor, $s_w = 1.25$ (see Table 5-1);
∴ bucket load, $V_b = 2/1.25 = 1.60$ cy [from Eq. (7-3)];
    cycle time, $CT_d = 32$ seconds (from Table 7-5);
then, using Eq. (7-4),

$$q_d = \frac{3600\,(0.99)1.60}{32} = 178 \text{ cu yds/hr.}$$

the current total hourly costs to use:
a 2-yd shovel might be $25 per hr, and
a 2-yd dragline might be $22 per hour.

These figures, which are reasonable and representative, favor the dragline on the basis of cost per hour. But the costs per unit produced are:

for the shovel,

$$c_s = \frac{25}{247} = \$0.101 \text{ per cu yd;}$$

for the dragline,

$$c_d = \frac{22}{178} = \$0.124 \text{ per cu yd.}$$

This comparison favors the 2-yd shovel.

These variations in the costs of draglines compared to power shovels could be shown graphically (Fig. 7-18). The above comparison is for a given set of job conditions—the material excavated, depth of cut, angle of swing, etc. It is based on peak efficiency, i.e., no delays in the work cycle. Other conditions lead to added curves, which could very possibly show a dragline favored over the same size shovel in cost per unit produced.

More often, these two types of excavators are not in direct competition to be selected for a given construction operation. For an operation dealing with hard, blasted rock or tough material that will stand in a steep bank, a shovel is more suitable. Certainly, that would be the case if the excavated material is to be loaded into haul units. On the other hand, if the operation is to excavate loose material and cast it to one side out of the way, a dragline is obviously the better choice. The selection of a clam-

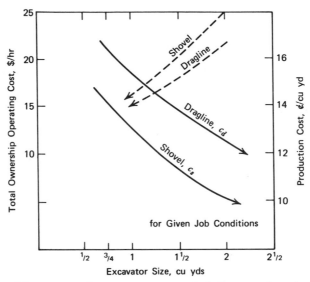

FIGURE 7-18 Comparison of $/hr and ¢/yd for power excavators.

shell or orange-peel bucket unit would be for more specialized operations, which were mentioned earlier in this chapter.

References

1. "The Functional Design, Job Applications, and Job Analysis of Power Cranes & Shovels," Technical Bulletin No. 1, Power Crane and Shovel Association (New York, N.Y., 1956).

2. "Cable-Controlled Power Cranes—Draglines—Hoes—Shovels—Clamshells," Technical Bulletin No. 4, Power Crane and Shovel Association (Milwaukee, Wisconsin, 1968).

3. Farrell, Fred B. "Some Equipment-Management Problems on Highway-Construction Jobs," reprinted from *Highway Research Abstracts,* Vol. 23, No. 10 (November, 1953).

4. Martinson, E. O., "Power Cranes and Shovels—Use and Application," reprinted with permission of Power Crane and Shovel Association by *Construction Methods and Equipment* (October 1951–May 1952).

5a. "Mobile Power Crane and Excavator Standards—PCSA Standard No. 1," booklet by Power Crane and Shovel Association (Milwaukee, Wisconsin, 1968).

5b. "Mobile Hydraulic Crane Standards—PCSA Standard No. 2," booklet by Power Crane and Shovel Association (Milwaukee, Wisconsin, 1968).

5c. "Mobile Hydraulic Excavator Standards—PCSA Standard No. 3," booklet by Power Crane and Shovel Association (Milwaukee, Wisconsin, 1969).

6. "Operating Cost Guide," Technical Bulletin No. 2, Power Crane and Shovel Association (Chicago, Illinois, 1965).

7. McClimon, Alan S. "Estimating Manual for Hydraulic Excavators," booklet by Koehring Division, Koehring Company (Milwaukee, Wisconsin, 1971).

Chapter 8

■■■■■■■■■■■■■■

■■■■■■■■■■■■■■■■■■■■■■■■■■■■■■■■■■■■
FOUNDATION
AND ERECTION
EQUIPMENT
■■■■■■■■■■■■■■■■■■■■■■■■■■■■■■■

Introduction to the Operations

A great deal of the building and heavy construction above and below ground level requires the use of lifting equipment. The foundation for a bridge, a building, a tower, a tank, or whatever, may require piles, caissons, or sand drains. These are generally vertical, load-carrying legs extending through weak earth into or onto firm material. The installation of these forms of foundation calls for the use of driving equipment. They are special pieces of equipment, which require the use of vertical lifting equipment to handle them (Fig. 8-1).

The work to erect the structural parts and operating units to be installed in a building, etc., also calls for heavy equipment. In that case it would be called erection equipment. This is also vertical lifting equipment, such as cranes or derricks.

In the case of foundations the design may call for timber, concrete, or

FIGURE 8-1 Equipment for pile driving (courtesy of MKT Division, Koehring Company).

steel piles of varying weights and lengths. For instance, a 12-inch-square concrete pile will weigh about 150 lbs per linear foot, and it might be 30 or more feet long. The 30-footer of this size, which is a modest reinforced concrete member, will weigh over 2 tons. Piles of that material could be much heavier as the dimensions increase. Another common heavy pile is steel with an H-beam cross section. That sort of pile may weigh 200 lbs per linear foot and be 100 or more feet long. A 100-foot H-beam pile of that weight totals 10 tons. When the planner adds the weight of the hammer and other accessories to drive such piles, it amounts to a sizeable load.

For erection work the individual pieces to lift into position some distance above ground will be of various weights. Ordinary structural erection will often have a piece weighing between 5 and 10 tons. Some special erection jobs have much heavier single loads to handle. All of these call for some sort of vertical lifting equipment. Some mobility or reach will also be required, depending on all the pieces to be erected. The planner of this type of lifting equipment must take into account the range of weights and heights, the equipment positions and reach needed, and several other variables. Now some details for foundation and erection operations will be discussed.

8.1 Pile-Driving Equipment

Piles are structural pieces driven into the earth to provide support for loads from bridges, buildings, and other constructed works. They extend through poor load-carrying soil at or near the earth's surface. Their length is more than 10 times the width. They are essentially symmetrical, with about equal widths at right angles to the longitudinal axis. So, piles are very much like columns in a building.

A pile may be driven with its final, full cross section as in the case of timber or H-beam piles. In some cases to save weight and damage to the structural strength, the outer shell of the final pile is driven. This is true with an open-pipe pile or several of the special patented piles, such as Raymond's step-taper pile. Any pile or driven member must have enough strength to withstand the column action of a force applied on its end in the direction of the longitudinal axis when it is not supported for a portion of that length. The solid cross-section, H-beam, or pipe piles have that strength in their own cross section. The pile weight to handle, in such cases, is the full weight per foot of the given cross section times the length in feet. For the special Raymond type of pile, a steel core or mandrel fits in the pile shell down to the driving foot. This heavy steel core may be almost 100 feet long, tapering from about 8″ diameter at the bottom to more than 16″ at the top. A lifting piece of equipment to handle that type of pile has to carry the mandrel's weight for each pile driven.

The available hammers to drive such different piles are of great variety. Hammers drive piles by the blows of their ram impacting on the pile top or by vibrating the pile while the hammer's weight rests on it. The total weight of a hammer must be lifted by the handling equipment for the driving of each pile.

In addition to the weights of the pile and the hammer, several other weights must be figured for the handling equipment. A key item is the set of leads required to guide some hammers, such as a drop hammer. Pile-driving leads are shown in Fig. 8-1. The leads may have a working platform at their bottom end. One or more men might be supported on such a platform.

The equipment to handle all these loads will have a maximum load in some common units of weight,

$$W = W_p + W_h + W_l, \qquad (8-1)$$

where W_p = weight of the lifted pile or pile shell and mandrel;

W_h = weight of the hammer; and

W_l = weight of leads with any attachments.

Unless the pile can be set up separately, this total load will have to be carried from the end of the handling equipment's mast each time it is moving into position to drive a pile. The handling equipment may be a crane. The crane's lifting capacity would have to be checked for the maximum load, W, at the necessary radius for good operating position. The check must include the working height as well. Such a check can be made on a curve similar to those in Fig. 7-13. If the pile-driving load is being handled by specially designed equipment, checks can be made with the specified limits for its lifting capacity.

8.1.1 Pile-Driving Hammers

The great variability of soil and job conditions leads to the variety of hammers for such an apparently simple operation as pile driving. The soil to drive into or through may be fine-grained silt, which is loose and saturated, or clay in a stiff, consolidated state. Or the earth material may be granular, such as sand and gravel which is highly compacted or loose and submerged under water. It may have boulders here and there within it. More than likely it is a combination of these or other variations in any stratum of similar material. A majority of pile driving is done through more than one material stratum.

While the material's composition may be inconsistent, the location of the water table or the water content of the material for a given site can be rather accurately found. This factor, along with the soil material, will govern the pile design being used. A pile is basically designed to be a point bearing or a friction pile. The bearing pile gets its load-carrying capacity mainly from the load it passes to the earth at its bottom end. In that case any driving hammer or method that gets the pile down and firmly embedded at its bearing level, without the pile being structurally damaged, will be satisfactory. The pile should be driven nearly to refusal, or a set (*s*) of about 10 blows/inch of movement.

For a friction pile to function as it is intended, there must be holding power from the soil along the designed length of the pile. That generally means the pile must be driven into the soil like a nail is hammered into wood. The wedging and friction built up along the length in the frictional material are important. The frictional resistance to drive against can be calculated, if one knows the unit pressure in the subsoil. It is assumed that the pressure at any point is equal in all directions. Then, if the soil pressure near the pile is found to be 80 psi, this is the normal (perpendicular) pressure on the pile's sides moving through the subsoil. With a coefficient of friction between the soil and pile material of 0.25, the skin friction is $F_s = 80(0.25)144 = 2880$ lbs/sq ft. A pile with many square feet of surface exposed to the frictional soil will need some heavy forces applied from an impact-type hammer. Otherwise, a vibratory hammer will have to be used to eliminate the frictional resistance to driving.

It must be noted that a friction pile will probably also be counted on for some point bearing. The frictional resistance of the soil may be developed after the pile is in place. That happens by the soil recompacting or consolidating around the pile after its disturbing action has stopped.

The selection of the best pile-driving hammer to install certain piles in a given situation presents quite a problem. Fortunately, some variation in the hammers that might do the job can be chosen and do it satisfactorily. Certain characteristics and physical quantities should be noted for a feasible hammer selection.

A vibratory or sonic hammer has its internal mechanism for vibrating the pile at or near its natural frequency. This causes the pile to slide through the soil like a hot knife through butter. The hammer attached to a pile is generally heavy enough with the vibrating mechanism to force the pile down to its designed depth. Obviously, the vibratory action breaks down the frictional resistance along the pile sides. So, this type of hammer is used more for point-bearing piles, particularly those driven through wet, fine-to-course sand and gravel.

Pile-driving hammers that are based on delivering an impact force have a ram-like part to contact the pile top. Such a driving part has to have sufficient weight and energy per blow. The basis for determining the required driving energy has for many years been the Engineering News (EN) formula. Actually, more than a half-dozen formulas are used. The wide publicity that the EN publication gave to that formula established its name. And its simplicity and reasonableness set it apart from others. It figures the safe load for a pile by

$$L = \frac{2\,wh}{s + 0.1} = \frac{2\,E}{s + 0.1} \tag{8-2}$$

where L = pile's safe load, pounds, based on a safety factor = 6;

    w = weight of striking part, pounds;
    h = fall of striking part, feet;
    E = energy for driving, foot-pounds; and
    s = pile movement with last blow, inches.

The first version of the formula is based on the striking part having a free fall. With the double-acting or diesel hammers there is additional energy applied downward at the top of the stroke. In that case the total energy, E, should replace the potential energy term, wh.

The EN formula with its s term presents a problem in its use. To help apply the EN bearing value determination, a set of curves is worthwhile. They might be exemplified by the curves of Fig. 8-2.

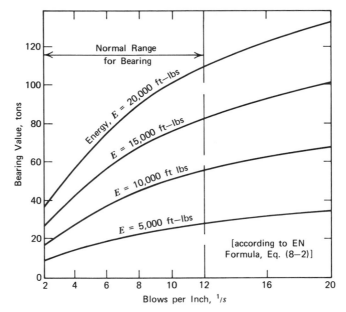

FIGURE 8-2 Pile-bearing capacity from driving energy.

The total energy of the striking part, or ram, available when it hits the pile is consumed in several ways. Part of the energy is used in deforming the pile top and generating heat. A part of it will be consumed by the rebound of a lightweight ram. It is the remaining, net energy that is actually useful to drive the pile further into the earth.

If the total energy is produced by a weight, w, falling freely a distance, h, numerous combinations of weight and height could be used. A lightweight drop hammer falling from great height will not drive a pile very well. Such a hammer will have a high velocity, $v = \sqrt{2\,gh}$ (where $g = 32$ ft/sec/sec), on impact. But its net energy for driving the pile will be low. It is better to have a relatively heavy striking part with a low velocity at impact. This is shown in Fig. 8-3 by the curve for net energy available to drive the pile.

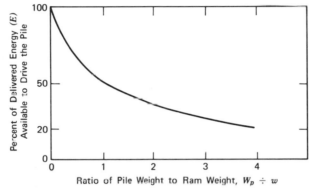

FIGURE 8-3 Effective pile-driving energy.

This curve shows that to have a relatively high proportion of net energy for driving the pile, it is best to select a hammer with a striking part which weighs about the same as the pile being driven.

Impact-Type Hammers. The pile-driving hammers of the impact variety are known as:

1. drop hammers;
2. single-acting hammers;
3. double-acting hammers;
4. differential hammers; and
5. diesel hammers.

The drop hammer is merely a specially shaped weight (w) that is hoisted to some height (h) above the pile top. When released, its free fall is guided by tracks on the leads to make sure it hits the driving block on top of the pile. The block reduces the damage to the pile top from the driving impact. As suggested, the drop hammer should have a weight about equal to the average pile to be driven. Its fall height, h, can be found from the pile formula if the required load capacity, L, is known and a suitable last set, s, is chosen. A value of $s = 0.1$ to 0.5 inch is common. Use of a drop hammer results in a slow operation because lifting the hammer is not automatic and fast. It is cheap without any extra power needed. It might be chosen if there are only a few piles to drive.

A single-acting hammer, which uses pressure in a cylinder to raise the ram, is shown in Fig. 8-4A. The force to raise the ram for each stroke is provided by compressed air or steam. The ram's weight (w) amounts to nearly half the hammer's total weight (W_h). A simple system of valves directs the pressure to lift the ram and then releases it at the top of the cylinder for a free fall (h). The ram travels up and down within the chan-

FIGURE 8-4A An impact type of pile hammer (courtesy of MKT Division, Koehring Company).

nel tracks, which are a part of the hammer. Generally, longer leads made out of timbers or steel members are used to guide the bouncing hammer on the pile as it is driven into the ground.

The single-acting hammers deliver approximately one blow every second, which is many times faster than a drop hammer. The rate will vary from 50 to 80 blows per minute, provided there is enough pressure and volume of the compressed air or steam power. The specific requirements for different sizes are shown in Table 8-1.

One notes that a lighter Vulcan hammer, with a ram weighing about 500 lbs, is designed for timber and sheet piles. Several heavier Vulcan hammers compete directly with the given two lighter MKT single-acting hammers. Though several other sizes are available, the biggest MKT hammer is given. As noted in the tabulation, there would be some problem to match satisfactory compressor equipment with this big hammer.

Some steam- or air-powered hammers are "open" to have the intake and exhaust work openly with the surrounding atmosphere. Others are enclosed hammers which use hoses and special port controls for inlet and exhaust. The closed type of hammer can drive piles under water as much as 80 feet below the surface. There must be hoses long enough to reach the surface and at least 1/2 psi of additional pressure per foot of water depth to keep it from flooding the hammer.

A double-acting or differential pile-driving hammer provides steam or compressed-air pressure for both directions of the ram's stroke by a more complex valve system. Thus, the ram does not get its force for impact on the pile just by the ram's free fall due to gravity. It has that and, in addition, pressure is applied for the downward part of the stroke. This means that the stroke can be shorter and the hammer more compact than a single-acting one. It also results in a faster rate of blows, giving it advantages somewhat like those of a vibratory hammer. That is, the double-acting hammer essentially keeps the pile continuously in motion during the driving. In such a state of motion the soil has less chance to develop frictional resistance along the pile's surfaces.

A differential hammer has upper and lower piston pressure chambers. It gets the added force of air or steam pressure from the differential of pressure between the two chambers. This type of hammer is a compromise between the double- and single-acting pile hammers. It has the frequency of blows like the double-acting hammer and a ram weight and stroke like the single-acting hammer. There is also a hydraulic pile-driving hammer that operates with the same principle as a differential hammer. That one operates with fluid pressure which is not as successful as one using air or steam pressure.

TABLE 8-1 Data for Some Single-Acting Pile-Driving Hammers

Specific Hammer	Vulcan No. 3	Vulcan No. 2	Vulcan No. 1	MKT S8	MKT S10	MKT S20
Weight, lbs	3700	6700	9700	18,300	22,400	38,600
Length, ft	9.5	11.5	13.0	13.3	13.0	15.4
Stroke (h), ft	2.00	2.42	3.00	3.50	3.50	3.42
Ram (w), lbs	1800	3000	5000	8,000	10,000	20,000
Energy rated, ft-lbs	3600	7260	15,000	26,000	32,500	60,000
Blows/minute	80	70	60	55	55	60
Pressure required, psi	80	80	80	80	80	150
Boiler hp	18	25	40	120	130	190
Volume of free air, cfm	220	336	565	850	1000	—

Data from Warrington-Vulcan and MKT Division, Koehring, specifications.

The double-acting type of hammer, like the original McKiernen-Terry (now known as MKT), is an enclosed hammer that is self-contained enough to absorb its own vibrations. While driving a pile, it will continue to hang directly under its hoist line and, so, does not need any separate leads to guide the hammer. The combination of a high rate of blows and self-containment means that the double-acting type of hammer can be used another way. That is as a pile extractor. An extractor needs to have a positive pull on the pile through the hammer to the hoist line. The crane must be able to pull with 50 to 100 tons tension. There are also specially designed extractors with physical dimensions as shown in Table 8-2. The B-3 hammers in Table 8-2 can be outfitted to drive piles underwater. That is, the hammer can operate submerged.

A diesel pile hammer is another impact-type with energy added to the force of gravity on the ram. The additional force is a combination of compressed air for preloading the pile before the ram hits and a diesel fuel explosion which occurs at impact. This amounts to a three-part blow for each stroke of the ram.

The diesel hammer has a simple mechanism for operation. It is started merely by lifting the ram by a hoisting line to trip it for its first downward stroke. In that descent it actuates fuel and lube pumps and air intakes. These provide the air that is compressed between the ram and anvil block and the mix for the diesel explosion at the bottom of the ram's stroke. The explosion not only helps to drive the pile downward but also lifts the ram for its next stroke. As the ram travels upward, it compresses air at the top of the cylinder which adds to the downward thrust for the next stroke. To turn the diesel hammer off is merely a matter of pulling a slack rope which disengages the fuel pump.

The three-part blow of a diesel hammer causes some difficulty in finding the energy delivered to drive the pile down. This is complicated by the fact that greater pile resistance causes the stroke and compressed-air pressure at the top of the cylinder to increase. That means more energy will be delivered by a given hammer on a pile when it gives more resistance to driving. The diesel hammer is outfitted with a pressure gauge on its side to show the air pressure being exerted on the top of the ram at each stroke. From this the total energy per stroke can be determined. The maximum striking energy is often increased 50–75% above the average potential energy of the ram falling freely in the cylinder's height. The specifications for a diesel hammer should give the available energy in foot-pounds converted from the psig pressure shown on the gauge. Some new

TABLE 8-2 Data for Some Double-Acting and Differential Hammers and Extractors

Specific Hammer	MKT Extractor E2	MKT's (Double-Acting)				VULCAN'S (Differential)	
		No. 3	No. 6	9-B-3	11-B-3	30C	50C
Weight, lbs	2600	675	2900	7000	14,000	7040	11,800
Length, inches	100	58	63	98	133	—	—
Stroke, inches	3	5¾	8¾	17	19	12½	15½
Ram(w), lbs	200	68	400	1600	5000	3000	5,000
Energy, ft-lbs	700	—	2500	8750	19,150	7260	15,100
Blows/minute	450	400	275	145	95	133	120
Boiler hp	30	15	25	45	60	40	60
Volume of free air, cfm	400	110	400	600	900	488	880

From MKT Division, Koehring, specifications and **R. L. Peurifoy's** *Construction Planning, Equipment and Methods*, 2nd edition.

diesel hammers give more than 90,000 ft-lbs maximum energy in a single blow.

For the equipment planner, who must select a suitable pile-driving hammer for his construction job, a comparison of the impact-type hammers will be helpful. Since the pile's capacity is directly proportional to the energy delivered from the hammer, the available energy will be the basis for comparison. Table 8-3 gives data on some representative hammers of the impact-type.

Vibratory Hammers. A vibratory hammer is contrasted to an impact type by the speed of applying its load. It does not use the forcefulness of separate, distinct blows, but the vibration set up in a pile by rapid succession of load variation. Where one or two blows might be applied every second with an impact hammer, the vibratory will cause 15 to more than 100 load variations every second. At the highest speeds, or frequencies, the hammer can put a pile in the resonant vibration range as does the Bodine Sonic Pile Driver.

The regular vibratory-type of hammer weighs three to five tons. Its purpose is to get the pile vibrating so the soil around it is kept in a fluid state. Then the pile slides down through such material, without much frictional resistance, under the combination of the pile's weight and that of the hammer on top of it. Ideally, the vibratory hammer could get the pile to vibrate at its natural frequency. However, that frequency will vary as the pile has different amounts of free length above the earth to be penetrated. Also, the surrounding soil material will have an influence on the pile's natural frequency. So, the object is to vibrate the pile at a high enough frequency for it to slide through the earth.

Basically, a vibratory hammer has a series of horizontal shafts with eccentric weights on them. The hammer has a power unit, which may be off to one side with just feeder lines to the moving hammer on the pile. It delivers the power to rotate the weighted shafts at high speed. This power is most generally hydraulic or electric. The shafts operate in pairs, with one revolving clockwise and the other counter-clockwise. The eccentric weights are attached so that their effects add together in the vertical direction but cancel each other horizontally. To be effective, the vibratory hammer must be firmly clamped to the pile being driven. Then the vertical effect of the eccentric weights amounts to either a downward compressive push on the pile or an upward tensile pull. By Poisson's lateral effect in the pile due to longitudinal compressive and tensile forces, there will be horizontal movement of the pile side surfaces. When the pile is in compression, its sides move outward; and when in tension, they move

TABLE 8-3 Comparison of Some Impact-Type Pile-Driving Hammers

Hammer	MKT DE-10	MKT 9-B-3	Vulcan No. 1	Vulcan 50C	MKT 11-B-3	MKT DE-30	Vulcan No. 0
Type	Diesel	Double-Acting	Single-Acting	Differential	Double-Acting	Diesel	Single-Acting
Energy (E) rated, ft-lbs	7,500	8,750	15,000	15,100	19,100	22,400	24,400
Ram(w), lbs	1,100	1,600	5,000	5,000	5,000	2,800	7,500
Hammer's (W_h) weight, lbs	3,900	7,000	9,700	11,780	14,200	9,400	16,200
Weight ratio $\left(\dfrac{w}{W_h}\right)$	0.28	0.23	0.52	0.42	0.35	0.30	0.46
Energy to ram $\left(\dfrac{E}{w}\right)$	6.80	5.47	3.00	3.02	3.82	8.00	3.26
Energy to wt. $\left(\dfrac{E}{W_h}\right)$	1.93	1.25	1.55	1.28	1.35	2.38	1.50
Blows per minute	48	145	60	120	95	48	50
Air power req'd, cfm	—	600	565	880	900	—	840

inward. With a high frequency of load variation, the outward-inward movement of the sides is very rapid. It amounts to a vibration of the sides which keeps the soil from getting any friction or grip on the pile. The vibrations from the hammer are almost isolated from the crane hoisting equipment by a spring support system.

A vibratory hammer is most effective in water-saturated, noncohesive soils that it can keep in a fluid state. Its usefulness is questionable in a heavy, cohesive clayey material. The outward-inward vibrations of the pile sides is very positive in metal piles, such as H-beam, pipe, and steel sheet piles. Under the right conditions a vibratory sonic hammer can put down a pile in one-hundredth the time it takes a single-acting steam hammer.

A vibratory hammer is also useful for extracting piles, particularly sheet piling. In fact, most such hammers are designed to work as either a pile driver or extractor. As an extractor the MKT model V-10 hammer requires a maximum of only 20-ton line pull from the hoisting equipment. This is much less than the 50- to 100-ton maximums for impact-type extractors. Their frequency of about 400 blows per minute is only one-half to one-fourth as fast as the cycles of a vibratory driver-extractor. One hydraulic vibratory unit is a compact piece of equipment used strictly for extracting sheet piling (Fig. 8-4B). This is cleverly called the Alli-up and operates at the fast pace of 2300 cycles per minute. Its effectiveness probably comes mostly from a 15,000-lb impact force. This is caused by hydraulic pressure up to 200 psi delivered by a flow of at least 28 gpm.

8.1.2   Pile-Driving Leads

A guiding frame known as leads is used with some pile-driving operations. Leads are necessary when using a drop-hammer or an open-type, single-acting pile hammer. They may be helpful when driving vertical piles with other types of hammers. It is essential that leads be used for driving batter piles. Such piles are installed at an incline from the vertical which is held by the leads.

Pile-driving leads may be simply two heavy timbers which form the tracks for a hammer. The timbers would be held in a given spacing by a cross piece at the top, which also serves as the pickup point, and horizontal U-brackets at several places along their length. The weight of timber leads can easily be figured from their length and make-up. They are generally made by the user as swinging leads to hang freely and be held by a single crane line.

Other leads are manufactured of metal beams or angles and other

FIGURE 8-4B   Vibratory hammer on sheet piling
(courtesy of MKT Division, Koehring Company).

shapes to form an open frame with tracks. These are swinging or under-hung leads with the basic top section as short as 15 feet long. With inter-mediate extension sections and a bottom brace or supporting section, the total length of ready-made leads may be 50 to 100 feet long. Such open-frame leads will probably weigh between 60 and 90 pounds per linear foot. If a bottom brace frame is used to tie the leads to the crane hoisting equipment, that will probably add 1,500 to 5,000 pounds.

Underhung leads are arranged so that a crane's boom point can be a part of the top cross support between the lead tracks. This means that the leads are supported directly under the end of the boom. Then the crane's hoist line is needed only for the hammer and perhaps to set the pile to be driven. The other common type is swinging leads. As their name suggests, these leads swing freely and require a lifting line from the crane to sup-port them. With this arrangement the leads can be lowered as the pile is driven and the hammer follows it. Thus, the leads may not have to be as long as the underhung type. Of course, these swinging leads cannot be lowered much if a bottom brace ties them to the crane's platform.

Some special long leads are held below their top end by the supporting boom. Such an arrangement is particularly good for long piles, especially batter piles. Manufactured lead sections can be used for this with a sliding and pivoting boom connector. Just as with other pile-driving operations, and particularly with long leads set for batter piles, a care-ful check must be made for the safe use of the hoisting equipment.

8.1.3   Hoisting Equipment for Pile Driving

The hoisting equipment for a pile-driving operation must be able to handle the extreme loading and reach needed. Checking reach will mean both horizontal and vertical distances. A combination may govern the equipment to be used. The maximum vertical height will have to include the greatest amount of pile extending (before driving starts) above the ground or surface supporting the hoisting equipment. Total height means adding the hammer on top of the maximum pile height. With underhung, swinging, or no leads, the upper boom point must be above that total height. If the hung leads are supported by a sliding boom connector, so they can extend above the boom point, there is not so much concern for the hoisting boom's length.

The horizontal reach of the pile driver is a matter of concern in check-ing load capacity. It must be remembered that a crane loses lifting capacity quickly as the load radius increases. The same load-radius relation applies to derricks or similar boom-type hoisting equipment. For this reason, there is an advantage to the use of a mobile crane which can move about and get closer to the places for each pile. However, if there are many piles to be driven in a small area, the use of a mobile crane may be more expensive than necessary.

Where many piles are to be located, say, 5 feet on centers, maybe the driving for one or more days can be handled with the hoisting equipment in practically one position. In that case it would probably be more economical to use a simple derrick or special pile drivers moved on large rollers, such as the equipment made for the Raymond International company. If such equipment is available, its load-carrying capacity per dollar spent is worth the contractor checking for comparison. Many piles are driven along a waterfront or into the water. If they can be driven from a barge, it should be more economical to set up a derrick with movable boom of substantial size and load-carrying capacity than to use a movable crane on the barge. To drive many piles in a large cofferdam, contractors have in some operations made use of the reach and load capacity of a hammerhead or tower crane. This type of hoisting equipment will be identified later in the chapter. Their use for pile driving was generally with an enclosed hammer and no leads. Of course, the choice of what to use will depend on the availability of equipment, as is true with most selection problems.

For driving batter piles, the calculations to check the hoisting equipment will be somewhat different from those discussed previously for a crane's capacity. The crane's capacity is checked for a particular load, L, at a given lifting radius, r, from the crane's center of rotation, as explained in Sec. 7.2.2. In the case of a pile being driven on a batter, the loads of the hammer, the pile, and the leads are at different distances from the crane's center of rotation. These different distances can be covered by summing their separate moments about the center. An example will be used to help explain the procedure. This case is the extreme one of an aft (pile pointed toward the driver) batter and the hammer at the top of the leads (Fig. 8-5). The moments of these pile-driving loads can be summed as follows:

$WR = W_h r_1 + W_b r_2 + W_l r_3 + W_p r_4 + W_b r_5$ giving the equipment load radius,

$$R = \frac{W_h r_1 + W_b r_2 + W_l r_3 + W_p r_4 + W_{br} r_5}{W}, \tag{8-3}$$

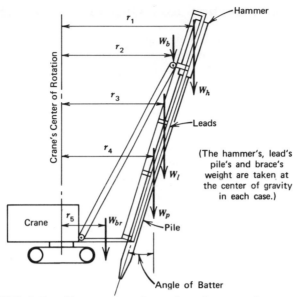

FIGURE 8-5 Crane loading for driving batter piles.

where W_h = hammer's weight, lbs;
   W_b = weight of lead connector and half of crane's boom, lbs;
   W_l = weight of leads, lbs;
   W_p = pile's weight, lbs;
   W_{br} = weight of bottom brace, lbs;
   $W = W_h + W_b + W_l + W_p + W_{br}$;

and the distances, r, are as shown with the same units as the radius, R. Now the crane's load-carrying capacity can be checked with these values of total load, W, and equivalent operating radius, R. Use would be made of a load curve such as those shown in Fig. 7-13 for the given crane. If the crane can handle the total load, W, at a radius equal to or greater than R, it is safe enough. If not, a larger crane should be selected.

In a pile-driving situation with limited headroom, a crane or other regular hoisting equipment may not be usable. Motorized fork lifts have been modified for such operations. The fork lift is equipped with hammer rails on the tracks for the fork. No leads are used in this sort of pile driving. Generally, the pile is driven in short lengths to be welded together or otherwise attached when the newly driven section is at the lower limit of the hammer's reach.

8.1.4 Selection and Cost of Pile Drivers

The job of selecting pile-driving equipment must account for the piles to be driven, the jobsite conditions, and the availability and costs of equipment for the operation. More specifically, the steps in the selection procedure might be listed as:

1. *determine the type* (material and form), *size, weight, and length of pile* to be driven;
2. *identify the special job conditions* influencing the pile driving—under water, batter piles, in limited headroom, etc.;
3. *determine the choices of hammers* that could drive the piles to the designed load-carrying capacity and to any specified depth; if there is a specified penetration into point-bearing earth material, *determine what air- or water-jetting equipment* might be used to reduce the driving;
4. *select the most feasible, economical, and available pile-driving hammer* for the operation as determined and identified above and, if needed, *select the external power generator* (air, steam, or hydraulic) and its distribution lines to the hammer;
5. *if leads are needed, design or select* type, spacing, and rail size for hammer, and length for hammer and piles to be driven;
6. *select hoisting equipment* that can adequately and economically handle the largest pile, hammer, and accessories for the above determined operation; and
7. *check the loading extremes* for the hoisting equipment to be sure it can handle the operation safely.

The good planner for this equipment will naturally be cost conscious, as well as concerned for the physical dimensions of the job. He will want to do the job satisfactorily for the lowest total cost. The construction costs include all the equipment and manpower to operate it as long as the pile-driving takes. Time is the biggest variable in this type of operation. Since the piles are driven into somewhat unpredictable subsurface conditions, the time will necessarily be unpredictable. So the total cost, which is time dependent, will be quite variable. One method to help get the pile driving done in the least time is to use the largest feasible hammer that will not damage the piles. This is desirable because it provides a reserve of forcefulness to overcome the unpredictably tough instances in the driving. Also, the larger hammer with a heavier ram will have a more efficient energy transfer, as shown in Fig. 8-3.

In the determination of a feasible hammer for the operation, it is desirable to know how the different hammers vary in cost. Of course, a simple drop hammer will be the least expensive, but it results in the slowest pile driving, so the total cost of operation may be high. A single-acting air- or steam-powered hammer may be ten times more expensive than the drop hammer. And the compressed-air-power system will add a cost about equal to or higher than that of the single-acting pile hammer.

A double-acting or differential pile-driving hammer will cost 20 to 30% more than a single-acting one with approximately the same energy rating (see Table 8-3). The fact that the frequency of blows by the former is higher should mean a faster operation, so the higher hammer cost will be minimized in the total cost. But the double-acting or differential hammers require air compressors 30 to 50% larger and more costly than required for a single-acting hammer of about the same rated energy.

A diesel hammer may be 50% more expensive than the single-acting air- or steam-power hammer of about the same energy rating. However, the diesel hammer has its power built into the driver, so there is no additional expense for power equipment, such as an air compressor or steam generator. A diesel hammer is also economical in operating-fuel expense. However, because it has a low frequency of blows, the total time and cost of operation may be high.

Other advantages in selecting a diesel hammer for driving piles are that it is comparatively lightweight and operates well in cold weather. The main disadvantage with a diesel hammer is the difficulty in determining the energy per blow. As explained in Sec. 8.1.1, that is because the energy varies with the resistance offered by the pile. However, that provides a safety factor in that the diesel hammer will automatically stop if there is no pile resistance under it.

8.2 Pier and Caisson Drilling Equipment

The construction of piers and caissons for the foundations of buildings and other land-based works call for a major equipment concern. Piers and caissons are vertical foundation elements constructed with the use of augur-type drills, open pipes or casing, and buckets for excavation and backfilling. These tools all require hoisting equipment to handle them.

A pier or cassion is not driven like a foundation pile but is located on a stable subsurface stratum in the earth. By "located" one means it is built in place. It is used where the firm stratum is too deep for a footing to be constructed in an open excavation. On the other hand, the point loading of piles might lead to uneven settlement and damage. The cross-sectional area of the pier or caisson is generally several times greater than that of a bearing pile usable in the same location. If more bearing area is needed, this type of foundation element may be belled out at the bottom. Building it in place allows the constructor to shape the foundation like he would a footing. However, if the hole for a building pier or caisson is opened in unstable ground or soil that is submerged under the water table, it will have to be lined during construction to keep its shape.

The holes drilled for building piers or caissons can be one to 20 feet in diameter and extend down a half-dozen to as much as 200 feet in depth. In addition to the opportunity for inspecting the subsoil and varying the dimensions to suit the soil conditions found, there are several other advantages with this type of foundation. They can be constructed very close to adjacent buildings without much danger of subsidence such as might be caused by an open excavation for footings. There will not be the noise nor possible damage from a pile-driving hammer close to an existing structure. Constructing piers or caissons for a building's foundation can be the most economical in certain soils and job conditions. The drilling equipment and tools are generally fairly simple and economical. Basically, a crane-type unit is enough for handling everything without auxiliary power or other equipment needed.

8.2.1 Drilling Methods and Equipment

Various methods exist, with as much variety of equipment, to construct building piers or caissons. Commonly, they involve a crane or similar hoisting unit to handle a drill shaft known as a Kelly Bar. The vertical shaft is operated through a horizontal supporting rotary "table" on the frame connected to the crane equipment. A large horizontal gear on this table rotates the Kelly Bar for drilling. It could be operated off the crane's secondary drum, while the main hoisting drum handles the vertical support for the drill shaft and digger. To operate the drill directly from the crane's line could prove to be entirely too slow. Its speed would be governed by the line speed of only a couple hundred feet per minute at best.

A specially manufactured drilling unit will have a built-in power unit to rotate the drilling mechanism. This can apply considerable twisting force, more than 1/3-million ft-lbs at speeds up to 90 rpm. The arrangement of specially designed drilling equipment mounted on a crane is shown in Fig. 8-6. Two basically different processes used to drill the hole for a caisson are the "dry process" and the "wet process." In the dry process the drill may be an augur, as shown in Fig. 8-6, or a bucket. An augur

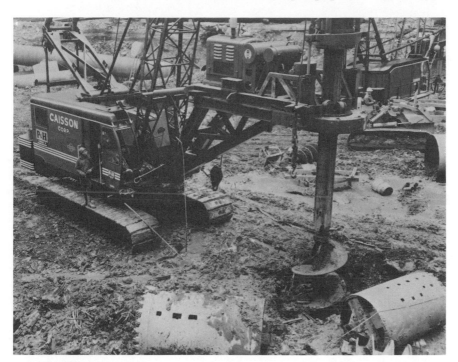

FIGURE 8-6 Caisson drilling equipment (by Calweld Drilling Equipment Division).

drills by its rotary action as it moves downward into the earth under its own weight. It removes the dug earth material by the same rotating movement as it is pulled out of the hole. A bucket drill for the dry process can be used somewhat like a post-hole digger. It is dropped into the earth, and its shovel- or knife-like leading edge cuts the material. When it stops penetrating, the bucket is closed on material within its outside faces. Thus, the bucket is closed enough to bring material up in it. A bucket has the advantage over an augur drill of being able to move the dug material well away from the hole. This is done by removing the bucket from the hole and dumping its contents to one side.

Buckets for the dry process are of several varieties. The grab with two jaws and the many-leaved, orange-peel bucket, operate like the post-hole digger as discussed above. Another variety of bucket is more like a barrel with its axis an extension of the Kelly Bar. Such a bucket has a part of its bottom open to receive the dug material. The opening extends part way up the side with a sharp cutting edge. This bucket cuts through the earth by being turned or twisted like an augur. It has the same advantage of other buckets over the augur. However, bucket drills cannot operate as fast as the augur type, nor drill effectively after the bucket is filled.

The wet process for drilling in caissons is used where there is a good water supply and the subsurface is fairly saturated. As the name implies, the material is removed from the hole in a water solution, or slurry. The wet condition of the material and the process itself generally require lining the hole for the caisson from the start of its construction. A pipe of the caisson's diameter is used and driven down, like a pile, as the hole deepens. The drilling pipe is a smaller diameter inside the caisson and carries water down to the cutting head. From there it circulates back up between the drill pipe and the caisson sides. The water flowing up carries the loose, cut material with it. Water overflowing the dug caisson hole is collected in a sump to be pumped back through the wet process. This amounts to a recirculating system, using a diaphragm or other type of sump pump discharging into the drilling pipe.

Another variation on these equipment setups is used for drilled-in caissons. For that type the caisson lining is the drilling tool with a sharpened, bottom leading edge. This liner of steel or reinforced concrete is rotated to churn into place in the ground. The crane-type equipment to handle drilled-in caissons is some of the largest for installing columnar foundations.

Sometimes the caissons need to have larger bottom bearing area than the necessary column section. It would be wasteful and perhaps difficult to drill in or drive a caisson of the required, enlarged bottom diameter. So, a fairly small-diameter caisson is put down to the depth from which the enlarged, belled-out base can be constructed. This can be done by a special, rotary cutter with knife edges that can be flared out when they are down at the right depth. Another way to cut the belled-out base is by hand tools. For that case and also to handle the removal of excavated material in the belled portion, the upper caisson with vertical sides must have a diameter of several feet, at least. Obviously, if laborers are to work in the caisson, it will have to be in dry, stable ground material.

8.3 Equipment Used for Erection Operations

The erection operation in construction is basically a matter of vertically lifting loads, reaching out with them, and holding each in a set position until the load is safely placed and can be released. The lifting step has to be in a controlled fashion so that the load maintains the horizontal, or whatever, orientation required for its placement. This will generally require a two-point, or more, pickup with hooks, slings, or a bridle arrangement. The load must be moved both vertically and horizontally in most cases to reach its point of connection to the permanent structure. Generally, the load must be held at that point of connection during its erection until it is firmly attached by pinning or bolting. It is for this reason that the productivity in tons per hour is not determined by the erection equipment but more by the manpower making the connections.

Various types of equipment can be used for the erection operation. These will generally fall into two categories—cranes or derricks. The cranes are movable erectors, introduced in Sec. 7.2. For erection purposes there are mobile cranes, locomotive cranes, revolvers, or tower cranes. Derricks are stationary erectors because their base does not move when loaded, except with a derrick boat. There are stiff-leg derricks, guy derricks, gin poles, derrick railroad cars and boats, and tower derricks.

All of these types of erection equipment have certain basic components. The common parts for erection are the boom as a main supporting member and the ropes and tools to pick up the loads. The boom has its supporting lines and those to lift and move each load. The basis for the boom's load capacity is like that for the crane discussed in the previous chapter. Some general understanding of ropes and tools for erection will be covered shortly. However, it must be understood that there is a great variety of these erection elements, and their safe and economical use should be determined by specialists for any involved operation. The main differences between the cranes and derricks for erection are noted in their general use. Cranes have a self-contained mechanism with counterweights, all for the sake of the mobility of the equipment. On the other hand, derricks use a mast and structural frame or guy wires tied to solid supports beyond the derrick's parts for stability. A fundamental difference is found in the boom's topping-lift lines and angles, as will be discussed shortly.

For any piece of erection equipment the wire ropes or cables used for lifting are designed with the same principle in mind as with mobile cranes. That is, they are of a size that is balanced with the structural parts of the erection equipment. The ropes that are actually working directly to pick up loads are frequently called the "load falls." There may have to be two or more lines, called "parts," of the rope to pick up a given load. In that case sheave wheels in pulley blocks are used. In any extensive erection operation, more understanding is needed than can be discussed in this book, and an erection specialist should be consulted. However, some fundamental understanding by the general planner of construction equipment will be helpful.

The wire rope used for erection operations is commonly an improved plow-steel rope. This variety is more expensive but stronger and safer than crucible cast-steel rope. The rope or cable used for holding lines such as guys, where flexibility is no problem, can be the stiffer and less expensive seven-wires-per-strand rope. Rope used for running in lift lines must be flexible and may be the 6 by 19 rope, where there are 6 strands of 19 wires per strand. This common lifting line has the breaking strength of about 100,000 lbs/sq in (psi). Usually, a safety factor of 4 will be used, giving a working limit (s_w) of 25,000 psi. Then the rope's capacity to take a load is found as

$$L_w = As_w = \left(\frac{\pi D^2}{4}\right) s_w. \tag{8-4a}$$

where L_w = load is weight or force or both, lbs
A = cross-sectional area of rope, sq. in.
D = diameter of rope, inches
π = 3.14
s_w = working stress limit, psi

In tons this is

$$L_w = \left(\frac{\pi D^2}{4}\right) \frac{25000}{2000} = \frac{3.14}{4}(D^2)(12.5) \approx 10D^2.$$

This means that a 1-inch-diameter rope will be good for approximately 10 tons and a 1/2-inch-diameter for about 2 1/2 tons.

To get a two-point pickup, it is frequently necessary to use several hooks or slings with the ropes in a bridle-type arrangement, as shown in Fig. 8-7. In this case the sling ropes take the lifting tension at an angle. If there are two parts of sling lifting line, the load (L) is taken by $L/2$ vertically in each line. But the pull along the sling's rope is increased as in Eq. (3-12). Therefore, if the angle Θ is 30°, the rope takes a tension, $T = (1.16) L/2 = 0.58L$.

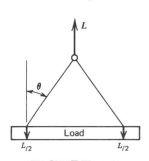

FIGURE 8-7
Lifting sling.

As mentioned previously, the heavy loads frequently require running the lifting line through sheaves or pulley blocks to distribute the load to two or more parts of the line. Basically, the load is divided by the number of the

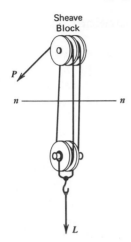

Sheave
Block

P

n —————— n

L

FIGURE 8-8
Multipart lift line.

moving parts of lift line cut by the section n-n shown in Fig. 8-8. That would suggest that the tension in these parts of line is L/n (in this case, $L/3$). Actually, there is a difference in the pull of the line either side of a sheave due to friction as it moves. For this reason the pull, P, in the lead line to the hoist drum is found by

$$P = L \frac{K^n (K - 1)}{K^n - 1}, \qquad (8\text{-}4)$$

where L = total load picked up;

n = number of parts of line, as explained above; and

K = ratio of stress in side of rope unwinding from sheave to that about to wind around the sheave, which ranges from K of approximately 1.05 for metal-line bushed sheaves to 1.025 for ball- or roller-bearing sheaves.

An example of the use of this formula will be helpful. For roller-bearing sheaves and a three-part lift line ($n = 3$), the lift in the running line is found as

$$P = \frac{(1.025)^3 (1.025 - 1)}{(1.025)^3 - 1} L = \left(\frac{.0269}{.077} \right) L = 0.349L,$$

compared to stationary $T = 0.33L$ tension.

The choice of equipment for erection depends on such important factors as the job layout, site conditions, and the loads to handle. Advanced planning is very important with drawings for the methods and sequence of steps in the operations. The question of productivity in tons per hour for the equipment is generally meaningless, except for building erection with many small pieces. The following sections will have no detailed analysis but some fundamental points for the sake of generally understanding erection equipment.

8.3.1 Cranes for Erection Operations

The use of crane-type equipment for erection work takes advantage of their mobility. The mobile cranes will be mounted on crawler tracks or truck mountings, as described in the previous chapter. More specialized cranes, such as the revolving-gantry type or tower cranes, are mounted on steel wheels which roll on tracks or on special wheel mountings without tracks. The arrangements and use of this type of equipment for erection will be discussed in the following paragraphs.

The crawler- or truck-mounted crane was originally introduced in Sec. 7.2. These are the most maneuverable of all erection equipment, which has only to find sound footing to support its heavy loading. The truck-mounted crane will use outriggers to improve its footing. A crawler-mounted crane may increase its stability for very heavy loads with a loading ring, like a hoop skirt, that can be blocked against the ground for wider support points. At least one manufacturer has a rigging arrangement to convert a mobile crane into a guy derrick for erection work.

As with all erection equipment, the upper end of the boom must have supporting lines running back to some other part of the equipment's structural frame. These are attached to the top end of the boom, called its point, and the line and its accessories are known as the topping lift. In the case of the mobile crane these run back to an A-frame on the crane's superstructure. This is the line that goes to point G in Fig. 7-12. As can can be seen there, the angle between the boom and its supporting line is small and provides a minimum of direct help to the boom in lifting the load. The only way to take a heavy load with this equipment is to have the boom nearly vertical so that there is only a small angle between the load and the boom. This means that the mobile crane is limited in its reach for heavy loads to somewhere between a 12- and 25-foot radius from its center of rotation.

A mobile crane is the most expensive erection equipment in terms of the cost per pound lifted. This is because of its load limitations and the expense of the base mounting and superstructure, which is more elaborate than with some other erection equipment to be discussed shortly. The standard mobile crane's erection use is for occasional load pickups or for low industrial buildings that cover a lot of area and need the crane's mobility. It is helpful to handle long, awkward loads that are not necessarily heavy, such as roof trusses. When the load is too long or heavy, it may be handled by two cranes working in tandem at either end.

A new rigging arrangement can convert some standard mobile cranes to tower cranes. This is done by using an additional counterweight to support the load lines. The original boom is replaced by a vertical mast topped by a boom that can be pivoted from a horizontal to almost vertical position on top of the mast. The boom is only slightly shorter than the mast, giving almost as much reach horizontally as a regular tower crane, discussed in Sec. 8.3.3.

Revolving-Gantry, or "Whirly," Cranes.   The revolving-gantry-crane type of lifting or erection equipment has a superstructure basically like the mobile crane mounted on a turntable for revolving. It is familiarly known as a "whirly" crane. An example is shown in Fig. 8-9. As the picture shows, this crane's superstructure is supported on top of a high gantry frame. The gantry raises the hoisting equipment, generally, with four legs spread nearly equally apart to straddle work area and give more horizontal stability to the equipment. The wheels under the legs run on rail tracks, frequently attached to a temporary work bridge. This équip-

FIGURE 8-9   Revolving-gantry cranes in action (courtesy of American Hoist & Derrick Company).

ment is used for a construction project with more length than width, such as concrete dams and locks. Each wheel is commonly powered by a separate electric motor, but operated together by a control console in the crane's cab. This crane equipment has a long reach to cover the main construction site with a boom 100 to 200 feet long that can be operated almost horizontally. This is possible with a high A-frame or tower structure that reaches nearly as high as the end of the boom.

The lifting capacity of a gantry crane is much greater than that of a mobile crane. This is partly because its topping lift has a multipart line in sheave blocks all balanced with a sturdy boom to handle the heavy loads. Another reason is that the topping lift's angle with the boom is greater than with a mobile crane. Cranes of this sort have a load capacity several times that of a mobile PCSA crane at the same radius. Some have the same capacity but do not lose their load ability as rapidly as a mobile crane does when increasing its reach. This is shown in Fig. 7-13. In summary, the heavier-load-lifting ability of a gantry crane is due to its greater stability and higher supporting A-frame, but these are gained at the sacrifice of less mobility than the mobile crane.

8.3.2 Derricks for Erection Operations

The derricks used for erection operations in construction are built to handle heavy loads from a fixed position. The support frame—consisting of the base, mast, A-frame, and all other than the boom—remains in a stationary position relative to the ground, or other support medium, during its lifting operation. The boom is generally rigged with sheave blocks and fall lines to handle a load just like a movable crane.

The basic difference between a derrick and crane is in the topping-lift connection to support the upper end of the boom. On a derrick the lines go to a mast or A-frame at a point anywhere from mid-height to higher than the highest point of the boom's vertical reach. In this way the topping lift makes a much larger angle with the boom axis than in the case of the crane. This allows a derrick to carry much greater load than a mobile crane can without requiring many parts of topping-lift support lines (see Fig. 8-10). If there are not too many parts of the lift line to move, the angle can be changed much more rapidly. Details of the dimensions, rigging, and operation of derricks can be found in a manufacturer's specifications. The following discussions will give an indication of the main differences in derrick-type equipment.

Stiff-Leg Derricks. The stiff-leg derrick is one equipped with a vertical

FIGURE 8-10 Mobile crane converted to a derrick called
the Sky Horse (courtesy of American Hoist
& Derrick Company).

mast which is shorter than the boom. This support member is tied into
framework of other members which are the angling legs extending from
the peak of the mast away from the boom to form a triangle with the
mast's base at the bottom frame level. A plan view is shown in Fig. 8-12,
where the triangle mentioned is $R_1 R_2 R_3$.

This derrick is a piece of erection equipment to consider where there
is a strong base level to support the entire horizontal under-frame. It
may be used on level ground, a boat deck, or the more-or-less-horizontal
chord members of a bridge. In the latter case the derrick's width (d) would
have to equal the bridge width between trusses, and a cross member would
have to support the reaction at R_1. A stiff-leg derrick could also be used
for higher lifts by having it supported at a higher level on a triangular
tower with vertical legs under R_1, R_2, and R_3. This would be called a

FIGURE 8-11 Twin stiff-leg derricks for bridge erection
(courtesy of American Hoist & Derrick Company).

tower derrick. Generally, the whole supporting frame for the derrick is
moved on rollers by winching under its own power along beams or rails.

In order to avoid taking too much area at the base level and to keep
the supporting legs relatively short, it may be necessary to use counterweights or ballast on the base frame between the back support points of
the stiff-leg derrick. This is somewhat similar to the arrangement for a
mobile crane. It is done to avoid negative reactions at R_2 and R_3, which
would be an indication that the derrick is about to tip over on its boom.
The reactions for the supports of the derrick can be found simply from
Fig. 8-12 as follows.

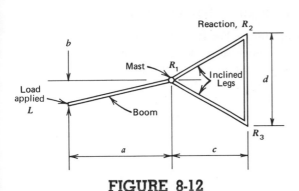

Due to the load, L, reactions are found as:

$$R_1 = + \frac{L(a + c)}{c};$$

$$R_2 = - \frac{La}{2c} - \frac{Lb}{d}; \text{ and}$$

$$R_3 = - \frac{La}{2c} + \frac{Lb}{d}.$$

FIGURE 8-12
Plan of a stiff-leg derrick.

The negative reactions must be counteracted by the derrick weight and counterweights on frame between R_2 and R_3.

These reactions act as vertical load on the bearing surface. They are vertical forces, and the axial force in the legs will be larger. In the case of a tower derrick the vertical legs will have to support these reaction loads as their axial force.

Guy Derricks. A guy derrick is an erection piece of equipment with a mast taller than the boom so that the boom can turn under the guys supporting the mast. These guys are somewhat stiff cables that run from the top of the mast to solid hold-down points on adjacent buildings, bridge points, etc. There will generally be about six guys to give the mast stability in all directions. The base for the mast takes very little space and is supported on a "bull wheel" turntable powered by cable turning the wheel in a horizontal plane.

The guy derrick has the largest angle of all erection equipment between the topping-lift line and the boom. This means that it has the best load capability for the lightest topping-lift lines. Guy derricks also represent the least equipment structure for erection, so they should be the most economical in cost per pound lifted. They are used generally for tall buildings where there is plenty of opportunity for tie-down supports of the guys surrounding the building to be erected. The guy derrick can move up the building as it is erected, using the derrick's own power and parts—i.e., its mast and boom—like a caterpillar moving up a tree.

8.3.3 Climbing and Tower Cranes

Another type of erection equipment takes the advantages of both revolving cranes and derricks into its operation. These are the climbing and tower cranes, which are generally stationary for the lifting of any heavy loads, but are mobile for getting to and handling their lifts. Several notable manufacturers of this sort of equipment have their own designs. Equipment of this sort commonly has a lightweight, vertical, latticed mast of heights that range from 90 feet to 200 feet. It is used for erection and material handling on tall buildings and other structures which can be handled from one setup area with a relatively short horizontal track.

A climbing crane may work within the outside lines of a tall building (Fig. 8-13). It can be set up in the location for a set of elevators or stair wells. These facilities are erected later, after the climbing crane has used their central location to advantage.

FIGURE 8-13 A tower crane in action.
(courtesy of Buck Division, Desa Industries, Inc.)

Design of Climbing and Tower Cranes. A basic design of this type of equipment has a horizontal boom and a counterweight jib opposite to the boom. The smaller units have a boom ranging in length from about 60 to 120 feet, and the jib is 40 to 50% of the boom length. The larger climbing tower cranes have booms from 120 to 175 feet long with jibs of the same relative length as the smaller unit. The boom and jib are supported by lines over the top of an A-frame projecting 20 feet or so above the boom support point. The load is supported from the boom on a trolley that can be run back and forth along the boom's length. Consequently, the load for each ten feet of greater reach is reduced only 10 to 15% as compared to the rapid decrease of load in a mobile crane when the boom is lowered for greater reach. An effective design to minimize the structural material in this crane's boom is to have a counterweight moving on the jib to always keep the rig in nearly perfect balance.

All of the horizontal moving and lifting parts of the tower crane rotate about the mast structure like a swing bridge on a turntable on the main mast. This turntable is just under the control cab, which swings with the boom and counterweight structure. The climbing crane moves up a structure for higher erection work by using a hydraulic jack system. This enables the mast to telescope upward with its several concentric sections moving within itself.

Another noteworthy design of this type of equipment is the so-called tower crane. This crane has a boom which moves in a vertical plane like mobile PCSA cranes. The topping-lift lines go back to an A-frame and then vertically down to a weighted base with the lines running parallel to the mast of the crane. This type of unit has booms that range from 50 to 180 feet long. The load is handled through sheaves at the boom point just like on a mobile crane. However, in this case with the high A-frame the boom angle can be anything from nearly horizontal to more than 60° upward. This gives the tower crane a maximum height of nearly 350 feet from the track, or about half that above rubber-tired supports where they are used. The load variations on this type of erection equipment are similar to those for the PCSA crane discussed in Sec. 7.2.2. The control cab for this unit is near the top of the mast under the boom. The tower crane can move on the track with a minimum of curvature and rotates at the bottom of the mast.

Load Capacities and Operations. The climbing and tower cranes have much less lifting capacity than the bulky revolving cranes described in Sec. 8.3.1. However, they are generally much more versatile. For instance, the maximum loads for the Buck climber range from 4 to 12 tons at a

60- to 70-foot radius. The capacity reduces to one-fourth as much at its maximum radius. All loads are handled with a 2- or 4-part line with the opportunity to convert from one to the other quite simply. The lifting speeds vary with the load level either by a direct-drive or torque-convertor type of transmission. The line speed varies from 160 feet per minute (fpm) for the heaviest load up to 900 fpm at the highest speed. This means that loads can be lifted from the base of the climber crane in 10 seconds to one minute. The trolley of a Buck climber moves horizontally at a maximum speed of 150 mfp. The swinging of the boom and jib for either the climbing or tower crane is at a speed of 1 to 1 1/2 rpm.

For the World Trade Center in New York City, tower cranes, each with 55-ton capacity, were used. The Pecco tower crane can take maximum loads from 6 to 40 tons at a 30-foot radius, reducing to one-sixth to one-tenth as much at the maximum radius. This means that its capacity is in the order of a half or more of the load capacity for a comparable PCSA crane, but it has twice as much reach as that more mobile type of erection unit. The heaviest loads move at less than 100 fpm, so they take several minutes to lift to the maximum height, but light loads can be moved at more than 300 fpm. A modern method to move loads quickly to high or inaccessible places is by use of heavy-duty helicopters as the erecting equipment.

Power and Costs. In any construction-erection operation requiring a single setup or short track, it is probably most economical to use electric power. This will certainly be the case for a limited-area construction site in a metropolitan setting where the electric power lines are available for tapping. The electric power for climbing and tower cranes is required for three functions. The first, of course, is for hoisting power, and a 60- to 125-horsepower (hp) electric motor should be sufficient. The power for this can be obtained from available lines or a DC generator to provide 75 to 150 kva. A review of Sec. 3.1.3 on electric power sources and characteristics may be helpful here. The second requirement for electric power is to operate the trolley or travel motor which would be satisfied by a 5- to 12-hp electric unit. The third power requirement is for a swing motor which might be a 6- to 20-hp unit. The drive components for powering the climbing or tower cranes are located at a level with and in back of the control cab high on the supporting mast. However, they can be operated with a remote-control panel from ground level.

Tower-crane-type equipment is rather specialized and has no set pattern of costs. However, a rough estimate might be made, based on assuming $500 per foot of mast, boom, and jib for the unit equipped with its basic components. This includes the essential rigging and electric power units. Additional mast or boom sections will involve a cost of several hundred dollars per foot of length. The climbing accessories for those units that can climb up a structure will add 10 to 20% to the original cost of the erection equipment. These are extremely rough figures for only "ballpark" estimating. Costs for any erection work must be handled as a unique operation figured on the basis of whatever special equipment is to be used. However, erection specialists can generally give a good estimate without too much detailed planning.

References

1. Riker, Warren N., Chapter 22, *Handbook of Heavy Construction*, Frank W. Stubbs, Jr., Editor-in-Chief, McGraw-Hill Book Company (New York, N.Y., 1959).

2. "Catalog S-55," Raymond Concrete Pile Company (New York, N.Y.).

3. "Single Acting Steam Pile Hammers," Warrington-Vulcan, bulletin 68A, Vulcan Iron Works (Chicago, Illinois).

4. "MKT 'McKiernan-Terry' Pile Driving Equipment," MKT Division, Koehring (Dover, New Jersey).

5. Housel, William S., "Michigan Study of Pile Driving Hammers," *Journal of the Soil Mechanics and Foundations Division*, ASCE, Vol. 91 No. SM5 (New York, N.Y., September 1965).

6. " 'Sonics' Drive a Pile 71 Ft, While Steam Drives Another 3 In.," *Engineering News-Record*, McGraw-Hill Publications Company (New York, N.Y., November 9, 1961).

7. "Hydraulic Pile Hammer Works," *Engineering News-Record*, McGraw-Hill Publications Company (New York, N.Y., June 7, 1962).

Chapter 9

■■■■■■■■■■■■■■

■■■

MATERIAL-HANDLING
EQUIPMENT
■■■■■■■■■■■■■■■■■■■■■■■■■■■■■■■■

Introduction to the Equipment

In the first chapter it was noted that most construction operations deal
with and involve equipment handling material. We have previously discussed compressors handling air and pumps for water. Equipment for
excavating, trenching, tunneling, etc., deal with natural earth material,
and pile drivers deal with finished foundation piles. The equipment to be
discussed in this chapter has a somewhat different purpose in that it
carries construction material. The material-handling equipment loads
and carries the construction object it is handling, moving it to a new location, where it is deposited in essentially the same condition as before the
move.

The variety of equipment to be considered moves material both vertically and horizontally. It includes equipment to handle solid objects
such as blocks, palletized material, pipes, and structural members without
trying to connect them as the erection equipment does. This material-
handling equipment includes fork lifts, lift booms, and mobile gantries.
Another group of the material-handling equipment deals with loose, bulk

material. These are conveyors, dump trucks, and wagons. A third category of equipment that would satisfy the above definition of material-handling in its broadest sense is the equipment used for moving other equipment. This includes the flat-bed trucks and trailers.

The basic considerations for material-handling equipment are:

1. the motion required—from where to where;
2. the material, loads, and quantity to move;
3. the time available, because this equipment generally should not govern the construction operation's timing; and
4. the space needed for the material to be moved and available space for the equipment to handle it.

The important points in applying these basic considerations will be evident as the variety of this type of equipment is discussed in the chapter.

9.1 Fork Lifts and Related Equipment

The equipment to move solid objects of enough size, or bundled, packaged, or palletized material, does not have to have its own container. Fork lifts are the prime example; one is shown in Fig. 9-1. The fork-lift

FIGURE 9-1 A fork lift in action (courtesy of Pettibone Corporation).

type of equipment is most known for its industrial use but is equally applicable in some forms in a construction yard, where it generally must be mounted on pneumatic tires for load distribution and maneuverability.

9.1.1 Design Features of a Fork Lift

The main identifying feature of a fork lift is its two horizontal arms that support the load and raise or lower it along a vertical mast. The mast is held at the front end of a two-axle tractor unit which serves as the counterweight to the load projecting out front of the equipment. The tractor is somewhat unique in that on a majority of models, its front wheels closer to the fork are often larger than the rear wheels, which are primarily for steering. This tractor unit has also been used by one tractor manufacturer interchangeably with a front-end-loader type of equipment.

The fork-lift tractor may have two- or four-wheel drive with a turning radius within two times its wheel base for great manueverability. That wheel base may be as short as 6 feet or up to almost 20 feet. The relative lifting capacity of the fork lifts is increased by increasing the wheelbase for greater counterweight, as shown in Fig. 9-2. A fork lift will generally have 4 to 6 forward and one or more reverse gears for travel at speeds of 1 1/2 to more than 20 miles per hour. A desirable feature for a fork lift is to have a torque-converter power transmission so as to provide smooth travel and speed changes. This will avoid shaking a load off the equipment due to sudden starts and stops.

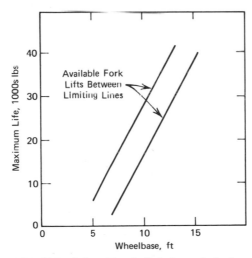

FIGURE 9-2 Fork lift loadability.

A more complete understanding for the use of this type of equipment will involve other basic design features. The fork arms are cantilever beams 3 to 8 feet long, supported on a cross frame that rides up and down the mast. The arm spacing is changeable to a limit of 3 to 5 feet apart, but its maximum load capacity is found at the 2- to 3-foot (24″ or 36″) spacing. The mast is primarily a structural frame with two vertical members to form the track for the fork to ride up and down on. Some masts can be extended upward by telescoping sections of progressively smaller pieces of mast that are hidden in the basic height of mast. This allows the mast to extend from an 8- or 9-foot height to around 30 feet maximum from the surface supporting the fork lift.

All fork-lift masts can be tilted 10° to 25° forward. The smaller angle may be the limit when the mast is fully extended. It can be tipped backward only 8° to 16° because of the operator's cab behind the mast. Lifting speeds for the fork range from 40 to 100 fpm up and down with the slower speeds for a loaded fork. These speeds result in lifting or lowering the loads the full height in 5 to 30 seconds. However, this timing will generally not affect the work cycle significantly.

The load-lifting capacity of the fork lift depends on the spacing of the fork arms, the height and tilt of the mast, and the surface on which the equipment is supported. Obviously, a fork lift that is fully loaded and transmitting, say, 30 psi on the tire bearing area cannot carry its load safely on a soil that has only perhaps 10 psi bearing capacity. That fork lift should operate fully loaded only on a paved or compacted surface. The lifting capacity is generally shown with the fork arms at 24″ or 36″ apart and ranges from 2,000 to 70,000 lbs. This compares with one-half to equal the weight of the unloaded fork lift. Its lifting capacity will vary with the height of its telescoping mast. The higher the mast, the less load it can carry with safety. This is shown in Fig. 9-3.

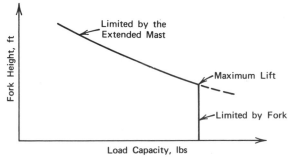

FIGURE 9-3 Variation of fork lift capacity.

9.1.2   Design of Other Mobile Lift Equipment

A type of equipment similar to the fork lift is known as the high-lift loader. It has a fork arrangement for holding a load just like the fork lift, but it has much greater lifting range from one position of the base equipment. This range of distance is both vertical and horizontal. It is made possible by an elevating boom like a crane's boom to support the mast from its top point.

As seen in Fig. 9-4, the elevator has a fork-lift-type mast on top for a strictly vertical lift of from 10 to 15 feet. The boom that pivots in a vertical plane might cause trouble in keeping a load on the lift arms of

FIGURE 9-4   A high-lift loader in action (courtesy of Lull Engineering Company, Inc.).

the fork. However, Lull's design has an automatic self-leveling device for holding the fork level, at any height, even with the equipment carriage on a sloping ground. This automatic leveling is by a parallelogram type of design between the mast for the loading fork and the ram supporting the mast. The fork and its mast can be tilted 13° forward or backward. The high-lift loader will carry loads up to 5000 lbs, and at its maximum height of 40′ it can carry a 3000-lb load. Furthermore, it has a 4- to 5-foot horizontal reach in the mechanism of the mast and fork load-handling members. With these features it is obvious that the high-lift loader combines the material-handling capabilities of a modest-size fork lift and the reach of a small crane type of equipment.

Another type of self-contained unit for material handling could be called a mobile gantry. This has a lifting frame high enough above its wheels to pick up loads between its support legs. The load pickup is generally by a block and tackle or a chain-hoist type of mechanism, which is one part that governs the equipment's lifting capacity. The other key part governing the load capacity is the wheel units at the bottom of the four legs. The most common type of this equipment used for construction is known as a lumber carrier. A carrier can straddle a stack of lumber, pick up the stack with its lift lines wrapped around it, and carry the load to a new location. This type of equipment is also used for precast concrete structural members.

Other mobile gantry forms of equipment are custom-built to the given job dimensions. If there is a lot of handling for a particular size and weight of loads, a custom-built gantry can be the most economical material handler. An example is a pipe jumbo or gantry for perhaps 10- to 40-foot-long sections of maybe 15-ft-diameter pipe. Equipment for these large diameters can be designed to run through the pipe and then pick it up by jacking their lift legs at either end of the pipe. Therefore, the pipe is carried surrounding the central part of the gantry frame while the extendable legs at either end have pneumatic wheels to move the material.

Another type of special material-handling equipment that is self-contained in its operation is called a sideloader. This is a truck-mounted unit with one side of its bed opened to pick up material with a crane arrangement, as shown in Fig. 9-5.

9.1.3 Costs of Mobile Lift Equipment

The fork lift used for a short vertical lift of less than 25 feet fills the gap under the smallest crane for material handling. At the point of load and

FIGURE 9-5 A sideloader in action (courtesy of Three Crown Corporation).

lift where the smallest mobile crane of 12 1/2-ton capacity could match the fork lift, the cost of the two would be about the same. For smaller loads to be lifted within the vertical range of the fork lift, it can be used at less cost than a crane-type unit. This is shown in the sketch of lift capacity compared to cost (Fig. 9-6). The same Figure compares the cost

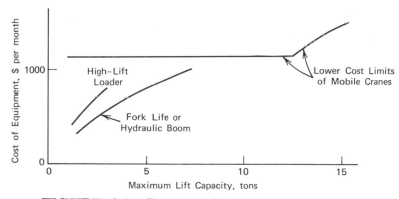

FIGURE 9-6 Relative lift capacity to cost.

of a high-lift loader with those of a fork lift and a small crane. As can be seen there, the high-lift loader is more economical than a crane for loads of generally less than 1–1 1/2 tons. The fork lift is the cheapest form of self-contained lifting equipment for lifts less than 20 feet high.

The mobile gantry form of material handler is designed for special, awkward loading that does not have to be lifted more than 10 feet up and where the loading is repeated many times. This is generally a more expensive piece of equipment than the fork lift for the same load capacity.

9.2   Portable Material Booms

Another type of material-handling equipment is the portable boom, which generally attaches to or works on the flat bed of a truck, for handling solid material. This piece operates like a crane boom, taking about 24 inches of the flat-bed length for its supporting mechanism.

The material booms are of two general varieties. One type is a multi-jointed boom primarily for horizontal movement, with a modest vertical change for the materials handled. This is called a loader boom and operates very much like the boom arrangement of a backhoe-type crane unit discussed in Chap. 6. The other variety is a high-lift boom for maximum vertical lift obtained with a telescoping boom. The higher and smaller sections of the boom extend out from within the basic boom length. This loading boom has comparatively little horizontal reach flexibility but does swing on a turntable support mechanism at its base. It should be recognized that these loader and high-lift booms are not designed for erection purposes but for material handling where the individual loads are not very heavy.

9.2.1   Design Features of Material Booms

A loader boom is made of several segments attached by rotating joints like the human arm, hand, and wrist, as shown in Fig. 9-7. The loader boom's action is generally gained from hydraulic cylinders which give it a positive control. The outer working end may be outfitted with a lifting fork of 3- to 4-foot-long tines, which can have a twisting "wrist" action in the horizontal plane. This rotation is as much as 200° for a hydraulically powered unit and a full 360° with electric power.

The lifting forks and arm action are supported on a mast with a base which rotates through a full horizontal circle. The multidirectional action makes this boom good for handling small, packaged material. The

FIGURE 9-7 Loader boom's working setup (courtesy of HIAB by Stanco Mfg. & Sales, Inc.).

footing of the loader boom may be attached to the truck bed, or it may have the ability to travel along a flat bed or platform to reach materials and deposit them with greater horizontal range. The loader itself weighs somewhat more than one ton. Its load capacities range from a maximum of about 7500 lbs with a 5-foot boom to one-third as much with a 16- to 18-foot-long boom. The largest units made have somewhat greater range in capacity, approaching a 10-ton limit.

The high-lift boom is generally a telescoping piece of equipment attached to a truck bed or other wheel-vehicle frame. At the end of the telescoping boom is a loading gooseneck or jib section. The telescoping sections of the boom and the rotating gooseneck are hydraulically controlled for more precise load handling. This type of equipment is designed to fill the load capacity gap below the smallest mobile PCSA crane. The high-lift boom unit is built with jacking feet to spread the load from the truck or vehicle it is mounted on to give greater lateral stability perpendicular to the supporting vehicle's length.

The lift boom gets its extreme height by extending, generally, three sections of 20-foot boom plus the jib for a total of 80 feet of height above the surface supporting the vehicle. The 20-foot basic length of the boom section allows it to be folded onto a truck bed for traveling between uses. The boom can assume various angles in the vertical plane from a high of 80° above the horizontal for maximum height and lift down to 15 or 20° below the horizontal for a reach below the support level.

The load capacity of a high-lift boom ranges up to 23,000 lbs but reduces with increasing height and radius of the boom just like any crane-type equipment. The high-lift boom, with its load capacity and working radius up to 24 feet or so, is useful to handle sections of reasonably large-diameter pipe, steel highway forms, castings for sewer work, and small building trusses. Each of these types of material could be loaded on the truck bed with the boom. Then the boom is retracted and loaded on the truck for travel to the place where it would be unloaded for the construction.

Several material-handling booms are useful for a variety of construction operations. One design is a telescoping boom with only two sections, which can be mounted on top of a tractor's mid-section. It can handle material such as sections of pipe on the job site. This sort of unit telescopes to around 15 feet of height for a total above the tractor base of more than 20 feet. It could handle a maximum load of somewhat over a ton, but only perhaps a half-ton at its maximum height.

The equipment that has been discussed in the first part of this chapter is for handling solid materials without the need for a container. Material-handling equipment for loose, bulk materials will be discussed next.

9.3 Equipment to Handle Loose Material

Loose or flowing material that needs to be moved from one location to another generally is carried on a conveyor or in hauling equipment, such as dump trucks or wagons (generally called haulers). They provide, if well-planned, a regular succession of containers to move the material. It may be uniform material, such as sand or gravel; or it may be inconsistent, such as earth with stones or rooty material mixed in it. Haulers can move the material over a roundabout or zigzag route, though a straight line between the loading and deposit points would be ideal, if possible.

The major problem in moving material in haulers is that of interruptions at the loading and unloading points. Equipment that handles the loading or unloading part of such an operation should not have to wait long, if at all, for a hauler. A related concern is that a hauler should not have to wait long to be loaded or unloaded. This problem with haulers, and the ways to solve it, will be discussed later in this chapter.

Conveyors provide the means for continuous movement of loose, flowing material on one or more flights of the conveyor system. A flight is one continuous line of conveyor belt, chain of buckets, helical screw, or enclosed pipe conveyor. To make a continuous flow of material possible, each flight of the conveyor system must continue to move, even though material may not be fed to it continuously. A noncontinuous feed will be wasteful of conveyor power, but so would stopping and starting the motion frequently. Aside from the power waste, there is no serious problem with irregular loading of the conveyor system, and it has that advantage over the use of hauling units. On the other hand, a conveyor should be fed fairly uniform material and move it in a straight line on each flight. More specific points in the planning for conveyors are covered in the next sections.

9.4 Conveyors

The youngster who has played on top of a sand pile, or anyone climbing a steep hillside of loose stones and rocks, knows how easy it is to move loose material downhill. The sand or rock slide, though dangerous, represents the easiest and most economical means for moving material. It moves under the influential force of gravity and it may or may not seem controlled. Some construction operations can take advantage of gravity to move material down an inclined surface in a controlled fashion. To be under control, the material must not be "snow-balling" down the incline. It should be moving as a steady mass, either as one body on a sort of lubricated surface or as a stream of uniformly displaced loose material in the total mass, like sand in the pile.

The concept of a body of loose material moving under the influence of gravity as a mass has been used to advantage by contractors in recent history. One ingenious contractor had the operation to move the material from the top of a hill to its base. He set up a conveyor belt parallel to the slope of the hill and loaded the excavated material on it at the top end. The loose material tended to fall downhill but had enough interlocking within itself to move as a mass and enough friction on the belt to drag the belt downward with it. This equipment setup not only moved the material down to the base of the hill, but also enabled the belt to generate electrical power through its roller supports for other uses.

Conveyors are commonly used to move material horizontally, on the incline, or vertically upward with the use of a power unit to drive the moving mechanism. These could be belt conveyors, bucket conveyors, or screw conveyors, all of which will be described in the next paragraphs.

9.4.1 Bucket Conveyors

A bucket conveyor consists of a series of buckets of uniform size and spacing supported on a chain or chain-like linkage or on heavy-duty conveyor belting. This linkage is looped around sprocketed wheels at the ends of the conveyor equipment. At least one of these end wheels serves as the driving force to move the conveyor around the looped circuit.

The capacity for a bucket conveyor to move material is dependent directly on the volume of each bucket, the material being moved, and the speed of the chain linkage along its circuit. The peak productivity, q_{pk}, for a bucket conveyor is found as

$$q_{pk} = \frac{60 \, V_k \, \delta \, s}{2000 \, d_b}, \text{tons/hr}, \tag{9-1}$$

where V_k = a bucket load, in cubic feet;
   δ = density of moved material, in pounds per cubic foot;
   s = speed of bucket travel, in fpm; and
   d_b = spacing between buckets, in feet.

If the material being moved is essentially liquid and seeks a level as water does, then the amount of material carried in a watertight bucket is equal to its struck capacity found by outside physical dimensions. If the material is granular solid particles, the bucket may carry a heaped load of the loose material. Then the volume in a loaded bucket will have outside dimensions which figure more than the struck capacity. However, the swelling of the material in loading leaves voids in the bucket's load. Unless the actual heaped capacity of loaded solids can be determined, it would be safer to calculate the productivity based on the struck capacity of the buckets.

The bucket conveyor is the only type described in this chapter that can move material vertically, or even on a steep incline, upward. It also can move liquid materials. This assumes that the buckets do not leak. The buckets can be supported on gravity pivots so that the open end is always facing directly upward except when the bucket is tripped to unload.

This type of conveyor may be found as a material transfer component in an aggregate- or concrete-producing plant. In that case it probably serves ideally as it was designed and no economical comparison with some other arrangement is practical. A bucket conveyor has low installation and operating costs. Its best advantage probably is the minimum of ground space required. However, it has a low capacity for moving aggregate material compared to the belt conveyor to be discussed shortly.

The ordinary bucket elevator is designed for granular material, not liquids, and has its buckets fixed in a set position with the continuous chain or belt. This allows it to discharge loaded material by centrifugal action. Its speed is carefully selected so that the material is thrown from the bucket as it passes over the head sprocket. The designed speed of the bucket conveyor should be used. Otherwise, its productivity may be reduced by waste of material improperly discharged. Using a higher speed will also lead to excessive maintenance problems with the bearings, etc.

9.4.2 Other Conveyors for Fine or Fluid Material

A screw conveyor also moves material of a granular or fluid consistency with a moving physical part that serves as an impeller. The impeller is a helical screw shaft which turns concentrically on the axis of a hollow pipe tube or open trough. The material should be fine-grained or liquid, to move along the tube by the helically shaped vanes on the screw shaft. The screw conveyor operates most effectively moving material horizontally or on a downward incline. It can move material up an incline of 15° up to maybe 45° with the horizontal. This will require extra power or a reduced flow of material. The screw conveyor transmits its power through torque in the screw's shaft. If the material load is reduced, the conveyor can be operated at higher speed with the same horsepower applied. The screw conveyor is used to move material such as finely powdered cement in a cement mill or concrete-mixing plant. In use with material as sensitive to moisture as cement, it is advisable to be sure that the closed tube is airtight to the surrounding atmosphere. An open-trough screw conveyor is commonly used in a sand-washing plant. Generally, screw conveyors have only two major troubles. They may jam with lumpy material. And there is an appreciable amount of noise when the screw operates without a self-lubricated bearing.

Another type of conveyor for moving cement or finely ground dust particles is the pneumatic air pump pipe or hose fed by a cement hog or from an aeration tank. The tank gets its material fed by gravity. The cement hog is like a vacuum cleaner that chews into loosely packed cement in a barge's storage bin or other such container. It has a short screw conveyor that serves to charge its air pump. Either of these pieces of equipment has an air pump which activates the cement particles into suspension when it discharges the dust-like material in dry air under pressure through a discharge hose or pipe to the place for its storage or deposit. It is essential to have a screened air vent in the discharge tank to release the air pressure.

These pieces of equipment for handling material are highly specialized and do not belong with so-called basic equipment, except that they are akin to the screw conveyor. They will be used where the material and conditions are ideal for their use. Generally, the pneumatic type of conveyor is more expensive in both installation and operating costs. The contractor or engineer who has a construction operation involving fine, dust-like material to be moved several hundred to a few thousand feet should consult the manufacturers of the special air-pump-type conveyor equipment. There may be several equipment setups that could do the job and, so, competitive bids might be obtained to insure reasonably economical operation.

9.4.3 Belt Conveyors

The most commonly used conveyors to move construction materials are belt conveyors. They are operated in a horizontal or somewhat inclined direction. They can handle granular, bulky materials and solid chunks that are small enough to fit and hold on the belt. The maximum inclination for handling these loose materials on the belt depends on the angle of repose (α) for the particular material. As long as the angle of inclination, i, (in Fig. 9-8) is less than α, the material can be carried by the belt.

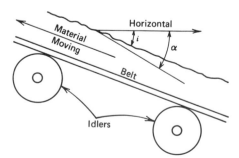

FIGURE 9-8 Section at conveyor belt trough.

With most materials of construction it is best to think of the maximum conveying angle (i) being half as great as the material's angle of repose (α). That is on a regular, plain belt. If the conveyor has cleats bolted to the center third of the belt width and spaced 6″ to 10″ apart, the maximum conveying angle can be increased 50%. Some represented values are shown in Table 9-1.

TABLE 9-1 Material Properties Compared For Conveying

Material	Avg. Weight, lbs/cu ft, δ	Maximum Conveyor Incline Angle, i		Angle of Repose, α
		Regular	Cleated	
Cement, Portland	90–100	23°	—	40°
Concrete, 4″ slump	110–150	20°	—	—
Concrete, 6″ slump	110–130	12°	—	—
Earth, loam, dry	70–80	20°	30°	30°–45°
Earth, moist	90–110	23°	—	45°
Gravel, blended	90–100	18°	27°	30°–40°
Sand, bank, dry	90–110	17°	—	30°–35°
Sand, bank, damp	110–120	22°	—	35°–45°
Sand, saturated	110–130	15°	—	15°–30°
Slag, crushed	80–90	10°	—	25°
Stone, crushed, unsized	85–90	20°	30°	—
Stone, 4″+, sized	90–100	15°	—	—

Note that rapid acceleration of a loaded belt moving upward will, in effect, temporarily reduce the angle of repose for the material. This may cause angle i to be greater than the temporary angle α, and slippage of material will occur until the belt arrests the differential movement of the material. It is best to avoid a sudden start of a loaded, inclined belt conveyor.

The essential components of a belt conveyor system are the continuous belting, the idlers, a driving unit, pulleys and take-up units to maintain tension in the belt, and the supporting structure. The belting is continuous around a circuit of minimum distance from one end to the other. It has enough flexibility to go around 12″ to 36″ pulley wheels at the ends to turn the belt without producing excessive wear by flexing. The belt is supported by idlers. These are anti-friction bearing rollers long enough individually, or in groups of three for the troughed belt, to support the full belt width. It would seem desirable to have a maximum spacing of the idlers. Then there would be fewer idlers required for a given length of belt and perhaps less friction at the idler bearings to resist movement. On the other hand, with maximum spacing of idlers the belt will sag more between idlers. To counteract this will require more tension in the belt with the resultant greater bearing pressure on the end, or turnaround, rollers or drive wheels.

With years of experience the manufacturers of belt conveyors have optimized the size and materials for the belting, the design and spacing of idlers, and other features of belt conveyors. They can be driven by electric, gasoline or diesel engines. The power drive is generally mounted at the head end, as shown in Fig. 9-9, to keep it freer from dust and dirt and put it in the most effective position.

Portable belt conveyors in lengths from 20 feet up to 80 feet long can be operated with inclination up to 25° from the horizontal. They generally have a lightweight truss frame and legs supported on wheels for portability between jobs. The wheels, swivelled to be perpendicular to the belt direction, help to pivot the conveyor about its tail or foot end for radial stacking.

Another type of semi-portable belt conveyor is the tower or mast-type

FIGURE 9-9 Radial stacker conveyor (courtesy of Barber-Greene Company).

radial stacker. This conveyor is very similar to the fully portable belt conveyor, though with lengths up to 150'. The chief difference is in the central support system, which is sturdier and capable of rapid adjustment in the tower stacker. The adjustments are made possible by a separate engine for the mast hoist. Also, the horizontal travel, pivoting around the tail end, is often done on an installed track. More permanent conveyor setups for commercial material plants, etc., make use of parallel-track stackers, shuttle belts, and trippers. A manufacturer of such equipment, such as Barber-Greene, is a good source for literature about them.

Capacity and Productivity of a Belt Conveyor. A belt conveyor is a high-capacity conveyor when properly operated, simply because the carrying belt is continuous and can be loaded for its full moving length between end supports. The load capacity of the belt conveyor depends on its cross-sectional area and the material it is carrying. If the material is small-grained material, such as sand or gravel, the belt's capacity would be found from the dimensions shown in Fig. 9-10.

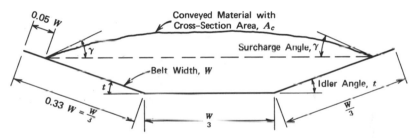

FIGURE 9-10 Cross section of loaded belt.

The belt is generally loaded from the feeder to 90% of its full width. The cross-sectional area is found by assuming that the material forms an arc with a tangential angle at each end of the belt, called the surcharge angle (γ). This angle is normally 15° less than the material's angle of repose (α). The angle (t) for the troughing idlers might vary between 15° and 30°, or approximately equal to the maximum inclination for the belt. However, with more flexible belting to carry material that is not too abrasive, the troughing idler angle may be set at 35°, 40°, or 45° maximum.

The appropriate troughing idler angle (t) should be based on the physical characteristics of the material, particularly its surcharge angle. A reasonable selection of the troughing shape can be determined by following the rules given below:

20±° troughing idlers used with
  heavy material (δ = 100 lbs/cu ft or heavier);
  surcharge angle (γ) of 20° or steeper.
35° or 40° troughing idlers used with
  heavy material with γ = 15° or less;
  light material with γ = 20° or steeper.
maximum 45° troughing idlers used with
  light material with γ = 15° or less.

For commonly assumed values of the surcharge angle for bulk materials and several standard belt widths, the cross-sectional area of the material carried is shown in Table 9-2.

TABLE 9-2 Cross-Sectional Area of Belt Load (based on angle of troughing idlers, t = 35°) A_c in sq ft.

Surcharge Angle, γ	Cross-Sectional Area, A_c	For 18″ Belt	For 24″ Belt	For 30″ Belt
10°	$.075W^2$	0.169	0.300	0.469
15°	$.087W^2$	0.196	0.348	0.545
20°	$.099W^2$	0.223	0.396	0.619
25°	$.111W^2$	0.250	0.444	0.694
30°	$.123W^2$	0.277	0.493	0.769

The areas and, consequently, capacities are decreased by 15 to 30% or increased by 5 to 13% for 20° or 45° troughing idler angles, respectively. A wider belt leads to a smaller capacity increase with the steeper idler angle.

The productivity for a belt conveyor depends on the cross-sectional area of material moving on the belt, the weight of the loaded material per cubic measure, and the belt's speed of travel with its load. The maximum output for a belt conveyor is found as

$$q_p = \frac{60 \, A_c \, \delta \, s}{2000}, \text{ tons/hr}, \tag{9-2}$$

where A_c = cross-sectional area in sq ft (refer to Table 9-2);
  δ = density of material, lbs/cu ft; and
  s = belt speed in feet/minute (fpm).

Generally, the belt speeds will vary between 100 and 800 fpm. For an example, consider a 30″ belt moving at 200 fpm and carrying wet sand with an angle, $\gamma = 30°$, and density, $\delta = 110$ lbs/cu ft:

$$q_p = \frac{60\,(0.769)\,110\,(200)}{2000} = 520 \text{ tons/hr.}$$

In order for a conveyor to move this much material, it must have a power unit of sufficient size to overcome the resistance to this movement. The power requirements for operating a conveyor will be discussed next.

Power Required for a Belt Conveyor. The total power required to operate a belt conveyor is the sum of the requirements for:

1. moving the empty belt over all idlers;
2. moving the conveyor's load horizontally;
3. lifting or lowering the conveyor's load; and
4. turning the end pulleys and other moving parts under the load.

All of these requirements—except, possibly lowering the load—present resistances for a power unit to overcome. If the load is moving downward in a conveyor system, it will be giving the power unit an assist due to gravity. This is like the case of the contractor moving the material off the mountain top.

To analyze the power requirements one at a time, let us consider the empty belt shown in Fig. 9-11A. The power delivered to move the belt must produce the differential tensile force, ΔT_e. This force is exerted here to overcome the frictional resistance on the idler bearings. Thus,

$$\Delta T_e = b_f W_e, \tag{9-3a}$$

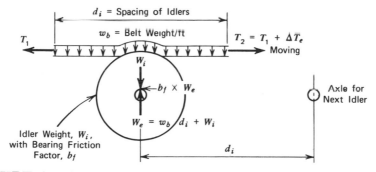

FIGURE 9-11A Factors governing the power requirements of a belt conveyor.

where W_e = weight of empty belt and moving parts (including idler) distributed on an idler bearing, and

   b_f = friction factor on idler bearing,

which varies between 0.015 and 0.035. The high value is for horizontal belting of a 500-foot, or shorter, conveyor. Longer conveyors in motion have an inertial advantage of less friction on the idler bearings. The b_f value of .015 is for the belt returning with slack side tension on the underside of the conveyor. For a conveyor on some incline the b_f value for the belt in tension will reduce from 0.035 to about 0.025 for small, short belts on a 33% slope. This frictional value will drop to 0.016 for wide, long conveyors at that steep angle (18°).

The view just given for the pull and, so, the power required to move a piece of empty belt is easy enough to visualize. However, it leads to some difficulties. An approach worked out by the Barber-Greene Company (the source for the above values) will be followed to find the power requirements for a belt conveyor. Reference should be made to Fig. 9-11A and following ones along with the symbols introduced with them.

To move the empty belt requires knowing the belt's weight. Then the required pull or tension for movement of the belt will be found from

$$T_b = (b_{ft} + b_{fr})\, w_b L, \qquad\qquad (9\text{-}36)$$

where b_{ft} = idler bearing friction where belt is in full tension (.035 to .016);

   b_{fr} = idler bearing friction where belt is on return idlers (.015);

   w_b = weight of the belt in lbs per foot of its length; and

   L = length of belt between head and tail pulleys, feet.

The belt's weight will vary from the smallest for conveyors of about 3 lbs/lin ft for an 18″-wide, light-duty (for 30–75 lbs/cu ft material) belt to the large 21 lbs/lin ft for a 60″-wide, heavy-duty (for 130–200 lbs/cu ft material) belting. This variation can be shown by the curves of Fig. 9-11B.

The other major consumer of power to move the main parts of the conveyor is the idlers. Rotating the idlers under a belt takes differing force depending on the idler's weight, its shaft diameter, and the weight on the idler. A basic equation similar to the previous one for an empty belt could be written. It would turn out to be

$$T_i = (b_f W_i)\frac{L}{d_i} = k_x L, \qquad\qquad (9\text{-}3c)$$

where k_x = an idler friction factor, lbs/lin ft, and other symbols are as defined with the previous equation and in Fig. 9-11A. This requirement

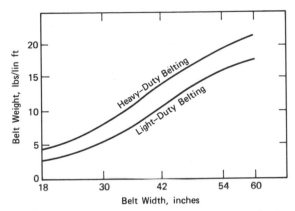

FIGURE 9-11B Weight of conveyor belting.

for belt tension and, so, power can be found using curves such as those in Fig. 9-11C.

Now the power required to move the main parts of the conveyor can be found. It will be

$$P_{eb} = \frac{[k_x + (b_{ft} + .015) \, w_b] \, Ls}{33,000},$$ (9-4a)

where P_{eb} = horsepower (hp) required for an empty belt;
s = belt speed in feet/minute (fpm);

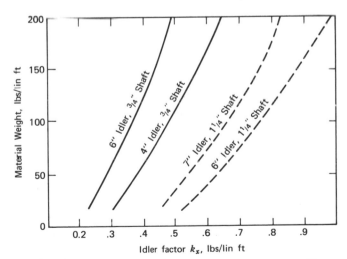

FIGURE 9-11C Belt tension to move idler.

and other terms have been defined above. It can be noted that the value of k_x may be twice as large as $(b_{ft} + .015)w_b$ for light belting, but as the belt size increases with increasing idler sizes, the second term becomes equal to or greater than k_x. Combining previous equations, this power expression can be simplified to

$$P_{eb} = \frac{(T_i + T_b)s}{33,000} = \frac{T_e s}{33,000},\qquad(9\text{-}4\text{b})$$

where T_e = the tension or pull in the belt required to move main conveyor parts, and other terms are as previously defined. The relative variation suggested in the above paragraph can be said for T_i (related to K_x) and T_b.

The power required to move a load horizontally on the belt is also applied by a ΔT, as shown in Fig. 9-11A, to overcome the frictional resistance to motion at the idler bearing. It can be found by using

$$P_{hm} = \frac{b_f w_m L s}{33,000},\qquad(9\text{-}5\text{a})$$

where b_f = idler bearing friction (see Fig. 9-11A);

w_m = weight of load per linear foot = $\dfrac{(33.3)q}{s}$, or found

using the cross-sectional area, A_c, from Table 9-2;
q = rate of conveying material, tons/hr.
L = length of loaded conveyor, feet; and
s = belt's moving speed, fpm.

Also, power to move a conveyor's load horizontally can be found for a given ton/hr productivity. This is found by substituting the above expression for w_m in Eq. (9-5a), which results in the equation

$$P_{hm} = \frac{b_f(33.3)qL}{33,000} = \frac{b_f qL}{990},\qquad(9\text{-}5\text{b})$$

Note that this expression does not have the belt speed (s) as a separate term. However, the speed for moving the material is in the q term.

Maximum belt speeds are generally recommended by the manufacturers of this equipment. For the small, 18″-wide belt this maximum is 250 to 400 fpm, depending on the material being moved. Loose stone without much fine material in it will tend to roll back on the belt if it is moving too fast. For the large, 60″-wide conveyor belts the maximum recommended speeds are 500 to 900 fpm. Maximum speeds for the belt widths between these extremes progress uniformly upward for increasing widths.

Another part of the moving conveyor equipment that consumes power includes several items called "accessories" to the system. These accessories require additional belt tension for:

1. pulley friction from non-driving pulleys, 30 to 50 lbs/pulley;
2. belt-propelled trippers, 70 lbs for 18″ belts to 180 lbs for 60″-wide belts;
3. belt plows, pull = 5 × (belt width, inches); and
4. skirt board friction, which depends on the material being conveyed.

This last accessory depends on the material, so the power requirement for that and some other losses will be added to the power for moving the material horizontally. That is, the power for accessories, P_{am}, is added to P_{hm} to give

$$P_h = P_{hm} + P_{am}. \tag{9-6a}$$

When the accessories for a given conveyor system have been determined, the P_a value is essentially fixed. It will not vary with the length of conveyor, belt speed, or material-moving rate (q). Therefore, P_h varies primarily with power required to move the material horizontally (P_{hm}). With set amounts of tension and, so, power to add for accessories, it should be observed that P_{am} will be more significant for short conveyors. In fact, for 25-foot conveyors the value of P_{am} may be about four times P_{hm}. With a 100-ft-long conveyor, $P_{am} \approx P_{hm}$. With the greater length of a 250-ft conveyor, P_{am} will add only 25% to the power needed to move the material horizontally. For a 400-ft conveyor or longer, the added power requirement for the accessories is negligible.

These comparisons suggest that the P_h power determination can be made by solving for P_{hm} from one of the (9-5) equations and applying a factor to account for the accessories and other power consumers. Thus,

$$P_h = f_m P_{hm}, \tag{9-6b}$$

where f_m = factor to cover the power for accessories dependent on the material being conveyed, taken from the curve in Fig. 9-11D. A similar adjustment is needed for the accessories and other power consumers to run an empty belt. This will lead to a modification of Eq. (9-4):

$$P_e = f_e P_{eb}, \tag{9-4c}$$

where f_e is also found in Fig. 9-11D.

The power required to move the conveyor's load vertically could be analyzed from a moving-force viewpoint. To see this for a belt conveyor,

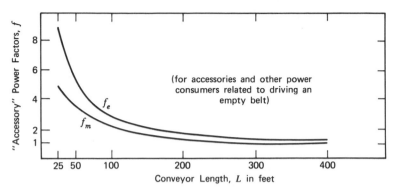

FIGURE 9-11D Adjustment factors for accessories and other power consumers.

we refer to a sketch such as Fig. 9-11A except on an incline. In this case the differential driving force, ΔT, applied by the power unit must overcome both the frictional resistance force (F_f) and the component of weight (W_d) opposite to the direction of motion. It is easier to determine the two components of power required by considering the kinetic energy (K.E.) of motion and the potential energy (P.E.) of position. The K.E. is based on the load moving at speed, s, and can be found by either Eq. (9-5a) or (9-5b) as for horizontal motion. The P.E. is based on changing the position of the load from one elevation to another a vertical difference, H. Determining the power in this way will make the expression readily usable for any form of conveyor—belt, bucket, or other conveyor.

The power required to provide P.E. work necessitates dealing with the total load or weight of material moved to a new elevation. Since power is work per unit of time—1 hp equals 33,000 foot-pounds per minute—it is

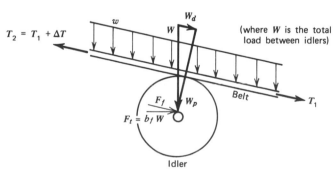

FIGURE 9-12 Inclined belt conveyor forces.

convenient to designate the load as so much moved per unit of time. Conveyors are generally described in terms of tons per hour. Therefore, the power for moving a load vertically will be expressed in such a quantity:

$$P_v = \frac{2000qH}{60 \times 33,000} = \frac{qH}{990},$$ (9-7)

where q = load, in tons per hour conveyed, and
H = total vertical distance moved, in feet.

Note that if the material is moving downward, this is not a power requirement but an advantage. In calculating the required power to drive a conveyor, the upward direction is positive (+) and downward is negative (−). The total requirement to be delivered by the power unit moving material upward is

$$P = P_e + P_h + P_v.$$ (9-8a)

The conveyor manufacturers have adopted a single equation to find the power requirement for a belt conveyor. In terms of the symbols and units defined in the preceding paragraphs, it can be written,

$$P = \frac{L[k_t(k_x + b_{ft}w_b + .015w_b)]\,s}{33,000} +$$

$$\frac{(Lb_{ft}w_m \pm Hw_m + \text{Accessories})s}{33,000}.$$ (9-8b)

The one new term introduced in this commonly accepted formula is k_t. This is known as the ambient temperature factor. If the temperature surrounding the conveyor is above the freezing point (32°F), then $k_t = 1$. At 0°F, k_t is around 1.2, and at −15°F the value is approximately 1.5. The person planning to use a conveyor in such cold conditions should consult an appropriate conveyor manufacturer or specialist.

As noted in the development for P_v, Eq. (9-8a) could be used for any type of conveyor. To find the components P_e and P_h for other than a belt conveyor, a modification will have to be made in the weight terms (w_e and w_m) to arrive at the appropriate weight per foot.

9.4.4 Power Unit to Drive a Conveyor

The power unit to drive a conveyor is generally positioned to operate on the upper, or head, end pulley, turn-around wheels. Its effective power is delivered for the required power found by Eq. (9-8) through the drive belt or chain. As mentioned previously, the power unit may be an electric motor or a gasoline or diesel engine. In any case it will be necessary to have a suitable power transmission system between the power unit's output and the delivered tension in the conveying belt or chain. This system may involve gears, a chain drive, a belt drive, or a combination of several of these. The required size of power unit will depend on the efficiency of the transmission system to deliver the required horsepower, P. It also should be noted that with a belt drive the friction on the drive pulley must be adequate to prevent unwanted slippage.

Even though Eq. (9-8a) looks simple enough, the valuation of the total power (P) requirement for a given conveyor system is complex. It involves using the various equations developed earlier along with the variable values that are required for them. The Barber-Greene Company has a set of tables that provide values for the separate power terms: P_e, P_h, and P_v. They also take into account power losses through the driving mechanism. These values are given in Tables 9-3, 9-4, and 9-5 for reference and handy use.

As can be noted in the above tables, the size of power unit to select depends greatly on the productivity (q) in tons per hour to be handled by the conveyor. Of course, the actual use and requirements of conveyor length, lift, belt speed, transmission of the power, and the like should be

TABLE 9-3 Horsepower at Headshaft to Drive the Empty Conveyor at 100 feet/minute[1] (For other belt speeds use direct proportion.)

Conveyor Centers, ft	Belt Width					
	18″	24″	30″	36″	42″	48″
25	0.44	0.53	0.62	0.72	0.82	0.98
50	0.47	0.57	0.67	0.77	0.89	1.06
100	0.52	0.63	0.76	0.87	1.02	1.21
150	0.57	0.69	0.85	0.97	1.15	1.36
200	0.62	0.76	0.93	1.08	1.28	1.50
250	0.67	0.82	1.02	1.18	1.41	1.65
300	0.72	0.89	1.11	1.29	1.54	1.80
350	0.77	0.95	1.20	1.39	1.67	1.95
400	0.82	1.02	1.28	1.50	1.80	2.10
450	0.87	1.08	1.37	1.60	1.93	2.25
500	0.92	1.15	1.46	1.71	2.06	2.40

TABLE 9-4 Horsepower at Conveyor's Headshaft to Move Horizontally Any Material at Any Belt Speed[1]

Conveyor Centers, ft	Capacity (q), in tons per hour							
	50	100	150	200	300	400	500	600
25	0.25	0.50	0.76	1.01	1.51	2.02	2.52	3.03
50	0.28	0.57	0.85	1.14	1.70	2.27	2.84	3.41
100	0.35	0.69	1.04	1.39	2.08	2.78	3.47	4.17
150	0.41	0.82	1.23	1.64	2.46	3.28	4.10	4.92
200	0.47	0.95	1.42	1.89	2.84	3.79	4.73	5.68
250	0.54	1.07	1.61	2.15	3.22	4.29	5.36	6.44
300	0.60	1.20	1.80	2.40	3.60	4.80	6.00	7.20
350	0.66	1.32	1.98	2.65	3.97	5.30	6.63	7.95
400	0.72	1.45	2.17	2.90	4.35	5.81	7.26	8.71
450	0.79	1.58	2.36	3.16	4.73	6.31	7.89	9.47
500	0.85	1.70	2.55	3.41	5.11	6.82	8.52	10.23

checked as discussed in this section for the given selection situation. A complete belt conveyor system is the most efficient and economical way to move loose materials.[2] This is dependent on having paths for the material to follow consistently during the total operation. An example of a complete conveyor system is shown in Fig. 9-13.

9.5 Hauling Equipment

A variety of equipment can be used to haul earth or loose bulk materials for a construction project. The earthmoving scraper is a hauler part of its working time. Other tractor-trailer combinations known basically as wagons are specially designed as earth haulers. The common dump truck traveling streets and highways is often a hauler of loose, bulk materials. In recent decades equipment very much like the dump truck but bigger has been developed for off-highway use.

The first mechanized haulers to be used on rough, construction haul roads were either crawler tractors pulling track-type wagons or conventional highway trucks. The former proved to be too slow, and the trucks were not rugged enough to take the rock loading and the twisting of their chassis. So with the initiation of big, joint-ventured projects such as Hoover Dam and Grand Coulee, it was necessary to develop bigger, faster,

TABLE 9-5 Horsepower at Conveyor Headshaft to Lift Vertically Any Material at Any Belt Speed[1]

Vertical Lift, ft	Capacity (q), in tons per hour									
	50	100	150	200	250	300	350	400	500	600
5	0.25	0.51	0.76	1.01	1.26	1.51	1.76	2.02	2.52	3.03
10	0.51	1.01	1.52	2.02	2.52	3.03	3.53	4.04	5.05	6.06
20	1.01	2.02	3.03	4.04	5.05	6.06	7.07	8.08	10.10	12.12
30	1.52	3.03	4.55	6.06	7.57	9.09	10.60	12.12	15.15	18.18
40	2.02	4.04	6.06	8.08	10.10	12.12	14.14	16.16	20.20	24.24
50	2.53	5.05	7.58	10.10	12.62	15.15	17.67	20.20	25.25	30.30
60	3.03	6.06	9.09	12.12	15.15	18.18	21.21	24.24	30.30	36.36
70	3.54	7.07	10.60	14.14	17.67	21.21	24.74	28.28	35.35	42.42
80	4.04	8.08	12.12	16.16	20.20	24.24	28.28	32.32	40.40	48.48

FIGURE 9-13 A conveyor system setup (courtesy of Barber-Greene Company).

and more rugged earth-hauling equipment. The possibility of bigger, more durable rubber tires paralleled this growth.

The variety of earth haulers that are available can be divided into two categories—over-the-road and off-the-road, better known as "off-highway," vehicles. The differences between vehicles in these two categories is based on the limitations imposed to travel on a city, county, state, or federal highway. These limitations involve maximums of load, width, and other dimensions, as well as measures to help maintenance and safety of the road. The result of these means that only haulers mounted fully on rubber tires and within certain size dimensions can be called over-the-road equipment. They include the smallest scrapers and wagons, as well as the highway dump trucks.

Equipment built strictly for hauling is frequently identified by its load capacity. A dump truck or hauler may be called by the tonnage it can carry. Thus, it may be a 15-ton truck. This is the maximum load that can be safely put on it or in its container bowl. Tonnage is the common way of showing capacity because it is a common measure to pay for what is hauled. On the other hand, excavated earth is generally measured in yardage to agree with the hauler's container size. Since the earth or loose, flowing material hauled for construction weighs between 2,200 and 2,800 pounds per yard in the hauler, conversion to obtain its limiting capacity in yards will be 75 to 90% of that stated in tons.

9.5.1 Design of Over-the-Road Haulers

Over-the-road haulers must be designed to operate within the restrictions set for the majority of highway use. Since no hauling vehicle will operate only on city and county roads, they will need to meet the limits for state and federal highways. These are generally set in the United States by the American Association of State Highway Officials (AASHO) in cooperation with the U. S. Bureau of Public Roads. The basic limits without special permit are a maximum vehicle width of 8 feet and a maximum axle load of 18,000 pounds or dual axle of 32,000 pounds. For safety's sake the over-the-road hauler for loose material must have a tailgate to prevent spillage and be able to travel at a speed of 40 or 45 miles per hour (mph), if expressways can be used.

Dump trucks of capacities up to 15 tons or 10 cubic yards are designed for over-the-road use. The trucks of the 1930's had an empty vehicle weight (EVW) slightly greater than their payload capacity. Now with larger sizes they keep within the AASHO maximum limits with an empty weight about 70% (for the largest ones) to 100% of their load capacity, which is a pay-load to empty weight ratio of 1.4:1 down to 1:1. They can travel easily at 35 to 50 mph with a total weight-to-horsepower ratio between 300 and 500. The container dump body is primarily of metal plate construction with dimensions to give the capacity specified with reasonable heaping of the load and not exceed the load limit at 3,000 pounds per cubic yard. The metal sides are somewhat shorter than the front or tailgate to make loading over the side easier. If the material to be hauled is lighter weight and the planner wishes to haul greater yardage, wooden sideboards can be added so that the top dimension is the same all around. The dumping mechanism is a hydraulic ram, or two, operated independently from the truck's driving mechanism for safety reasons. An important consideration in selecting dump trucks for continuous hauling operations is the dumping time. A slow dump time may be costly in the total hauling cycle.

The common dump truck has two axles but may have dual rear tires for better load distribution for rolling resistance and traction. This is particularly desirable where the truck must travel with its full load off a paved roadway. The specifics on tires, rolling resistance, and traction were discussed in Chap. 3. In cases approaching the load limits of highway travel on normal-size truck tires, the vehicles may be designed with two rear axles and dual tires. This amounts to a 10-wheel truck.

There are also bottom-dump, wagon-type haulers designed for over-the-road use. These meet the maximum axle load restriction, generally, by dual axles supporting the rear of the wagon and the front end supported on the rear of the tractor with dual axles. Such a combination amounts to a 10 x 4 wheel mounting. With a load close to 32,000 pounds on the rear, dual-axle drive wheels of the tractor part, there should be plenty of traction for the hauler to get on the road from a loading pit. If traction is not a problem, the load moved per trip can be increased by attaching another wagon with both front and rear axles to pull in train fashion.

9.5.2 Design Features of Off-Highway Haulers

Off-highway haulers, generally used to haul earth, are larger than the over-the-road vehicles described previously. They may be designed with maximum width and height of 12 feet to allow movement on a highway with special permit. However, to get higher production from this type of hauler off the highway, it will be designed for capacities giving 2 to 3 times as much load per axle as allowed on a highway. This is possible only because of recent, great improvements in oversize, low-pressure tires.

The off-highway hauler requires more power than its over-the-road cousins.[4] To get more, it sacrifices speed and settles for a maximum between 30 and 40 mph, unless its prime mover is a truck-type which can go faster. To achieve greater ruggedness and more power for the higher rolling resistance than an over-the-highway vehicle must overcome, an off-highway hauler frequently is designed with more empty vehicle weight. Its payload-to-weight ratio may be nearly 1 to 1, though some at the other extreme will be more than 1.5 to 1. The off-highway truck-type hauler has a rigid frame to take the twisting, rocking motion on its haul route. It also has a reinforced, double-bottom container bed to take the impact shocks in loading rocks and other hard material. The off-highway hauler does not have a tailgate and eliminates one mechanism, which results in improved dumping efficiency.

The off-highway haulers can be grouped into the following types: (a) rear-dump truck; (b) side-dump truck; (c) rear-dump tractor-wagon; (d) side-dump tractor-wagon; and (e) bottom-dump tractor-wagon. The rear-dump truck is basically an oversized, over-the-road dump truck with a stronger, rigid frame (Fig. 9-14). To have power for heavy loads and gain speed quickly, the engines are sized to handle 300 to 500 pounds of total load per horsepower. This off-highway truck may have a 4 x 2, 4 x 4, or 6 x 4 wheel mounting, the last being a dual rear axle with 4-wheel drive. A

FIGURE 9-14 A rear-dump truck in action (courtesy of WABCO, an American-Standard Company).

side-dump truck is designed to discharge its load by raising one side of the bed, instead of the front end, and dumping out the other side. In this case there is a high rear end plate as well as front of the bed. To gain the necessary stability for dumping, the side-dump truck has a dual rear axle mounting. Off-highway dump trucks have their hauling bed sides flared out more than over-the-highway trucks can be allowed because of the width restriction. The flared sides help to protect the tires against damage from rocks falling from the load or in the loading step. They also give the loading operator a wider target to hit from his bucket.

Manufacturers have been making bigger and bigger off-highway rear-dump trucks.[5] These are up to 200-ton capacity units. This growth in size is outstripping the development of suitable diesel engines. At first, the increased power was gained by using two engines in tandem. Now a few have more than 1400-hp in a single engine, which results in about 300 lbs/hp. More of these large haulers have weight-to-horsepower ratios higher than normal for a dump truck, so they have less speed or gradeability. To get the higher power outputs needed, the manufacturers have resorted to electric motorized wheels powered from a diesel engine generator. The need for higher power output has spurred the development of gas turbine power units which can produce considerably more than the 1000 hp threshold previously thought to be the limit.

The tractor-wagon haulers may have a single-axle tractor like the earth scraper. In fact, the same prime-mover may be used for either equipment application. This type of tractor unit has advantages in gaining the maximum possible traction and having a minimum turning radius. It has a 4 x 2 wheel mounting arrangement. Another common mounting is with 6 x 2 wheels or 10 x 4 with dual rear wheels, where the tractor has two axles—the rear one having the drive wheels. Generally, the wagon body is designed to be wider and deeper in the front to put more than half of the wagon's weight on the drive axle of the tractor. This helps gain traction by putting about 40% of the equipment weight on the drivers for benefit of the engine's power ability. Also, with that shape of the wagon body there is less spillage from a rear-dump without a tailgate.

Nevertheless, the 6 x 2 or 10 x 4 tractor-wagon has been limited to long hauls without much vertical lift. Tractive effort for tremendous future haulers is a problem. There can be an advantage for the tractor-wagon over an off-highway truck. The largest sizes of either type will probably have three axles to spread the load. In the case of trucks the back two are dual axles which will generally both transmit power. The weight on these two axles with the truck loaded is about 67% of the total weight of the hauler. That governs the maximum tractive effort possible. In the case of a 3-axle tractor-wagon, the two back axles will be at either end of the loaded wagon. That gives them about 80% of the total loaded equipment weight. To take advantage of that weight for maximum traction means that both back axles must have power wheels. This makes it a 10 x 8 tractor-wagon and seems to be the direction of development with four-wheel electric drive for the larger-than-100-ton haulers in the future.

All the tractor-wagon haulers are obviously articulated pieces of equipment because of the hitch-type connection between the tractor and wagon parts. This makes them highly maneuverable in a forward direction. Their turning radius, with a 90° tractor-to-wagon turn, is much shorter than for an off-highway truck of equal load capacity, which can turn its wheels 40° to 45° from a straightforward position. A reasonable comparison of this factor is shown in the set of curves of Fig. 9-15. Such a comparison is more reasonable between rear-dump trucks and rear-dump tractor-wagons. The bottom-dump tractor-wagon is built with greater length and a narrow body because it must dump the load between its

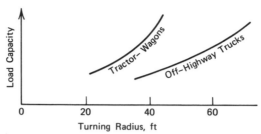

FIGURE 9-15 Comparison of turning abilities.

wheels. Nevertheless, the articulation makes even this tractor-wagon very maneuverable.

The bottom-dump tractor-wagon is designed to handle free-flowing materials such as loose earth, sand, gravel, and crushed stone, but not rock. Of course, this equipment has gates which open the bottom of the container body. It should be able to dump its load in 3 to 5 seconds. The sides of the body all slope inward like a hopper to help the dumping of the total load into a receiving hopper underneath the wagon or into a windrow between the wheel tracks while the tractor-wagon moves forward. The latter dumping step would generally be done on a fill area and take longer than the dump time given above. The fill will be uncompacted during this part of the operation, so a bottom-dump wagon is designed with large, single tires to carry relatively low pressures for better floatation. The use of single tires is also necessary to give more dumping width between the wheels on an axle.

The total weight (GVW)-to-power ratio for a bottom-dump tractor-wagon is generally between 300 and 600 lbs per hp. This means the equipment handles a lot of load for its power, which is satisfactory for travel on the level or modest grades and a slower gain of speed, than a truck, for long hauls. The rear-dump tractor-wagons are designed for a somewhat lower, 350:1 to 475:1, GVW-to-hp ratio. This is still higher than the off-highway dump truck, but it gives the rear-dump wagon, for instance, the ability to pull a load up steeper grades than bottom dumpers.

Though the rear-dump tractor-wagon does not have quite the gradeability nor the top speed that an off-highway dump truck has, it does have a great advantage in maneuverability. A rear-dump wagon is designed to raise the front of its body to dump its load in 12 to 15 seconds. It can even shorten its wheel base by a "jackknife" type of action for unloading or turning in cramped quarters. Another design of this type of equipment uses an ejector mechanism, like a scraper's, for precision dumping while in forward motion. And the articulated feature between the tractor and wagon means the equipment can operate on rough topography where the axles may not be in the same horizontal plane.

9.5.3 Use and Productivity of Haulers

The selection of suitable haulers for a given material-moving operation depends on a complete job analysis. This means every part of the work cycle for the pieces of equipment. For a material-moving operation the cycle includes load, haul, dump, return, and spot for the next load.

Taking each component of the cycle separately, one can discern the factors influencing the selection of haulers. The total cycle for an earth-moving operation was discussed in Sec. 5.2. An outline will serve to review the key points:

1. Loading requires knowing:
   a. The size and type of loading machine—Is it a continuous loader like a belt loader, or does it have positive bucket control like a shovel or front-end loader, or does it have a swinging bucket like a dragline?
   b. The type and condition of material to be loaded—Is it free flowing like gravel, or moist and sticky like clay, or large and chunky like blasted rock?
   c. The capacity of a hauler.
   d. The skill of the operators.
2. Haul or travel requires knowing:
   a. The distance each load is to be moved, broken into continuous straight and curved stretches.
   b. The condition of the haul route—Is it paved or surfaced? With what and for how much of the route? Where there is a dirt road, what are the traction and rolling resistance?
   c. The grades to travel on and for how much of the route.
   d. Miscellaneous conditions affecting haul speed and movement— what are the direction and grade changes that cause acceleration/ deceleration and braking? Is the route well drained, or might traction and rolling resistance be variable? Are there bridges or underpasses to navigate through on the route?
   e. The hauling equipment's ability to perform with the road conditions, grades, and other adversities in traveling.
3. Dumping affects the hauler selection by:
   a. The type and condition of the material just as with loading.
   b. The way the material will be handled at the dump—Is it to be

dumped down the slope of an embankment, or is it to be spread for compaction, or is to to be dumped into a hopper?

    c. The type and maneuverability of hauler for a restricted dump area.

4. Return travel must be covered like the haul travel with load (see point 2 above). This component of the haul cycle generally will not govern the selection of hauler. An exception might occur where the hauler must return empty up a much steeper grade or poorer route than it had to travel over with a load.

5. Spotting the hauler for loading requires knowing:

    a. The type of loading machine as in point 1a.

    b. The various positions the loader will take to load out the material.

    c. The maneuverability of the hauler to get into good position for loading.

For an operation to move loose bulk material, where the one-way haul distance is more than several thousand feet long and other factors do not make scrapers or conveyors more economical, haulers filled by a power shovel or other loader will probably be best. If a significant part of that haul distance can be effectively run on existing roadways, then over-the-road haulers will be planned. Bottom-dump, truck-type tractor-wagons, with light enough axle loadings, should be considered if the material to be hauled is free-flowing and to be spread out as it is dumped. The dumping time will be a minimum in this case. It can be included in travel time that must show the speed slowed down to perhaps 10–20 mph for getting in and out of the dumping area. For other parts of the haul distance this hauler may reach top speeds of 40 to 60 mph. The use of a bottom dumper will require having ample turning space in both the loading and dumping areas for trailing equipment that will be 40 to 60 feet long overall and, practically, cannot operate in reverse.

In the more common case of hauling loose material a significant distance over-the-road, with some variation in grades and variable dumping, an ordinary dump truck likely will be planned. The maneuvering to dump, generally, by backing up to a designated spot, will take time from that for traveling. Raising the bed and dumping a load will consume about 1/4 minute. The maneuvering in the dump area may run the total time to be planned for dumping (DT) to 1/2 minute. The time to allow at the other end of the cycle, i.e., loading time (LT), will depend primarily on the loading equipment. Between these end points of the hauler's cycle is the time for traveling. This variable time (VT) includes both traveling in one direction with a load (VTL) and returning empty (VTE).

To find the traveling times, the haul route for the dump truck must be broken down into lengths of common grade and rolling resistance. Then the gear to give the necessary power and, consequently, the top speed for each length of the haul can be determined. At this point it is necessary to recognize that the haul vehicle will not travel at the top speed for the full haul length. The planner must be able to estimate the average speeds for each moving part of the haul cycle. Table 9-6, a table of factors for converting top speeds or maximums to reasonable ⟨ erage speeds, will be helpful.

TABLE 9-6* Factors to Convert a Maximum to an Average Speed (of a vehicle with 300 to 400 lbs/hp**)

Length of Haul Road Section, feet	Unit Starting from Stop	Unit in Motion when Entering Section
100– 350	.25–.45	.50–.66
350– 750	.45–.55	.66–.74
750–1500	.55–.68	.70–.88
1500–2500	.68–.78	.78–.93
2500–3500	.78–.84	.87–.95
3500 and up	.84–.92	.90–.97

* Modified from booklet, "Production and Cost Estimating of Material Movement with Earthmoving Equipment," by TEREX Division, General Motors Corporation, 1970.
** Note: Factors are slightly higher for lower lbs/hp ratios and lower for higher ratios.

In the earthmoving operations to haul material without using existing roadways, the same basic question of scrapers vs loaders with haulers occurs. For earthmoving with a one-way haul of more than 3,000 feet, the haulers may be more economical. Assuming this to be the case, the plan ner has to decide between using off-highway dump trucks, rear- or side-dump tractor-wagons, or bottom-dump wagons.

The rear-dump truck will be more effective than the others for han-dling any loose material and hauling it where there are some steep grades and where higher speeds are desired on level, straight runs. The rear-dump tractor-wagon has a low end opening for lip loading of larger rock by a front-end loader or power shovel. It can maneuver in relatively tight quarters and dump quickly. So it is the best hauler for a tunnel excava-tion operation or others with narrow haul routes. It has moderate grade-ability and speed, which makes the rear-dump wagon more suited to shorter hauls than the off-highway truck. The less common side-dump truck or wagon is used where the operation will permit dumping material over the side of a stable embankment or on the run. This will reduce the dumping time by eliminating the need for maneuvering and backing up to dump.

A bottom-dump wagon is more frequently used in construction for dumping free-flowing material into windrows on the run. This would be likely in an operation to build up an earth embankment or the base course for a highway pavement. Travel through the loose fill area may require the high floation of a bottom-dump wagon. This equipment may also be used effectively to haul free-flowing material over and dump into the ground-level receiving hopper for a material-crushing or mixing plant. In any use of bottom-dump wagons the haul route should not have an adverse grade of more than 3 or 4% for an effective operation.

Determining Loader-Hauler Production. The determination of productivity for an earth hauler can be easily figured. This will require calculating the heaped capacity, load limit, or efficient load from a loader and the time it will take to put this load in the hauler. To balance a loader and its haulers, the haul container capacity should be an integral number of loader bucketfuls. For example, 6-yard trucks would be balanced with a 1 1/2-yard shovel but not with a 1 1/4-yard shovel.

For an efficient operation with good balance between the loader and its haulers, the loader should need between three and six bucketfuls to fill a hauler. If it takes less than three, too much time will be spent spotting haulers at the loader. The loader will undoubtedly not be working to its expected productivity. At the other extreme of more than six bucketfuls, the hauler and its operator will be sitting idle too long and not using the expected production in travel time to move the material.

Now the determination of productivity for a loader-hauler operation will be discussed. With the many selection factors and possible choices of loaders and haulers, numerous combinations might be used. In any case the construction planner will follow a logical approach for his situation. A typical example will be used later to illustrate the suggested procedure.

The approach will obviously (1) start with the construction operation to be done, then (2) consider the site conditions to plan on, next (3) decide on the feasible choices of equipment to do the operation, then (4) determine feasible equipment combinations and their productivities and costs for the operation, and finally (5) select the equipment combination to use for efficient and economical operation, all factors considered. An outline to expand on these steps should help to clarify the approach and suggested procedure:

1. Given operation for loader-haulers:
   a. Quantity of material—How much is to be moved?
   b. Material's natural condition—Can it be handled as is, or would it be more manageable if worked on before moving?
   c. Distance to move material—Is it to be moved to a specific location for deposit, or is the material to be wasted where the planner chooses?
2. Site conditions for loading and hauling:
   a. Terrain between loading site and the place for depositing the material—What are the variations of grade and the alignment or curves on feasible routes?
   b. Surfaces to travel on—What traction and rolling resistances can be expected? Can these be improved to advantage? (See Sec. 5.6.4 for an example.)
3. Feasible choices of equipment:
   a. Loaders—What types and sizes can be considered for the given operation?
   b. Haulers—What types and sizes can be used with the loader choices and for the site conditions?
4. Determinations for each loader-hauler combination chosen in point 3:
   a. Find the maximum productivity for the loader [as with Eq. (7-1) for a shovel or Eq. (7-4) for a dragline].
   b. Calculate the loading time (LT) for haulers from the loader's cycle time or from its maximum productivity (q_{max}) by

$$LT = \frac{V_h}{q_{max}}, \tag{9-8}$$

   where V_h = pay-yards/cycle.

   The loader's maximum productivity is used because of the short duration and belief that no interruption should occur between the start and finish of loading a single hauler.
   c. Find the haul route, grades, and weight of loads for hauler travel.
   d. Calculate the hauling resistances, speeds, and travel times over the various segments of the haul route both for traveling loaded and empty.
   e. Calculate the total cycle times for a hauler, taking into account the allowance for acceleration-deceleration-braking and turning (by using "fixed times" or converting to average travel speeds) and dumping (DT). Referring to Eq. (5-6), the hauler's best expected time is written

$$CT_h = LT + VTL + DT + VTE, \text{ in minutes,}$$

and the normal cycle time allowing for the hauler's wait in line (queue) for the loader and the driver's working efficiency (f_w) can be found as

$$(CT_h)_n = \frac{CT_h}{f_w}. \tag{9-9}$$

f. Find the maximum (q_h) and normal ($q_h)_n$ productivities for the haulers as follows:

$$q_h = \frac{V_h}{CT_h} \times 60, \text{ pay yards/hr, and} \tag{9-10}$$

$$(q_h)_n = q_h f_w \text{ pay yards/hr.} \tag{9-11}$$

g. Decide on the number of haulers needed for each loader, based on a previous decision with point 1 above determining whether the loader or haulers should govern the production for the operation; the theoretical number (N) required is

$$N = \frac{(CT_h)_n - LT}{LT} + 1, \tag{9-12}$$

which generally does not result in an integral number: so one must choose a number of haul units, N_h, as the next higher whole number over the theoretical,

$$N_h \geqslant N \text{ for the loader to govern,}$$

or choose the next lower whole number,

$$N_h \leqslant N \text{ for haulers to govern.}$$

h. Calculate the cost per unit of material moved by each loader-hauler combination, using direct equipment and operator costs and indirect costs with allowance for estimated delays (calculations of costs will be discussed in Sec. 9.5.4).

5. Select the most appropriate loader-hauler combination from the results of point 4 based on:

   a. An economical choice when considering these combinations for this operation only.

   b. The equipment combination that is economical and most readily available for this operation or will be most economical when other related operations are considered with it.

An Example of Loader-Hauler Determination. To illustrate the approach and procedure outlined, an example for the determination of a loader-hauler combination will be shown now. This will find a feasible selection of loader and haulers for a given construction operation. It will not take the steps which relate to costs (points 4h and 5 in the above outline). The special points about costs for hauling equipment will be discussed in Sec. 9.5.4.

Given: an earthmoving operation to move 40,000 cu yds of rocky earth soil to be spoiled (wasted) by dumping down an embankment about 4000 feet away to fill part of a ravine.

the cut bank averages seven feet high and the cut area drains well;

—cut area is open enough for ample maneuvering room;

—hauling possibilities are:

(i) haul route all on earth surface with no bad turns—600′ down 2% grade, 1200′ up 3% grade, 2000′ practically level, and 300′ down 4% grade to dump—the return on essentially the same route; or

(ii) haul 700′ up 4% grade to highway, 4000′ on highway with an average of zero grade, and 800′ down 2% grade to dump—return over same haul route reversed;

—rolling resistance of the earth haul roadways can be readily maintained for no more than 70 lbs/ton under rubber tires.

Now, following the approach discussed to solve this loader-hauler operation, the outlined procedure will be used. The given information provides answers for points 1 and 2. Feasible choices to make for point 3 might include:

(i) a 2-yd or 2 1/2-yd shovel (quantity docs not justify higher-production loader) loading into rear-dump, off-highway haulers; or for alternative,

(ii) a front-end loader or 1- or 2-yd shovel with dump trucks for travel on the highway.

With these sizes and types of equipment numerous choices of specific units and combinations might be considered. This is the first point where the equipment available to the construction planner is obviously recognized, but that should not overly prejudice the following selection. It might be profitable to let go of older, less suitable equipment for a new set to do this operation.

Taking one feasible combination for the off-highway (i) alternative, it is desirable to have a balance of sizes between the loader and haulers. Referring to Fig. 7-6 to estimate bucket efficiency, $E \approx 80\%$, a 2 1/2-yd shovel will handle a payload of $0.8 \times 2\ 1/2 = 2.0$ yards. This loader will work $(4 \times 2\ 1/2 = 10.0)$ with a hauler of 9-yd struck and 11-yd heaped capacities. Such a hauler might have specifications showing:

> Empty weight = 28,000 lbs;
> Maximum load = 30,000 lbs;
> Engine = 140 bhp at 2100 rpm;
> Max. governed speed = 30 mph; and
> Tires = 12.00×25 front and drive wheels.

Now continuing to follow the outlined procedure with point 4:

(a) An estimate of the shovel's average swing must be made from the cut site layout—we will assume a 90° average swing;

> estimate 2 1/2-yd shovel's optimum depth of cut $(D_o) = 9$ feet (see Fig. 7-5),

$$\text{so } \frac{D}{D_o} = \frac{7}{9} \times 100 = 77\%,$$

and $(A{:}D) = 0.97$ (from Table 7-1).

The 2 1/2-yd shovel can load, at best, using Eq. (7-1),

$$q_s = \frac{3600(2\tfrac{1}{2})0.8(0.97)}{29} = 241 \text{ cu yds/hr.}$$

(b) Haulers will carry a pay load of cut bank material received from 4 full shovel dipperfuls, say, $V_h = 4(2\ 1/2)0.8 = 8.0$ pay yards. Then, using Eq. (9-9),

$$\text{LT} = \frac{8.0}{241} \times 60 = 2.0 \text{ minutes.}$$

(c) The information for the haul route was gathered with the given operation data and site conditions:

assuming material weight, $\delta = 2800$ lbs/cu yd, a load, $W_L = 8.0 \times 2800 = 22{,}400$ lbs $< 30{,}000$ max. Therefore, loaded hauler's total wieght,

$$W = \frac{(28{,}000 + 22{,}400)}{2000} = 25.2 \text{ tons.}$$

(d) Dividing the haul route into sections of uniform grade and travel conditions, we can show the hauler's travel loaded:

for 600 feet down a 2% grade, GR = -40 lbs/ton;
$\Sigma(\text{RR} + \text{GR}) = \Sigma R_1 = (70-40) = 30$ lbs/ton; required tractive effort, $(\text{TE})_1 = 30 \times 25.2 = 756$ lbs. Using Eq. (3-8) modified, we can assume mechanical efficiency of 70%:

$$\text{max. } v = \frac{33,000(0.7)140}{756 \times 88} = 48.6 \text{ mph} > \text{limit of } 30.$$

Using Table 9-6 to find an average speed, assume 0.50 factor, average $V_1 = (0.5)30 = 15.0$ mph; then

$$\text{VTL}_1 = \frac{600}{15.0 \times 88} = 0.45 \text{ minute.}$$

for 1200 feet up a 3% grade, GR = 60 lbs/ton; $\Sigma R_2 = (70 + 60) = 130$ lbs/ton; required $(\text{TE})_2 = 130 \times 25.2 = 3280$ lbs;

$$\text{max. } V_2 = \frac{33,000(0.7)140}{3280 \times 88} = \frac{36,800}{3280} = 11.2 \text{ mph;}$$

assume 0.75 factor, av. $V_2 = (0.75)11.2 = 8.4$ mph;

$$\text{then, VTL}_2 = \frac{1200}{8.4 \times 88} = 1.62 \text{ minutes.}$$

This procedure will be followed for each section of the haul, traveling both loaded and empty. It might be handier to carry these calculations in a tabulation. If there are to be many such calculations, a computer program would be useful and quickest.

To tabulate the calculations:

Haul Section	G, %	ΣR, lb/T	TE, lbs	max V, mph	speed factor	avg. v	VT, min
Travel loaded, $W_L = 11.2$ tons (T) and $W = 25.2$ tons:							
600	−2	30	756	30	.50	15.0	0.45
1200	+3	130	3280	11.2	.75	8.4	1.62
2000	0	70	1760	20.9	.80	16.8	1.35
300	−4	—	—	30	.30	9.0	0.38
					total travel loaded, VTL =		3.80
Travel empty, $W = 14.0$ tons:							
300	+4	150	2100	17.5	.35	6.1	0.56
2000	0	70	980	30	.80	24.0	0.95
1200	−3	10	140	30	.75	22.5	0.61
600	+2	110	1540	23.9	.40	9.6	0.71
					total travel empty, VTE =		2.83

(e) All parts of the cycle time (CT) have been determined above, except the time to dump a load. Referring to the discussion in Sec. 9.5.3, we can assume that, with maneuvering into position, the dump time, DT $= 0.5$ minute.

Therefore, $CT_h = 2.0 + 3.80 + 0.5 + 2.83 = 9.13$ minutes;

assuming a working efficiency, $f_w = 0.7$,

$$(CT_h)_n = 9.13/0.7 = 13.05 \text{ minutes.}$$

(f) To find a hauler's maximum productivity, use Eq. (9-10):

$$q_h = \frac{8.0}{9.13} \times 60 = 52.6 \text{ pay yards/hr;}$$

but its normal productivity is

$$(q_h)_n = 0.7(52.6) = 36.8 \text{ pay yards/hr.}$$

(g) For this example, the theoretical number of haulers is found, using Eq. (9-12):

$$N = \frac{13.05 - 2.0}{2.0} + 1 = 6.53 \text{ haulers.}$$

In this operation it is believed to be more important that the loader cut the material out of the way for construction that follows. Therefore, the shovel should govern the productivity for the operation. So the planner will choose a number of haulers,

$N_n \geqslant 6.53$. Therefore, he will use seven haulers.

The excess of hauling equipment over the loader's rate of production has another advantage. When it appears that haulers will bunch up in a long queue waiting for the loading, the dispatcher can wave a hauler out of his cycle to refuel or make any necessary repairs.

This same precedure can be followed to determine the equipment combination for alternative (ii) or other sizes and types of loader-haulers for either hauling alternative.

Haul Fleet Size. To plan an efficient and economical material moving operation with a combination of loader and haulers requires careful coordination of the dependent pieces of equipment. The loader is dependent on each hauler and vice versa. Early in this section the question of balancing sizes was discussed. This is one point for coordination. Another, probably more important, point for coordination is in the production rates of the dependent equipment. The haul fleet should be of a size that will allow the loader to reach its normal productivity at all times and its maximum production rate at least part of the time. However, this must not be at the expense of having too many haulers available for the entire moving operation. On very large operations to move millions of cubic yards the plan may call for standby equipment to be available.

As one of its series of field studies, the Highway Research Board sponsored a study to find how many trucks were needed for each shovel. The study involved shovels in the 3/4-yard to 2 1/2-yard sizes loading truck sizes from 4 to 14 cubic yards struck capacity. It observed the time lost by a shovel waiting on trucks and also the time lost by the trucks waiting in the queue at the shovel. These times were expressed as a percentage of the available working time, which is the total clock working hours less every delay of 15 minutes or more in duration. The observations found that the shovels waited from 4 to 25% of their working time. The dependent trucks waited from 25 to 4% of their available working time. An interesting empirical relationship was concluded from those observations. That is, the sum of the time losses by a shovel and its hauling trucks adds up to 29% of the available working time.

The Highway Research Board (HRB) published a set of curves from that study relating the total of hauling capacity to the time lost by the shovel. Hauling capacity was expressed in the shovel dippers to fill the fleet of trucks. For example, in the illustration worked earlier in this section it was planned that four dipperfuls would fill a hauler, and with seven of them the hauling capacity of the fleet is 28. The HRB curves are shown in Fig. 9-16. If the observations of that study can be applied to the example given earlier in this section, the 2 1/2-yd shovel will lose between 12 and 15% of its available working time with the seven haulers planned. With the shovel governing the maximum production for that hauling operation, the likely working efficiency is $f_w = (1.00 - 0.15)100 = 85\%$. The best productivity that can be expected throughout the working hours would be (referring to point 4a of the previous outline)

$$(q_s)_n = 0.85 \times 241 = 205 \text{ pay yards/hour.}$$

If it is assumed that the average available working time after delays for bad weather, major work stoppage, etc., is 60% of the total clock working hours, then $f_a = 0.60$ and $(q_s)_a = 126$ pay yds/hr. This would mean that it will take $40,000 \div 126 = 317$ clock hours or about 40 eight-hour work days to move the material with the suggested loader and haulers.

A method proposed recently for selecting the haul fleet size is based on

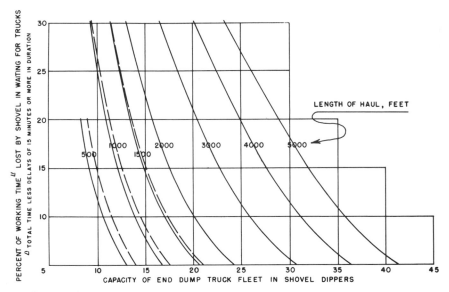

FIGURE 9-16 Relation of truck fleet capacity to shovel delays for varying lengths of haul.

queuing theory. This was proposed by Griffis[7] and developed from the background of work done by Shaffer, et al.[8] The method starts on the basis that the loader is the primary equipment and will govern production for a material-moving operation. The objective is to determine the likely productivity that can be expected for the loader. That must be done allowing for times when there may be no hauler in the lineup (queue) for loading. In other words, one must recognize the loader's lost time as observed in the Highway Research Board study.

The queuing theory method is based on the probability of having at least one hauler in the queue waiting to be filled by the loader. Actually, there are two main alternatives for the sake of this method. Either there will be at least one hauler in the queue, or there will be no hauler in the queue. The sum of these alternate probabilities must be equal to one. This can be shown by

$$P[n \geqslant 1] = 1 - p[n = 0] = 1 - P_o,$$

where n = number of non-productive haulers in the queue.

The article on this method by Griffis uses a whole different set of symbols for the material-moving determinations. To help relate it to the earlier discussions in this chapter, some modifications from the symbols in the article will be made.

Using the probability, $P[n \geq 1]$, the loader's expected productivity is

$$q = (1 - P_o) f_w V_h \left(\frac{60}{LT} \right) \text{ in cu yds/hour.}$$

All the terms have been used earlier, except the probability, P_o, that there will be no hauler in the queue to be loaded. The value of P_o *is* found by using an exponential summation equation which can be solved readily with a set of Poisson probability distribution curves or standard statistical tables.

The Poisson curve or table to use depends on a ratio shown by Griffis as μ/λ, where $\mu = 60/LT$ haulers loaded per hour, and

$$\lambda = \frac{60}{(CT_h)_n - LT}$$

haulers arriving in the queue per hour. Putting the ratio in these terms, one gets

$$\frac{\mu}{\lambda} = \frac{60/LT}{\dfrac{60}{(CT)_h)_n - LT}} = \frac{(CT_h)_n - LT}{LT},$$

which is one less than the theoretical number of haulers required by Eq. (9-12). The probability value is found as $P_o(k,x)$, where k is the number of haulers in the fleet and x is the Poisson ratio closest to μ/λ. Griffis showed an example using this method and basing his selection of fleet size, k, for the minimum unit cost to move the material. For his example with a haul of 1.3 miles one way, using a 3-yd shovel with 15-yd, bottom-dump wagons and other given or assumed data, he found $P_o = 0.208$ and $k = 5$ wagons to be the recommended selection. The equipment planner may wish to use statistical analysis for his haul fleet determinations. If so, it would be wise to study the references given at the end of this chapter on this subject.

9.5.4 Costs of Hauling Equipment

The costs to the contractor for having and using his own off-highway haulers are very similar to those of any other equipment. There will be the "standing," or ownership, expenses to cover the original purchase price, freight charge to have it delivered, and the expenses for insurance, storage, taxes, and interest on the money used. Also, allowance must be made for overhauling the equipment at appropriate times to keep the investment in good order for use throughout its expected lifetime. That may be 4 to 8 years. To estimate the maintenance or overhaul expense, 30 to 60% of the original price can be expected as reinvested in the hauler to maintain it in good working order. The total cost of repairs for an off-highway hauler should not be much more than $1 per hour or a few cents per yard hauled. Such an estimate does not include the cost of tires, which is a major expense for haulers and, so, takes special attention. That will be discussed shortly.

The other major charges for the use of hauling equipment are known as the "running," or operating, expenses. They will include the cost of fuel, lubrication, operator wages and fringe benefits, as well as the expense of minor repairs and adjustments for preventive maintenance. The extent of that maintenance will directly affect the lifetime and overhaul expense limits mentioned above. Also, these variables will most positively be influenced by the severity of the hauler's use.

The methods for determining the ownership and operating expenses were discussed in the first several sections of Chap. 2. A review of them should be made if they are not well understood by now. The one major variation in the case of rubber-tired haulers involves the tires and their maintenance and replacement. A first set of tires is purchased with the hauler and so is part of the original investment and might logically be an ownership expense item. However, the tires are not expected to have nearly as long a life as the original equipment. For that reason and because their cost is so directly dependent on the miles the hauler travels and, consequently, the hours it is used, the tire costs are generally considered to be an operating expense.

To account for the tire costs in a reasonable way, their original cost is deducted from the purchase price of the hauler. That cost is then divided by the expected life of the tires to get an hourly expense for the investment. Then a reasonable amount is added to that expense, generally a percentage increase, to allow for tire repairs during the hauler's operating time. A simple example with realistic, assumed values will show this.

Given: hauler purchased for $30,000, including freight (about 2%),
              with tires worth   −3,600

net purchase price, $E_o = $ $26,400.

For an expected life of 5 years and used 2000 hours each year, the straight-line depreciation for 10,000 hours, $e_d = $ $2.64/$hr.

To this ownership expense, the others mentioned above must be added.

Now, if the expected life for the tires

is 3,000 hours of use, $3,600/3,000 = $1.20/hr,

plus adding 15% for tire repairs   0.18

total tire (operating) expenses $1.38/hr.

This high expense for tires compared to that for the remainder of this equipment must be expected for heavy-duty haulers.

The expected life of tires varies with the job and haul route conditions, as well as the way the hauler is operated. As indicated in Sec. 3.2.2, tires show variation in wear based on the level of their inflation, the extent of the loading on them, and the speed at which they are driven. The job conditions affect the tire life expectancy by the kind of haul route surface. Is it very rocky, or is it compacted earth and well maintained? Most jobs will have a combination of conditions along the various parts of a total haul route. In spite of the many variables, it can be noted that rear-dump haulers, with their use for high power, steep grades, and rugged conditions, will have tire life expectancy of probably 2,000 to 4,000 hours of use. Bottom-dump wagons for more level, uniform travel without rocks may get their tires to last 3,000 to 8,000 hours at best for soft soil without much spinning.

Of course, the biggest operating expense, except perhaps with the large wagons approaching 100-tons capacity, is that for the operator. Then, after the tire expense, an important one is the expense for fuel. This will vary considerably, based on the condition and efficiency of the engine as well as the type of power unit used and the way the equipment is operated. For rough estimating purposes a hauler with an internal combustion engine should be expected to consume between 1 and 3 gallons per hour for each 100 bhp of the engine. Other operating expenses are of less importance, assuming that the operating maintenance which causes a shutdown of the job is part of the total ownership maintenance expenses mentioned earlier.

Costs for Highway Trucks.   The costs for over-the-highway dump trucks and other hauling units, which are expected to be on the go most of the time, will likely be handled differently from those of the off-highway units. Such hauling units may travel 15,000 miles and more per year, much more than most off-highway haulers. For that reason, their operating and maintenance expenses may be three to four times the original purchase price of the truck. Just the tires for such a truck may cost thousands of dollars per year. Therefore, it is very important that the truck is carefully selected or engineered for the jobs it will handle. It is probably better to purchase a slightly more expensive truck with a little more than enough load capacity, ruggedness, etc., than an undersized one which may require a lot of maintenance with the concurrent down-time. The use of long-tested, standardized units and durable parts is recommended.

It is generally recognized that a reasonable, economic life for hauling trucks is five years. However, the Automobile Manufacturers Association found that the average age of trucks on the road in 1967 was 7.8 years. These were predominantly owner-used trucks. This shows a poor regard for the costs of operating trucks as noted by a leading transportation research center. A study it made found that the cost of operating average, under-maintained vehicles can be as much as 50% more than normal costs. These revelations combine with the fact that private truck operations were never closely checked by their owners according to a University Research Center study.[9]

It might be suggested that a contractor, who does not carry on hauling operations primarily, should subcontract such work or resort to "for-hire" trucks. That probably would be more expensive because the trucking subcontractor will charge a rate to include his profit on top of all the equipment expenses. However, sub-contracting would eliminate the need for the private owner/contractor to have his own servicing and maintenance arrangements for the trucks.

Another alternative gaining favor is the use of full-service leasing of the required hauling units. This arrangement is one involving a contract with a firm in the business to provide one or more fully equipped and operated trucks for the user. The contract covers the procurement and maintenance of the best-suited trucks for operations anticipated by the user over the years of contracted time. It also covers all licenses, insurance, fuel, as well as repairs and parts. There is the added advantage that the leasing company can provide an additional truck or two, at practically the same cost per truck, on relatively short notice to have just the right number for a hauling operation.

The company that is in business to provide the full-service leasing, obviously, will charge to cover all its costs plus a profit. It is able to make a competitive charge because the company deals with a large fleet of vehicles for leasing. The leasing company buys its trucks at discounts of 20 to 25%, whereas the best a contractor might do in the purchase of his smaller fleet would be a 15% discount. Also, the leasing company will generally charge for a truck based on the net value for depreciation. The net is figured as the discounted purchase price less a reasonable salvage value for the truck after five years.

The previously mentioned study by the University Research Center found that the dollar advantages mentioned above and others based on discounts for quantity purchasing show a real advantage to full-service leasing. A user of leased trucks can realize a definite economic saving for his hauling operation. It was also found that the saving was not by a contract based on an average user. Generally, a full-service leasing agreement is custom-made to the user and his needs and operations. This is, at least partly, due to the fact that such expenses as insurance, licenses, and operating costs vary between the states, localities, and even users. However, in the cases studied with the user requiring a fleet of eight trucks, light or heavy sizes, there was an economic advantage to using a full-service leasing arrangement.

References

1. "Conveyor Manual," Barber-Greene Company (Aurora, Illinois, 1965 revisions).

2. Low, W. Irvine, "The Portage Mountain Conveyor System," *Journal of the Construction Division—ASCE,* Vol. 93 No. CO2, (New York, N.Y., September 1967).

3. McClimon, A. S., Chapter 3, *Handbook of Heavy Construction,* Frank W. Stubbs, Jr., Editor-In-Chief, McGraw-Hill Book Company (New York, N.Y., 1959).

4. Neul, Milton C., "The Economic Effects of Weight to Horsepower Ratios on Off-Highway Haulage Vehicles," Preprint No. 66A014, Society of Mining Engineers of AIME (New York, N.Y., March 1966).

5. "E/MJ Survey of Off-Highway Trucks," booklet of Engineering and Mining Journal (New York, N.Y., December 1968).

6. "How Many Trucks You Need Per Shovel," *Construction Methods and Equipment,* McGraw-Hill Book Company (New York, N.Y., August 1956).

7. Griffis, Fletcher H., "Optimizing Haul Fleet Size Using Queueing Theory," *Journal of the Construction Division—ASCE,* Vol. 94, No. CO1 (New York, N.Y., January 1968).

8. O'Shea, J. B., G. N. Slutkin and L. R Shaffer, "An Application of the Theory of Queues to the Forecasting of Shovel-Truck Fleet Productions," University of Illinois (Urbana, Illinois, June 1964).

9. "Truck Costs—A Comparison of Private Ownership and Full-Service Leasing," pamphlet by University Research Center, Inc. (Chicago, Illinois, 1968).

Chapter 10

■■■■■■■■■■■■■

■■■■■■■■■■■■■■■■■■■■■■■■■■■■■■

AGGREGATE
PRODUCTION
EQUIPMENT

■■■■■■■■■■■■■■■■■■■■■■■■■■■■■■

Introduction to Aggregate Production

Aggregate material is used in construction to give bulk or wearing surface.
It may be mixed with other materials to give a solid filler and structure
to a constructed item. The aggregate is generally made from natural
earth rock or granular material. The plant to make aggregate has crushers
in it, which are the first type of equipment in this book used to truly
process material. The crushers change the natural material in its size and
shape as well as serve to clean and separate it in a way other than by the
natural separation or disintegration caused by previously discussed equipment. An aggregate plant does its processing by crushing the natural
material and other solid, bonded chunks. Then it separates the output
into piles or bins containing different sizes of material for gradation purposes. A complete plant is shown in Fig. 10-1.

The sources of material processed in an aggregate plant include rock

435

FIGURE 10-1 Portable aggregate plant in operation (courtesy of Pioneer Division, PORTEC, Inc.).

from a quarry, gravel from a natural pit, and slag or cinders from a material-deposit area. The processed aggregate material is used for concrete, roadway surfacing (graded rock, macadam, asphaltic, or concrete), railroad ballast, and filter material which is clean and well-graded for water drainage or treatment.

The basic steps in the operation for producing aggregate material are as follow:

1. remove the raw material from the source, generally with a power excavator, bulldozer, or front-end loader, but in a gravel pit it may be under water, requiring a dredging method;

2. reduce the material to a form or size that the plant can handle—by breaking or blasting oversized rock or washing sticky clay and other foreign material out;

3. transport the material to the plant—loading it by power excavator, or other loader, to a conveyor or haulers for delivery; and

4. feed the material to the equipment components of the plant for crushing and other processing to arrive at a finished aggregate product.

Specifications have been extensively developed to govern the aggregate as a processed material, frequently used in a closely controlled construction mix such as concrete or asphalt. These specifications have been set by the American Society for Testing Materials (ASTM), the American Association of State Highway Officials (AASHO), and other construction agencies. They have requirements to test for many of the following characteristics in the final aggregate product: hardness, toughness, resistance to abrasion, specific gravity, particle shape and fracture, fineness modulus, organic impurities, surface moisture, soft particles, and gradation. Not all of the aggregate materials produced need to be tested for all of these properties. The ones to be applied will depend on the use for the aggregate. Many organizations have supported research to help aggregate producers and users with the problems of specifications. These include the Portland Cement Association, the Asphalt Institute, the National Sand and Gravel Association, and the National Crushed Stone Association.

The extensive use for aggregate in construction requires the greatest care in specification control for concrete and asphalt mixes. These require a hard, durable stone free from excess flat, elongated, or soft particles. It must be free of dirt and other objectionable material. The aggregate also must have a well-controlled gradation to meet the specifications for the appropriate mix.

A variety of equipment has been designed for an aggregate plant to meet the various specifications. The equipment includes:

1. feeders and hoppers to receive the original raw material, or material at other stages of the processing;

2. crushers to reduce the size of the material—these will be called primary and secondary or reduction crushers for different stages in the process;

3. conveyors and bucket wheels to move and direct material in the processing from one unit to another or into stock piles;

4. screens for separating, grading, and redirecting material to the following crushers, bins, or stockpiles; and

5. bins and other hopper-bottom units for temporary holding, or larger containers for permanent storage, of the material in a stockpile.

The important, power-requiring pieces of equipment in this variety for the aggregate plant will be discussed and described in following sections of the chapter.

Any aggregate production plant has a process to reduce feed from given sizes to a product having specified gradation of sizes at a predetermined tonnage rate. Thus, conveyors and other material handlers, including crushers, must be gauged for the capacity to meet their parts of this production rate. The production rate for an aggregate plant is initially set by the job requirement for the material. This in turn sets the rate for the crushers, which then governs the material which must be handled by the feeders, screens, etc. Also, two stones from different sources do not crush alike, so adjustments in crusher settings must be made after the process is started. Since the crushers govern other equipment within the aggregate production process, this type of important equipment will be covered first in the detailed discussion to follow.

10.1 Crushers for Aggregate Production

Crushers in the aggregate plant are called primary crushers and reduction crushers. The primary crushers take original or raw material that needs reducing for a specified product and break it into smaller sizes. This is generally done by jaw crushers, gyratory crushers, impactors, or single-roll crushers, which are able to reduce the large rocks from, say, 3- or 4-foot maximum size into pieces which range all the way down to the smallest particles one-hundredth that size. Actually, a crusher could not be set to crush a piece to 1/100 of the original size, but the fragmentation in the crushing process will result in some pieces that are small. This will be explained later with the grid, or percentage, chart.

Reduction crushers work on smaller-sized material that has been passed to it from a primary crusher, or was taken from a gravel pit, and needs further refinement in its sizes. Generally, a twin- or triple-roll crusher, cone crusher, or hammermill will be used for this purpose.

A more specific way to speak of the reduction in material size caused by a crusher is to indicate the "ratio of reduction." This is the ratio of the maximum size of feed (F) the crusher can take compared to the maximum opening at its discharge, which is the crusher's setting (s). These dimensions are ones to be found perpendicular to crushing surfaces. In the case of most crushers the size of feed is governed by the top opening through which the material is fed. However, for a roll crusher the maximum size of feed is the largest stone that can be nipped between the rollers. The approximate ratios of reduction, F/s, can be indicated for the variety of crushers already mentioned (Table 10-1).

In an aggregate plant the crushing from one maximum size down to smaller sizes is done in one or more crushers, each representing a stage in the process. With each crusher there is a "stage of reduction," which is the difference of dimensions between the maximum size of stone actually fed

TABLE 10-1   Ratios of Reduction, F/s, for Crushers (with the smaller number for the largest crusher setting and the larger number corresponding to the smallest setting)

Crusher Type	Smaller Models	Larger Models
Jaw crusher	5 to 10	6 to 14
Gyratory crusher	3 to 6	6 to 8
Cone crusher	2 to 9	5 to 15
Twin-roll (smooth)	1½ to 3½	1½ to 9
Hammermill	6 to 24	5 to 48

to the crusher and the maximum size coming out of it. The top size of output will have at least one dimension practically equal to the crusher's setting. Thus, if 6″ rock is fed into the crusher with a 2 1/2″ setting, the stage of reduction is essentially 3 1/2″. The reason for the slight difference will be explained later with the grid, or percentage, chart.

Where there are several stages in the reduction process, the stage of reduction is not as important for the primary crusher. The opening size for the feed is more important there. However, the stage of reduction is very important for the secondary, or reduction, crusher because of its need to control closely the gradation of the finished product. A larger stage of reduction generally will require more mechanical strength, power, and probably maintenance for the crusher.

10.1.1   Principles of Rock Crushing

Rocks and stones are broken into smaller stones and particles by at least one but, more generally, several of the following actions: attrition, pressure or compression, impact, and shear. These will each be discussed briefly in the following list.

1. Attrition is rubbing or grinding down by friction. There is some of this action in all crushers. It is most effective with material that is friable, and not abrasive (low silica content). The attrition action is most beneficial when maximum fine material is wanted.

2. Pressure or compression is a squeezing action between two surfaces. This is most obvious in jaw crushers but also takes place in other crushers designed for it. The action should be specified when material is hard, tough, and abrasive and a minimum of fines is required. It is not useful or recommended for sticky materials.

3. Impact is the instantaneous, sharp blow delivered by a hammer

on the material, which shatters it into many smaller pieces. The hammermill and impact breaker are examples of crushers that apply this action. It should be specified when material is not too abrasive (not much over 5% silica) and contains a high percentage of soft stone. It also is recommended if more cubical particles are desired. Crushers designed for this action will give well-graded products from largest to smallest sizes in one stage of crushing operation.

4. Shear is a cutting or slicing action found as part of most crusher action such as gyratory and roll crushers provide. It might be specified when material is fairly soft and easily crumbled, and yet a minimum of fines is wanted. For a crusher to apply this action, the material should not be too abrasive.

Of the crusher actions described above, the attrition and shear are evident in almost all of them. The pressure-compression and impact actions are the more significant factors in the design of crushers. The crushers to be discussed next individually can be listed in the order from the greatest impact action and least compression as follows: impact breaker; hammermill (high to low speed); rod mill (high speed); ball mill; single roll crusher; cone crusher; gyratory crusher; jaw crusher; and twin- or triple-roll crusher. Reading them in the reverse order gives a listing of crushers starting with the one designed for the greatest compression with the least impact action.

10.1.2 Design Features of Crushers

All aggregate crushers are designed to provide either a strong impact force or pressure through a compressive force on the rocky material. As just indicated in the listing of effectiveness, both types of force can be found in each crusher as well as some attrition and shearing action. The object of the design of a crusher is to provide the most forcefulness to get the desired product with the least cost of the original equipment, power, and maintenance needed due to wear and tear on the parts. So far, not one design has been found most effective in all respects of these criteria. Furthermore, no two types of rocky material crush in the same way. So for any particular job, generally, more than one type of crusher may be suitable. The planner tries to select the one most adaptable to all of the materials and possible uses he will have for it.

Important dimensions in the design of a crusher are its top opening and the crusher's settings. The top opening is a horizontal area through which fresh material is fed. The setting is the adjustable opening through which crushed material of that size or smaller can pass out of the crusher.

The next sections will discuss each of the crusher types used frequently for aggregate production.

Jaw Crushers and their Selection. The jaw-type crusher has a large rectangular top opening between the tops of the two crusher plates, or jaws, and the end walls. Jaw crushers are identified by the top opening in inches, such as an 18 x 36 crusher, which has jaws that are 36 inches long. The jaws converge toward the bottom, leaving a long, thin opening. The small dimension of this bottom opening is the jaw crusher's setting. The general makeup of the jaw crusher is shown in the cross-sectional view of Fig. 10-2.

FIGURE 10-2 Cutaway view of jaw and roll crushers in operation (courtesy of Pioneer Division, PORTEC, Inc.).

In this crusher design one jaw is stationary while the other moves back and forth, pivoting about an almost fixed hinge with the force supplied for leverage at the other end of the jaw. Two designs of jaw crushers have proven to be effective. One is the Blake, or double-toggle-type, crusher which has the pivot on a fixed shaft at the top of the movable jaw. This gives the greatest leverage and compression force on a rock near the top of the jaws. It permits rapid elimination of the smaller crushed material through the bottom to get out of the way for crushing the larger rocks. The other design is the overhead, or eccentric, jaw crusher, which has a single toggle at the bottom. That is the hinge point for the moving jaw. Then a revolving shaft on an eccentric axis at the top moves that jaw top through a circle in the vertical plane, as shown in Fig. 10-2. This design, with the greater jaw movement at the top and gravity helping to feed the crusher, gives a sort of forced-feed action and the potential for higher output of crushed material than the Blake type. There is better attrition action in the overhead eccentric than with the Blake type. A jaw crusher has the largest feed opening of the crushers to be discussed and the highest productivity for investment in the equipment.

Gyratory and Cone Crushers. In this type of crusher the moving element is a conical mantle on a revolving, essentially vertical shaft, as shown in the cross-sectional view of Fig. 10-3. The revolving shaft with its mantle

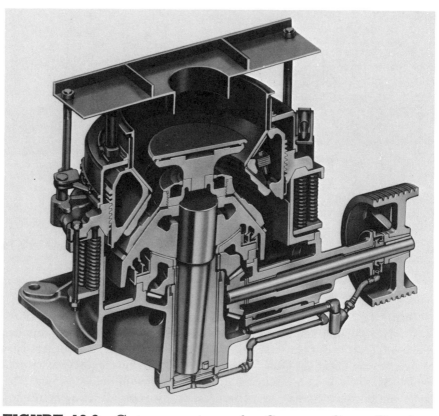

FIGURE 10-3 Cutaway view of a Symons Cone Crusher (courtesy of NORDBERG-Division of Rex Chainbelt, Inc.).

is on an eccentric bearing, giving a gyratory action to its rotation. The stationary crusher plates are vertically concave and horizontally circular to form a shell or bowl in which the mantle revolves.

The stationary shell of a gyratory crusher may be straight (smooth), a modified straight, or non-choking concaves. The latter types are helpful for use with sticky, moist, or dirty material that may stick to the inside of the crusher or in some way clog it. The nonchoking concaves are available only for the smaller sizes of crusher up to 18″ maximum feed. The modified straight concaves are a compromise between the straight and nonchoking variety. These various stationary plate designs do not affect the gyratory crusher's capacity very much.

The actions of the gyratory or cone crusher on the material are similar to those of a jaw crusher. The eccentric, gyrating motion means pressure is applied at different points around the circular shell crusher plates as the mantle revolves. The mantle revolves so that the bottom moves from a minimum to a maximum gap every half revolution. Generally, the maximum gap, known as the "open side," is the setting for the gyratory crusher. The material is fed through the top ring-shaped opening between the cylindrical shell and the mantle's smallest diameter. After being crushed, it leaves through the bottom gap where the mantle has its largest diameter.

A gyratory crusher is designated by the size of the receiving opening width, i.e., the width of the ring shape identified above. It has a high capacity for the size of material fed to it because of the full-circle discharge opening. That opening minimizes the production of slabby, elongated finish material, but it is not good for extra-hard material. The gyratory's capacity can be varied by changing the speed of rotation within reason.

A cone crusher is very similar to the gyratory one but is used as a reduction-type crusher. It has a modified mantle and bowl to better control the finished size of smaller feed material. Cone crushers are designated by the maximum diameter of the mantle in feet, which is different from its larger cousin, the gyratory. The crusher head, or mantle, is convertible to permit change for different material and sizes. This is an attempt to match the wide range of adjustment possible with other popular reduction crushers, such as the twin- or triple-roll crusher to be described next.

A disadvantage is that a cone crusher may pack or get clogged badly with sticky material. Otherwise, it produces about as well as the twin- or triple-roll crusher, producing good cubical particles. The cone-type crusher does require somewhat more head room and operating space than the roll-type crushers.

Roll Crushers. There are single-, twin-, and triple-roll crushers. These all take advantage of the compressive force on crushable material caught between a rotating cylinder (roll) and surface adjacent to it. The material is said to be nipped between the two like clothes in a wringer.

A single-roll crusher is generally a special-purpose one with a large-diameter, horizontal cylinder on which knob-like teeth project. The roll revolves over an adjustable plate or anvil as the other crusher surface. This crusher is identified by the diameter of the roll at the root of the projecting teeth and the length of the roll, both given in inches. The teeth in the single-roll crusher give a cutting and shearing action on the material. It is suited for crushing sticky or soft material that might pack and clog other types of crushers. It produces cubical particles without an excessive amount of fine material. It is used as a primary crusher and seldom set for less than a 2 1/2″ opening.

Twin- and triple-roll crushers are used as secondary, or reduction, crushers. In this capacity they are designed for a wide range of adjustment in their settings. They can be used for most construction aggregates down to the minus (less than) 1/4″ size. Sticky material may pancake between rolls, but it will not clog the crushers.

A twin-roll crusher has two horizontal cylinders rotating toward each other (one clockwise and one counterclockwise) as in the old-fashioned clothes wringer. One is shown in the plant arrangement of Fig. 10-2. It is designed so that crushable material dropping toward the gap between the rolls is pulled through or crushed on its way. The rolls are made with surfaces that can be described as coarse-corrugated, fine-corrugated, and smooth rolls. The two rolls in a crusher may be of the same surface or different, depending on their effectiveness with the material anticipated for crushing.

The twin-roll crusher is identified by two dimensions, its diameter and length of rolls in inches. One roll is "stationary"—i.e., its axis remains fixed—and is the powered roll for driving the crusher. The other roll is rotated by a star gear or chain drive from the first roll. This one is known as the "floating" roll because it is adjustable to change the gap between the two rolls and provide the crusher's setting.

The size of feed to a twin-roll crusher is governed not only by the gap between the rolls but also by the diameter of the rolls. Manufacturers recommend the maximum size of feed for a given roll crusher on the basis of the roll diameters and possible setting. Generally, this maximum size is not more than 8″ or 9″ for the largest rollers, which is about 1/6th of the diameter for course-corrugated rolls or 1/8th to 1/10th of the diameter for smooth rolls. The maximum size of feed for a roll crusher depends on the biggest stone that the rolls can nip between them. That dimension can be worked out by the geometry involved. Professor Peurifoy, in *Construction Planning, Equipment, and Methods,* shows the appropriate calculations.[3] He states that the angle of nip is 16°54′ for smooth rolls. The maximum size of stone that the rolls can pull through for crushing is found by the following formula:

$$F = 0.085R + s \qquad (10\text{-}1)$$

where F = the largest dimension of rock, inches;
 R = the radius of the rolls, inches; and
 s = the setting, inches.

If larger pieces than F are fed to the rolls, they cannot be nipped and will keep popping out.

The twin-roll crusher is one of the most popular secondary, or reduction, crushers designed to produce a uniform product of controlled gradation. To get the desired final product, the roll crusher is often set with a "closed" loop to recycle the largest material that passes through it. This is actually oversize, i.e., slightly larger than the crusher setting, as will be explained with the grid, or percentage, chart shortly. To avoid overloading a roll crusher, it is best to screen out of the feed any material which is below the setting and let it bypass this crusher.

A triple-roll crusher has a third roll almost directly above the stationary roll of a twin-roll crusher. Material enters this type crusher between the upper roll and the stationary, powered one and then passes for a second crushing between the two set like a twin-roll crusher, as shown in Fig. 10-4. Adding the third roll to this type of crusher increases the total possible stage of reduction for the feed material. However, its capacity is about the same as a twin-roll crusher of the same size and minimum setting. The power requirement for a triple-roll crusher is slightly higher than for the twin because it does more work in reducing the material's size.

The advantages of a triple- or twin-roll crusher are the following: (a) they have a large stage of reduction possible, so (b) they can take larger feed than other reduction crushers and, consequently, (c) they permit a primary crusher to have a larger setting ahead of the reduction crusher.

Impact Breakers and Hammermills. Some aggregate crushers use the age-old technique of breaking rocks with a sledge hammer, but in a controlled, high-production manner with mechanical, rather than man, power. The impact breaker is one of these which is used as a primary

FIGURE 10-4 Cutaway view of triple-roll crusher
(courtesy of Pioneer Division,
PORTEC, Inc.).

crusher. A hammermill is used more as a reduction crusher (Fig. 10-5). So these two are like the gyratory and cone crusher alternates.

The impact breaker, or impactor, is a large-size crusher that operates like the one pictured in Fig. 10-5 with one or perhaps two rotors, each equipped with three or more rows of projecting hammers around its circumference. It is designated by the feed opening like a jaw crusher's identification. The broken material hit by hammers moving at high speed is thrown against the sides of the crusher chamber, which are breaker plates, bars, or gates, and each impact breaks the material further. Unlike the hammermill pictured in Fig. 10-5, the impact breaker has no grate bars at the bottom so material can flow freely below the rotating breaker. The maintenance of the impact breaker's parts is a problem, especially when used with extremely hard or excessively abrasive material. It is used mainly for a single-pass crushing of stone or rock that is relatively soft.

A hammermill works like the impact breaker with the added crushing benefit from the hammers on the under side of the rotor grinding material down to fine material for finish size. This crusher applies attrition action

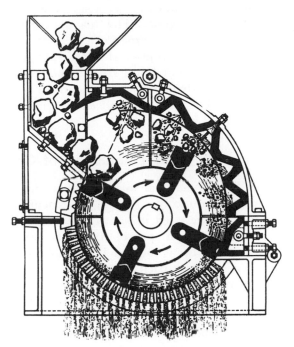

FIGURE 10-5 Cross-sectional sketch of the hammermill.

in the final grinding and so is used for sticky materials where a lot of fines are required. It is designated by the diameter to the hammer tips and the distance between side walls, both in inches. Its capacity, which will be noted in the following section, varies not only with the size of the hammermill but also with the size and properties of stone fed to it and the speed that the hammers are rotated.

Other Crusher Equipment. The types of crushers commonly used in the production of construction aggregate have been described. Others that are frequently found in custom-made plants are the rod mills and ball mills. These are all of the impact type used as reduction crushers to produce fine-sized materials. They will receive material previously crushed to about 1″ size and produce fine sand for construction aggregate.

A newer piece of equipment useful for this purpose is the centrifugal crusher. It takes minus 1″ stone in about a 5-foot-outside-diameter cylinder with an inner distributing table and shaft that revolves at speeds of 1500 to 1800 rpm. This throws the particles outward centrifugally against well-located breaker plates so that their impacts are generally perpendicular. Throw shoes and breaker plates are easily removed or reversed for maintenance purposes. The manufacturer, CIMCO Inc., claims the resulting product is sharp-edged, cubical fines suitable for finish asphalt or concrete mix. It consists of minus 1/4″ gradation material down to 100-mesh size. The centrifugal crusher produces 70 to 80 TPH with a 125-hp diesel or electric power unit. This whole crusher and power unit mounted on a suitable frame is small enough to be considered for a portable installation.

10.1.3 Capacities and Selection of Crushers

Aggregate crushers handle material to turn out a final product useful for some construction. Because this type of equipment is part of a production process, we are interested in knowing how much material each crusher can handle in a certain span of time. The usual term for this rate of production of a crusher is tons per hour—abbreviated TPH. This rate is known as the crusher's capacity.

The capacity of a crusher is dependent, first of all, on its setting, which is adjustable for each use. Each crusher has a range of settings. The limits of the range indicate in one way whether the crusher is designed to be fed large rock material, medium rock and gravel, on down to sand-size material. The crushers for large material have an input feed opening big enough to take the largest block of material it can handle. Maximum size of the material limits the capacity of a large crusher in that way. For the crushers handling small feed material, mainly the reduction crushers, the feed opening is dependent more on what equipment will charge the crusher with fresh material. It may be a conveyor belt, hopper spout, feeder chute, or screen. The crusher's capacity would be governed basically by what TPH it can handle at the given setting. It must not be fed material at a higher rate than its maximum TPH, or the crusher will be overloaded and become jammed.

The crusher size and settings are design features that govern its capacity. Other parts of a crusher, including its power, are designed to match, or balance, its dimensions and expected capacity. This is certainly true of jaw-type and gyratory crushers. In the case of impact-type crushers the speed of the hammers and the rate of feed also govern the capacity.

The capacity of aggregate crushers is dependent not only on designed features of the equipment but also on the material and how it is handled. Properties of the material, such as its hardness, stickiness, and shape, will affect a crusher's capacity. It is also affected by the way material is fed into the crusher. If the material comes in surges instead of an even flow, the crusher will likely have a lower capacity. Also, each type of stone or rocky material crushes differently.

With all of the variables affecting the rate of production of a crusher, it is difficult to give a positive number for its capacity. The manufacturers of this type of equipment provide tables on their crusher capacities. But they point out that these values may vary as much as 25% higher or lower than the actual production realized in use. Therefore, any capacity tables should be used only as a guide for estimating production. This emphasizes the need for checking an installed crusher to be more certain of its rate of production in operation.

Some further understanding of the variables and capacities can be gathered from the data provided by the manufacturers of crushers. Tables 10-2 and 10-3 are for typical examples of the crushers described in Sec. 10.1.2. The capacities are based on a full, continuous feed to the crusher. From these capacity tables several significant differences can be noticed between the two common types of primary crushers. A jaw crusher has a wider range of settings—generally, a maximum from two to three times the smallest setting. As noted in Table 10-1, a jaw crusher has a maximum ratio of reduction almost twice that for a gyratory crusher. From the tables above it can be observed that for a comparable maximum size of feed and setting, a gyratory crusher has much higher capacity than a jaw crusher. The gyratory type has this advantage without much increase in power to drive the crusher.

The selection of an appropriate primary crusher for a given use will have to be based on a variety of factors. These are not limited to the design features of the crusher. If the feed to the crusher is blasted rock from a quarry, the size and method for handling the feed are factors. For instance, a power shovel with its dipper has a limit to the maximum size of rock it can handle well. A 1-yard shovel cannot easily handle a rock over 2 1/2 feet on its maximum dimension. Therefore, a 1-yd shovel cannot be used to feed a jaw crusher larger than the 30 x 42 size; likewise, a 2-yd shovel can feed no larger than a 42 x 48 jaw crusher. If the largest 66 x 86 jaw crusher or 60″ gyratory crusher is to be fed from a quarry, where a shovel is loading the raw material, it will have to be at least a 5-yard shovel. It may be more economical to reduce the blasting, resulting in larger rock that can be handled by the loader-hauler combination and still fit in the primary crusher. Undoubtedly, a large ratio of reduction will then be required of the primary crusher.

If, for another job, the feed is from a gravel pit with relatively small maximum-size pieces, a large feed opening is not needed. It may be more economical to feed all of the pit run material into the primary crusher

TABLE 10-2 Capacity (TPH) and Horsepower for Jaw Crushers (a) Overhead or (b) Blake type

Type	Crusher Size, inches	Power Required, hp	\(\frac{3}{4}''\)	1″	1½″	2″	2½″	3″	4″	5″	6″	8″	10″
			\multicolumn{11}{c}{Setting of Crusher at Close of Stroke}										
a	10 × 20	20–30	5	10	17	24	33	50					
a	10 × 36	30–40	10	18	30	43	60	90					
b	15 × 30	60–80			33	43	53	62					
b	18 × 36	60–80			42	61	77	93	125				
b	30 × 42	100–130					125	150	200	250	300		
a	42 × 48	180–210								430	515	680	855
b	42 × 48	125–150								380	420	510	580
b	48 × 60	150–250									480	570	660

Note: The 42″ and 48″ crusher can be set as 7″, 9″, 11″, and 12″.

TABLE 10-3 Capacities in Tons per Hour (TPH) of Gyratory Crushers with Straight Concaves

Feed Opening, approx. inches	Shaft Speed, rpm	Power Required, hp	Open Side Setting at Discharge Point									
			1½"	2"	2¼"	3"	4"	5"	6"	7"	8"	10"
8 × 35	450	15–25	30	41	47							
13 × 44	375	50–75			85	133						
16 × 60	350	60–100				130	210					
30 × 98	325	125–175					310	390				
42 × 143	300	200–275						500	630			
60 × 196	250	225–300								900	1110	1530

Note: Gyratory crushers are designed to be set at 1¾", 2½", 2¾", 3½", 4½", 5½", and 6½", where these are within their range of settings.

rather than separate out the part that is already smaller than the crusher setting. That would call for a crusher with a higher capacity. There are many variations to these "if" situations, so the planner must select his crushers with the total operations and economics in mind.

The selection of secondary, or reduction, crushers is also a complicated determination. To help with the understanding of those crushers that handle smaller feed, some typical data from their manufacturers is given in Tables 10-3 and 10-4.

The data in the preceding tables are given to provide more understanding of the commonly used crushers for aggregate production. For any particular material-processing situation, it will be advisable to make preliminary determinations about the types of crushers needed. Then the planner should request suitable manufacturers of the equipment to make a detailed determination for his job and give quotations for an installation.

A few general statements will serve to summarize the main criteria involved in the selection procedure. The economic selection of any particular crusher is dependent on its ability to handle the maximum size of feed, reducing this at the highest possible ratio of reduction and at a minimum expenditure for original installation, maintenance, and power. In the case where most of the feed is coarse and stage crushing is required, primary crushers that meet the requirements of the previous statement and have straight crushing surfaces will be most economical. Where only a very small percentage of the feed approaches the crusher's size of feed opening, non-choking crushing surfaces in a high-capacity crusher may be advisable for overall economy's sake. If the total installation requires staging, and several different types of crushers could be used for each stage, it is necessary to do a cost analysis of each feasible combination to find the crusher plant with the lowest total cost. Such a cost analysis is generally a custom job to be done by manufacturers of the crushing equipment.

Grid Chart for Crusher Output. The previous section discussed crusher selection based on the maximum size of feed and the crusher's capacity or output by its setting. A given job's need for aggregate materials has more specific requirements than those. The aggregates are frequently specified by a full set of gradation limits, as well as properties of the material. These are given in the specifications published by ASTM, AASHO, various highway or airport building agencies, and other contracting authorities.

When rock is crushed, the product includes part at the crusher setting

TABLE 10-4 Capacities in Tons/Hour (TPH) of Twin- or Triple-Roll Crushers

Crusher Size,[a] inches	Maximum Feed Factor[b]	Power Required, hp	Width of Opening between Rolls, s						
			1/4″ with (F/s)	1/2″	3/4″	1″	1 1/2″	2″	2 1/2″
16 × 16	0.68	15–30	15 (3.7)	30	40	55	80	110	140
24 × 16	1.02	20–35	15 (5.1)	30	40	55	80	110	140
30 × 18	1.27	50–70	16 (6.1)	33	45	66	95	130	155
40 × 24	1.70	60–100	24 (7.8)	45	65	90	135	175	220
54 × 24	2.30	125–150	30 (10.2)	54	80	120	175	220	280

[a] The first dimension is the diameter of the rolls, and the second is the length of the rolls; the largest two listed roll crushers can be generally set at 3″ and 4″ also.

[b] Referring to Eq. (10.1), the value $0.085R$ is called the maximum feed factor here; the values in this table are for a twin-roll crusher and would be slightly higher for a triple-roll crusher.

TABLE 10-5 Capacities in Tons/Hour (TPH) of Hammermills (built by Allis-Chalmers Mfg. Co.)

Size Feed Opening, inches	Size Feed, in.	Shaft Speed, rpm	Power Required, hp	Capacities, TPH, for given Openings between Grate Bars, in.						
				1/8	3/16	1/4	3/8	1/2	1	1 1/4
6 1/4 × 9	3	1800	15–20	2½	3½	5	8	10		
12 × 15	3	1500	50–60	9	13	17	23	29	36	39
15 × 37	6	900	150–200	27	37	47	60	71	97	105

Data from R. L. Peurifoy.[3]

and others at sizes below that down to dust size. To help predetermine the full gradation of crushed material, a grid-type chart is useful (Fig. 10-6). It was originally developed using fairly ideal stone and is typical of

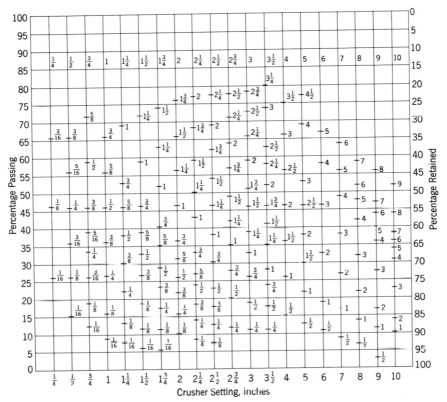

FIGURE 10-6 Percentage chart for crusher output.

a material breakdown. This gives the percentages passing or retained on a screen of each standard size for each given setting of the crusher. It was developed for pressure-type crushers, primarily the jaw and roll crushers, and cannot be used for impact-type crushers.

A characteristic in the design of the pressure-type crushers will be found on the percentage chart. The manufacturers of this equipment allow enough play or flexibility in the adjustable crusher face to let about 15% oversize material through. For example, when a jaw crusher is set at 2″, 15% of the material passing will not pass a 2″ screen opening. That material will be just slightly larger than the size of the setting. The crusher is designed this way so that an extremely hard, uncrushable piece will not jam the crusher.

The grid, or percentage, chart is read by entering on the vertical line given for the crusher setting being used. Then follow that line up or down to the size of opening or screen for which the percentage passing or retained is sought. With that size point found on the vertical line, read the desired percentage to the left or right. The point made about oversize material is found by noting points on each vertical setting-line giving the same size at the 15% retained level.

An example will serve to illustrate the use of the percentage (grid) chart. If a twin-roll crusher is set at 1″ to handle 40 tons of material per hour, (a) how much of the output will pass a screen with 3/4″ openings (i.e., minus 3/4″ material) and (b) how much of the material—percentage and tonnage—will be between the 3/4″ and 1/2″ sizes? The solution to each part is:

(a) on the 1″ vertical setting line, the 3/4″ reading is at 66% passing (34% retained). So the output will have (.66)40 = 26.4 TPH of minus 3/4″ material.

(b) on the 1″ setting line, the 1/2″ reading is at 46% passing. The amount of 3/4″ to 1/2″ is (66%–46%) = 20% of feed. The output in tonnage of 3/4″ to 1/2″ is 8 TPH.

10.2 Feeders and Related Plant Components

An aggregate production plant will have components such as feeders and scalping units to direct material to be crushed, separated, or stored for later use. The feeder components are used mainly to handle input material for the plant. They are of two types, called apron feeders and mechanical, or reciprocating-plate, feeders.

Apron feeders are generally used for quarry rock to be fed into a primary crusher. They are of a heavy-duty construction to take the shock from rocks dumped directly on them. To cushion the blow of material dumped on the feeder, a large hopper may be used to receive the dumped loads. The feeder is really a series of overlapping pans or plates forming a continuous chain like a conveyor belt. They will be on a slight incline and driven in the downward direction. Feeders are made with widths ranging from 2 1/2 to 8 feet and lengths two to three times their width. A feeder can be driven by a 5- to 20-horsepower motor, depending on the size and expected load on it.

The mechanical, or reciprocating-plate, feeders are generally used for material from a gravel pit. Frequently, the input has some material small enough that it does not have to be crushed, and it should be directed out of the main stream of material. The reciprocating plate is driven by an eccentric which is powered with a 3- to 20-horsepower motor to give the feeder a stroke for throwing the material along its length.

The main point to consider in the selection of the right size of feeder is to be sure that it can keep an even flow of material to the primary crusher. The feeder's size and speed should be enough to have 25 to 35% more capacity than the crusher. It may be necessary to incorporate hand controls to regulate the feeder in the case of irregular dumping of the raw material.

A scalping unit is frequently used in conjunction with the feeder. The scalper is either a stationary grate or grid, or it might be powered with vibratory motion. Its purpose is to remove some material between the feeder and the primary crusher. The scalping unit saves the plant from putting material that is too large in the crusher and thereby clogging it. Or the scalper makes it possible to by-pass a material that is smaller than the crusher's setting, thereby allowing the use of a smaller crusher with less capacity and more economy.

A grizzly bar design is most commonly used for scalping units. This design is made by evenly spaced, parallel, and inclined bars running in the direction of the feed to the crusher. Then the stone that is too large will pass along the bars and be discharged at their end while the smaller material drops through. Therefore, the spacing of the bars is set for a desired size of feed. Scalping screens may also be used after each crusher to remove or recirculate the oversized material. More will be said about this in the next section.

10.3 Screening and Screens

Any aggregate production plant or process has a vital need for screening to direct, separate, and control the material flow in the process. The screens for this purpose are used in connection with crushers and aggregate grading and washing plants. The two main purposes for screening in the aggregate process are (1) for "scalping," to remove oversize or undersize material in the crushing plant, or (2) to carry out a complete sizing of materials being produced. An aggregate production plant will have some of both functions as a part of the total system. An example to help visualize these uses for screens can be given in reference to the sort of portable plant shown in Figs. 10-1 and 10-2. A portable plant may have a 2 1/2-deck or layers of screens at the beginning of the process to catch the initial input on the top deck. The decks or layers that are parallel will be separated by enough distance to allow the materials to move between decks. The material retained on the top deck will go to the primary crusher. That which passes through the top and is retained on the second deck goes to the reduction crusher. The output of both crushers is recirculated to the top deck for repeat screening. Finally, all of the material that passes the second deck is ready for classification, with the excess sand removed through the bottom 1/2-deck. Now it will be helpful to have a brief discussion of screen designs and some idea of how they might be selected for a plant.

10.3.1 Design Features of Screens

A screen is generally made of interwoven wires forming a mesh with regular openings between adjacent wires. The three basic types of screens used for aggregate processing are known as (1) inclined vibrating screens, (2) improved horizontal screens, and (3) revolving screens.

The inclined vibrating screen has a flat plane to receive material on a slight incline. It is vibrated in a circular direction about an axis perpendicular to the plane of the screen. The circular motion is given to it by force from eccentric parts on the drive shaft. This makes the screen throw the material to advance it down the incline of the screen. Rubber mounts for the screen isolate its vibration from the frame supporting it.

The improved horizontal screen is a modernized version of the old shaker screen with improved effectiveness made possible by higher-speed motion with shorter stroke. This variety is similar to the inclined vibrating screen but needs less headroom. Since it is held in a horizontal position, this screen is not used for scalping.

The revolving screen is essentially a large drum with perforated sides or screening that has been shaped into the form of a cylinder. The drum revolves slowly on its inclined, longitudinal axis. Material passed into it from the upper end of the cylinder moves inside the screen drum until it passes through the side openings or out the lower end. The rate of material handled in this screen depends on its inclination and speed of revolution. The revolving screen may have increasing sizes of openings along the drum length. Or it may be designed with concentric screen drums of increasing diameter to give various separations with one rotating screen unit. The effective screen area for finding capacity of the revolving type is usually figured as one-third of the diameter (D) times its length (L), i.e.

$(1/3)D \times L$. If the revolving screen is a perforated plate, it is cheaper than either of the flat area screens and finds extensive use in washing plants.

For the effective use of screens in an aggregate production plant, it is well to give the design careful attention. This may require the services of the experienced manufacturers of this type of equipment for any given installation. Frequently, the screens are designed in combinations such as the parallel deck arrangement with the example of the portable plant suggested above. When more capacity is needed for any screen, they may be used in tandem instead of parallel to increase the total area and, so, capacity for each screen level.

The screens are fed from one of various types of plant components. They may be loaded from conveyors, bucket elevators, or some other type of feeder. If a conveyor is used to feed the screen, it should pass the material in a direction opposite to the incline for the screen to be most effective. On the other hand, a conveyor feeding a revolving screen will be moving in the same direction as that screen's downward inclination. In the case of any unit feeding a screen, it should not make the material fall freely on the screen any greater distance than necessary, to minimize wear and tear of the screen. To help prevent a screen from being clogged by sticky material, it may be desirable to have the screen heated. This can be done by an electrical hookup with current being passed through the wires of the mesh, heating them to perhaps 100° to 300°F. In the case of washing plants there is not much danger of the material clogging the screens.

Generally, the manufacturers who make screens will specify the capacity they expect in tons per hour (TPH) per square foot of screen area. This is for a fresh (unscreened) material coming directly on the screen with a good graduation of sizes, with the largest material not too much larger than the size of the screen opening. Certain key factors have to be taken into account in determining the screen area required, based on the variation in the material it handles and its location in the plant arrangement. One such factor might be called the deck correction. A top deck should have capacity as specified by the manufacturer, but each succeeding lower deck will have a 10% reduction due to the interference of the decks above it. For example, the third deck in a set would be 80% as effective as the top deck. Wash water or water spray will increase the effectiveness of screens with openings of less than 1″ in size. In fact, a deck with water spray on 3/16″ openings will be more than three times as effective as the same size screen without the water spray.

The other key factors for screen effectiveness have to do with the size or gradation of the material coming onto the screen. The manufacturer's screen capacity rating is generally based on 40% of the aggregate coming onto the deck being less than half of the opening size. If only 10% of the material is of that small a size, the screen will be only 50% as effective. At the other extreme, if 80% of the aggregate coming onto the deck is less than half of the screen opening, the screen can handle more than two times as much material as the manufacturer specifies. There are also corrections necessary in determining screen area for the case of a large amount of oversize material coming onto the screen. The manufacturer's capacity rating is generally based on 20 to 30% of the material coming onto the screen being of a size too large to go through the openings. If there were 50% oversize on the screen, it would be only 90% as effective; and 70% oversize would make the screen only 80% as effective as it is with the manufacturer's expected amount of oversize material on it. A full spectrum of these correction factors for screen area determinations may be obtained from the screen manufacturers. An aggregate equipment planner can find his own set of factors by trial and error with a set of interchangeable screens and material of various gradations.

10.4 Combining Aggregate Production Components

There are many ways and variations in putting together a total aggregate production plant. The variety of crushers discussed in Sec. 10.1 can be considered and lead to many combinations. Then the components that feed these crushers and take the material from them for recirculating or directing for grading the outputs can be arranged in a number of ways. Therefore, it is obvious that space in this book cannot be devoted to discuss typical setups. Some ideas can be gained from an example to show the approach to the solution of an aggregate equipment problem.

With the various alternatives and complexity of each unit, some simplification is needed for a clearer analysis. This is desirable in trying to solve any aggregate production plant problem. The simplification will be shown in the next section, dealing with flow diagrams.

10.4.1 Flow Diagrams for the Aggregate Process

An aggregate flow diagram is simply a line sketch of the various components in the system with arrows showing the direction of materials flowing through the plant. The simplification leads to just a sloping dashed line for a screen or scalping unit and pairs of parallel lines surrounding small circles at the ends to represent feeders and conveyor belts. The various crushers are also shown in simplified fashion. A roll crusher is simply two circles with a slight gap between them to show where aggregate material goes through it to be crushed. A jaw crusher is two nearly vertical lines converging toward the bottom where the crushing takes place. A hammer-mill can be shown by a larger circle with three or four radial lines from a central axis to show the hammers that revolve in it. These simplified representations for the various aggregate process components are shown in Fig. 10-7.

This example of a flow diagram shows a three-stage gravel plant, schematically. It lays out the various components to show the interrelationships and functions of the units. Obviously, the intricate details of each component and their support mechanisms are omitted for clarity. This is the sort of scheme that can be used to advantage in working out the solution of an aggregate plant problem.

10.4.2 Example of an Aggregate Production Solution

To give the construction equipment planner some scheme for solving a problem of aggregate production, a case requiring several crushers with feeders and screening will be shown. This is not intended to be a typical situation, though it does involve common crusher and screen units.

We will take the example of a quarry rock of 12″ maximum size being handled in a two-stage crusher plant at the rate of 70 tons per hour. The maximum size of output is to be 1 1/2″, and we will want a separation of materials over 1″ size and the minus 1″ output. The crushers that might be considered are those included in Sec. 10.1.2. For the screens to be considered, the manufacturers have given the information that 1 1/2″ screening has the capacity of 2.7 TPH per square foot and 1″ screening has 2.1 TPH/sq. ft. The solution to be found will include selecting adequate and economical-sized crushers for the two stages and the sizes of screens between them and below the secondary stage.

The solution to this problem with a suitable flow diagram follows.

In the selection procedure it was decided that a jaw crusher would probably be most economical as the primary crusher. The jaw crushers considered available are those listed in Table 10-2. As far as the maximum size of rock in the feed is a determinant, a 15 x 30 jaw crusher could be used. But it does not have enough TPH capacity at its widest setting (3″) shown. Also, a 3″ setting of the primary crusher would probably require too high a stage of reduction for an economical secondary crusher. So the selection is an 18 x 36 jaw crusher set at 2 1/2″, which has a capacity of 77 TPH—

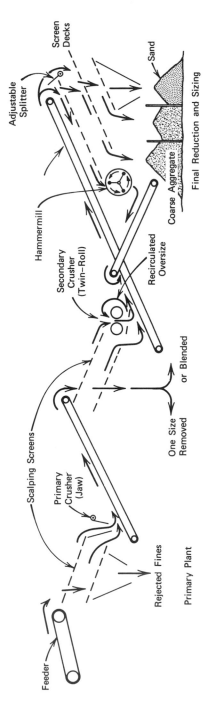

FIGURE 10-7 An aggregate process flow diagram.

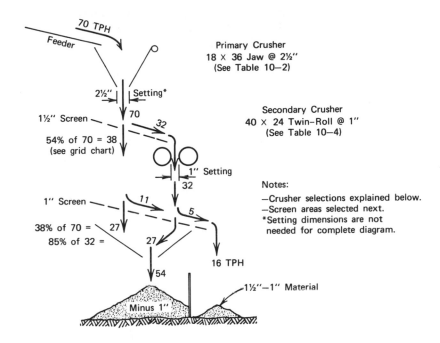

slightly above the expected input. A given condition was that there should be nothing larger than 1 1/2″ in the final material. Material this size and smaller can be screened out following the primary crusher to keep from overloading the secondary crusher. From the grid chart (Fig. 10-6) it is found that for a 2 1/2″ setting, 54% of the material will pass a 1 1/2″ screen, or 46% will be retained. This 46% of 70 TPH gives the 32 TPH fed to the secondary crusher.

For an economical secondary, or reduction, crusher a twin-roll crusher was selected from those given in Table 10-4. As far as capacity is concerned, the smallest 16 x 16 roll crusher set at 3/4″ might be used. But its stage of reduction limit is not high enough. With the given dimensions the maximum size of feed, F, is approximately 2 3/4″ (the increase over 2 1/2″ was explained following the grid, or percentage, chart). The 16 x 16 twin-roll crusher set at 3/4″ has a limiting $F = 0.68 + 0.75 = 1.43″$, according to Eq. (10-1). The stage of reduction at this setting would be $F/s \approx 2\ 3/4″ \div 3/4 = 3.66$. As indicated in Table 10-1, this would be high for the smaller models of twin-roll crushers. Basing the reduction crusher selection on the stage of reduction and, consequently, the maximum feed factor, the 40 x 24 twin-roll crusher in Table 10-4 will work with a 1″ setting. Its limit of feed size is $F = 1.70 + 1.0 = 2.70 \approx 2\ 3/4″$.

If the output of this crushing process should have less material of the larger-than-1″ size, the 40 x 24 roll crusher could have a closed loop. That is, it could have its oversize (15% of 32 TPH greater than 1″) recirculated through the roll crusher without exceeding its capacity. Then all material leaving that crusher with a 1″ setting would be of a 1″ size, or smaller.

Another possible solution to this aggregate process problem can be found with the data in this chapter. A 13 x 44 gyratory crusher with a 2 1/4″ setting has a capacity of 85 TPH. The maximum size of stone in the output is estimated from the percentage chart to be approximately 2 1/2″. All of the output from the primary crusher could be put through a 30 x 18 twin-roll crusher with a 1 1/2″ setting. (The same arrangement might have been planned for the first solution.) With 1 1/2″ the maximum size of final product, this roll crusher would have to recirculate its oversize material. Thus, it would be handling $70 \div 0.85 = 83$ TPH after the system is running for an hour. This is still below the 30 x 18 crusher's capacity of 95 TPH. The largest stone it can take, $F = 1.27 + 1.5 = 2.77″$, is larger than the 2 1/2″ largest size from the primary crusher. Output from this solution will show about a 33% greater amount in the 1 1/2″ size. Most crushing problems will be more specific about the proportions of the material output by sizes. The specifications, rather than economy, generally govern the crushers to be selected.

To find the required areas of screen, the planner needs to know the rate of material dropping on each as well as that material's consistency. The 1 1/2″ screen under the jaw crusher (see flow diagram) is the top deck. As explained in Sec. 10.3.1, no deck correction factor will be necessary. To find the relative amount coming on the screen that is half-size, make use of the precentage chart. With a 2 1/2″ setting the amount that is smaller than half-size on a 1 1/2″ screen will pass a 3/4″ screen. The chart shows this to be 30%. In that case the screen's effectiveness, based on the smaller-than-half-size material, is 80% of the manufacturer's capacity rating. The amount of oversize material dropping on this 1 1/2″ screen is all that will be retained on that size from a 2 1/2″ setting. The percentage chart shows that to be 46%. So the oversize correction factor should show an effectiveness of only about 90%.

Therefore, the corrected 1 1/2″ screen capacity is

2.7 (0.80)(0.90) = 1.95 TPH per sq ft, and for 70 TPH the minimum screen area required is 35.8 sq ft. Therefore, use a 4′ x 9′ screen to have width greater than jaw's 36″ length.

The 1″ screen is a second deck for the 38 TPH from the jaw crusher, so the deck correction factor is 0.90. Other screen corrections are found as discussed for the 1 1/2″ screen:

> half-size = 23%, so correction factor is 0.75;
> oversize = 11/38 TPH, or 29%, so factor is 1.0;
>
> therefore, capacity is 2.1 (0.90)(0.75)(1.0) = 1.42 TPH/sq ft;
>
> minimum screen required = 38/1.42 = 26.8 sq ft.

For part of 1″ screen below roll crusher:

> top deck, so deck factor is 1.0;
> half-size = 46%, so half-size factor is 1.1;
> oversize = 15 %, so oversize factor is 1.0;
>
> therefore, capacity is 2.1 (1.0)(1.1)(1.0) = 2.31 TPH/sq ft;
> minimum screen required = 32/2.31 = 13.8 sq. ft.

The design for the 1″ screening might use 4′ x 7′ under the jaw crusher and 3′ x 5′ under the 24″-long roll crusher. Or this could be one continuous 1″ screen 4 feet wide by 11 or 12 feet long.

As explained previously, the solution given above is just one of a number of possibilities. Each manufacturer of aggregate production equipment can probably arrange a workable solution. It might be best to have several manufacturers work out their solutions for the construction planner's problem. With each solution a projection of costs should be provided for the equipment installed. Then the planner for the aggregate production plant can make his analysis, based on how well the material specifications can be met and at what production cost. Frequently, an operating plant will have variation in its material output, so a certain amount of flexibility in the plant's production process is desirable. This may suggest to the planner that the best solution might be a somewhat more costly plant with desirable flexibility in its output.

References

1. "Producing Aggregates," A Reprint from *Construction Methods and Equipment,* McGraw-Hill Book Company (New York, N.Y., 1957).
2. Michaelson, Stanley and R. F. Friend, "Crushing Machinery . . . Its Selection and Application," article printed by Allis-Chalmers Mfg. Co. (Milwaukee, Wisconsin).

3. Peurifoy, R. L., *Construction Planning. Equipment, and Methods,* McGraw-Hill Book Company (New York, N.Y., 1970), Chapter 20, pp. 564–591.

4. Hansen, Hans I., Chapter 9, *Handbook of Heavy Construction,* Frank W. Stubbs, Jr., Editor-in-Chief, McGraw-Hill Book Company (New York, N.Y., 1959).

Chapter 11
■■■■■■■■■■■■■

■■■■■■■■■■■■■■■■■■■■■■■■■■■■■■■■■
CONCRETING
EQUIPMENT
■■■■■■■■■■■■■■■■■■■■■■■■■■■■■■■■■

Introduction to Concreting Equipment

The equipment for the production and placing of concrete deals with materials to be processed for construction use. The materials for concrete are aggregates—both fine aggregate, such as sand, and course aggregate, such as gravel or crushed stone—cement, water, and frequently some additives. Taken separately, some of these are natural materials; and the others, namely cement and additives, are processed materials themselves. When the right proportions of each of these materials are measured out and then mixed together in an appropriate way, they form the processed material known as concrete. This material is used for constructing foundations, structural elements such as beams, columns, etc. pipe and other conduit, and for paving on the ground.

Historically, concrete has been used for its various purposes throughout the 20th century. Until 1930 the emphasis was on controlling the mixing of concrete, and little attention was paid to the proportioning of the materials. The proportioning in those past times was strictly by volume, which could not be very accurate. The tools used to do that proportioning

467

were hand shovels and wheelbarrows to make a mix in the proportions of 1:2:4, 1:3:6, etc. The first ingredient in those three is cement, and the "1" would mean a sack of one cubic foot of cement material. Then the mix would include perhaps two cubic feet of sand and four cubic feet of gravel. The mixing in those early days was done so that the concrete could be controlled visually, and the man in charge would judge his mix by how it looked and its consistency. The mixes used were not too consistent, and when mixed by power equipment, it was by steam engines at first.

The early measuring of concrete materials by volumetric batching meant that each material had to be batched singly to insure a uniformity in the batches of concrete. To have a much better control of the concrete mix, modern batching has been done by weighing the various ingredients to make up the concrete mix. Weigh batching will be the emphasis of the following discussions on the various concreting equipment. The equipment to be considered in the following sections can be labeled separately as bins, batchers, mixers, concrete handlers, and certain special equipment for concreting.

11.1 Concrete Material-Handling Equipment

Generally, the concrete aggregates are moved from their shipping containers or storage to bins. The bins are used for temporary storage and, particularly, holding of the separate materials just before they are fed to the batchers to proportion the right amounts for the concrete mix. Aggregates are moved to these bins by various types of equipment discussed earlier in the book. These might be belt or bucket conveyors, a clamshell bucket operated from a crane-type unit, or front-end loaders used for low bins. The material from a bin is gravity-fed into the batcher directly underneath its hopper-type opening.

The cement material is held in an enclosed bin to keep moisture from prematurely reacting with the very small, active cement particles. Cement is delivered to its bin storage by special hopper-bottom railroad cars, tank trucks, or watertight barge units. The railroad hopper cars have about 400-barrel (bbl) capacity and the tank trucks about 250 bbl (approximately 100,000 lbs), whereas a barge tank can hold much more than either of these. A barrel of cement equals four cubic feet.

To unload a railroad car or truck into a hopper pit requires an enclosure to keep moisture from getting to the cement at this point. To take the cement from the hopper pit requires either a screw conveyor, enclosed bucket conveyor, or a pneumatic air-pump conveyor. These

FIGURE 11-1 Cement shipping and handling equipment (courtesy of Fruehauf Division, Fruehauf Corporation).

conveyors were discussed in Chap. 9. The screw conveyor would be used for a horizontal or inclined movement of the cement, whereas the enclosed bucket conveyor could be vertical or inclined. A combination of these might be used to move the cement material from the collecting hopper under the mobile shipping container to the bin storage. In the case of a pneumatic pump, its flexible pipeline can be used to make any horizontal or vertical movement necessary within reason.

For particularly small jobs which are inaccessible to a commercial ready-mix plant, old-time concreting equipment is still useful. This will include wheelbarrows and scales with a platform for the carrier and a vertical panel for weigh beams to proportion the concrete material. The wheelbarrows are loaded by shoveling from dumped piles of aggregate, and the material is weighed in the carrier on the platform scales, which have capacities running up to around 500 lbs. With such a simple concrete proportioning arrangement the cement will probably be used from sacks which weigh, generally, 94 lbs per sack. This sort of arrangement for very small jobs is useful in a remote, inaccessible area or on farm work with small yardage where it would be uneconomical to set up a portable concrete plant or transport from a ready-mix plant.

The various requirements for concrete material-handling equipment are generally expressed in terms of tonnage or barrels to meet a given rate of concrete in cubic yards per hour or pour. These capacities required for delivery of the materials and for the batching equipment are governed by the mixer to meet the set rate of concrete pouring. The meaning of such capacities will become more evident from the discussion in following sections.

11.2 Concrete Batching and Mixing

The separate granular materials to make up a concrete mix are fed from their bin storage into batching equipment. This is a component of concreting which gathers the correctly proportioned amount of each material by weight for a given concrete mix. That material may be coarse or fine aggregates, cement, and additives for air entrainment, controlled setting time, or other benefits. The water will generally be drawn from a tank and measured either by volume or by weight for its part of the batch.

The materials that have been properly proportioned in the batching equipment are then dumped together into mixing equipment to produce the wet concrete mix for delivery to concrete construction (Fig. 11-2). Since productivities are governed by the mixing equipment, batching and other equipment will be discussed after the mixers. The mixing equipment is designed to get the proper intermixing of the several ingredients with coating of all particles by the cement paste with whatever additives have been included. The mixers will include not only the enclosed container or drum with its designed features for mixing materials but also

FIGURE 11-2 A portable concrete plant (courtesy of the Koehring Road Division).

appropriate controls and recording devices attached to the mixer. These will be discussed in more detail in following sections.

11.2.1   Concrete Mixers

The concrete mixer is the piece of concreting equipment which generally governs the capacities required of all the interdependent concreting equipment. It has been found by extensive study that the mixer should be charged with all of its batched materials simultaneously, if at all possible. Other measures for getting the right mix of concrete material have been regulated in various ways over the years of developing this type of equipment. The mixers used to be standardized by the Mixer Manufacturers' Bureau[2] of the Associated General Contractors of America until that Bureau was disbanded around 1965. Now this equipment's performance is governed by various organizations, such as ASTM specifications for ready-mix concrete, the National Ready Mixed Concrete Association, and governmental agencies and other owner representatives.

A number of forms and sizes of concrete drums are used in the mixing equipment. These drums may be non-tilting and revolve on a horizontal axis (Fig. 11-3), or they may be tilting drums, the axes of which can be tipped from the horizontal. In Europe turbine mixers have been developed in recent years. They are cylindrical tubs rotating about their vertical axis with mixing paddles that revolve in a direction opposite to the tub rotation and inside its shell. Such a simple mechanism evidently produces good concrete. A variety of mixers are shown in Table 11-1. Other variables include the arrangements for loading and discharging the drums through openings designed for that purpose.

Recent studies have shown that introducing the water into the drum ahead of the solid materials makes a better mix. This apparently is due to the lubrication of the water on the sides of the drum preventing solid particle material from sticking and not becoming properly mixed. The water supply and measurement from a storage tank are best handled with a pressure of 150 psi to get positive delivery. The water tank capacity should allow mixing for 15 to 20 minutes without refilling to be more certain that the mixing cycles will not have to be stopped due to lack of water.

The drum volume is generally expressed in the model number for the mixer. That number represents the nominal capacity of the drum in cubic feet of mixed material. This is one-third to one-fourth of the actual gross inside volume of the drum. The reasons for this amount of extra space are for mixing the loose materials and because the drum must receive the

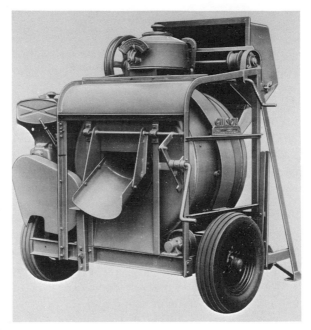

FIGURE 11-3 A small, portable, non-tilting concrete mixer (courtesy of Gilson Brothers Company).

initially unmixed material with a lot of void space in it. Standards have been followed in the past allowing a mixing drum to receive the batched material to produce a mix equal to the nominal cubic foot capacity of the drum plus ten percent. The use of this standard will be shown shortly.

The production rate of a mixer is governed by certain well-established standards. The ASTM specifications for ready-mixed concrete, designation C94-69, requires that each concrete batch be mixed at least one minute for the first cubic yard plus 15 seconds for each additional yard or fraction thereof. The mixing cycle is governed by the drum speed, and for best results that should be a minimum of four revolutions per minute (rpm) and a maximum of 18 rpm.

The maximum production rate for a mixer can be calculated in a simple way. The calculation merely requires using the above-mentioned standards, which can be applied to mixers remaining in one place to work. It can be formulated as

$$q_m = \frac{60(V)k}{27(c + m)} = \frac{2.22k(V)}{(c + m)},\qquad (11\text{-}1)$$

where k = amount of rated volume allowed—1.10 for S-type standard
      mixers and 1.20 for E-type paving mixers;

V = rated volume of the drum (its model number), cubic feet;

c = minimum time for charging and discharging mix, minutes;

m = specified minimum mix time, minutes; and

q_m = concrete produced, cubic yards per hour.

For example, with an estimate of three-quarters of a minute for charging
time and 1 1/4 minutes mixing time, a 28S mixer would produce

$$q_m = \frac{2.44\,(28)}{2.0} = 34.2 \text{ cu yds/hr}$$

assuming the charging is load and unload.

Data on the productivity and other design features of a variety of mixers
is given in Table 11-1. The 3 1/2S mixer is designed to take a mix with
one-half bag of cement, the 6S mixer is for a 1-bag mix, the 11S for a 2-
bag mix, and the 16S for a regular 3-bag or a rich 4-bag mix of concrete.
These are very portable mixers.

The E-type mixers, called "pavers" because of their mobility, used to
be very useful for pouring concrete for extensive ground slabs or highway
pavements. These have been the single-drum 27E and 34E pavers and
dual-drum 16E and 34E pavers. They are all non-tilting mixers that are
loaded at one end and discharged at the other, so that the drum does not
have to be reversed and can be highly productive. A paver-type concrete
mixer is shown in Fig. 11-4.

Dual-drum pavers have two cylindrical compartments end to end on
the same horizontal axis. Such a mixer can handle two batches at a time.
When the first batch has been mixed for half of its time, it is moved to the
second compartment, allowing the next batch to be charged into the
mixer. As the first batch is discharged, the second batch is moved into the
second drum compartment, and a third batch can be charged into the
paver. This process increases efficiency and reduces the time consumed
for each batch.

Since specifications now permit the hauling of wet-batched concrete to
the paving site, the pavers are not seen nearly so often as they were for the
two decades following World War II. Now a more familiar piece of equipment for concrete paving is the concrete spreader used in combination
with wet-batch hauling trucks. The concrete paving spreader is discussed
in Chap 12. The larger pavers have a boom to serve as the track for the
pouring bucket. This 25- to 40-foot-long boom can be swung horizontally

TABLE 11-1 Information on Concrete Mixers

Mixer Model (Size)	Design Features of Primary Interest						
	Charging Skip	Drum Tilting	Openings	Mount Wheels	Compartments	Max. Drum Vol.,[a] cu ft	Theoretical q_p,[b] cu yd/hr
$3\frac{1}{2}$S	no	some	1 or 2	2	one	16.5	5.7
6S	yes	some	1 or 2	2 or 4	one	27.4	9.8
11S	yes	no	1	2 or 4	one	48.3	17.9
16S	yes	no	1	4	one	68.4	26.0
28S	some	some	1 or 2	4 or *	one	114.	34.2
16E	yes	no	2 or 3	4	two	68.4	42[c]
27E	yes	no	2	crawler-truck	one	110.	47[c]
34E	yes	no	2 or 3	crawler-truck	one	137.	59[c]
34E	yes	no	2 or 3	crawler-truck	two	137.	90[c]
35S	no	no	2 or 3	4 or *	two	141.	85
56S	no	some	1 or 2	*	one	240.	61
84S	no	some	1 or 2	*	one	336.	82
112S	no	some	1 or 2	*	one	436.	100
larger	no	some	2	*	one		

* Mounted on skids or a supporting frame.

[a] Volume shown is for a single compartment of the non-tilting-drum type; the volume of one compartment in a single-opening, tilting drum will be about three-fourths of the volume shown.

[b] Based on charging and discharging time (c) of $\frac{1}{2}$ minute for up to 1-yard-size models and $\frac{3}{4}$ minute for larger sizes, and a mixing time (m) of one minute minimum up to 1 yard plus 15 seconds for each additional yard of fraction thereof, per ASTM designation C94-69.

[c] Three-fourths as much time as for the S-type is assumed for the total charging-mixing cycle time of paver-type (E) mixers.

FIGURE 11-4 The concrete paver, formerly a familiar sight (courtesy of Rex Chainbelt, Inc.).

through about a quarter circle to help pour a wide area. It can also be raised 30° to 45° above the horizontal to pour concrete from the bucket at a higher level or into a holding hopper.

The larger stationary mixers, such as the 56S, may have a non-tilting or tilting design. These are used in permanent concrete plants. Non-tilting mixers are simpler for low cost and dependability, using a relatively lightweight supporting frame. The tilting design can handle larger aggregates more easily and will discharge its mixed concrete more rapidly than a non-tilting mixer.

The mobile truck mixer is designed for different conditions of use from those for the paver or stationary mixer. It needs to be more compact because it travels over roadways while working. This means its drum will have less total volume for a given amount of mixed concrete. To get the necessary amount of mixing in less space for the materials will take more time. That is quite possible for a truck mixer which can be mixing the concrete while it is traveling.

A truck mixer is designed to mix a batch that takes up more than half of the total drum volume. The mixed batch may occupy slightly more than 60% of the gross volume (V) of the drum. With so little open space in which to turn over and mix the materials, it is necessary to revolve the drum more times. It has been found that between 50 and 150 revolutions at 8 to 18 rpm are necessary to get a well-mixed concrete.

Transit mixer trucks, as they are sometimes called, are standardized by the Truck Mixer Manufacturers Bureau (TMMB)[3] which is affiliated with the National Ready Mixed Concrete Association (Fig. 11-5). Such a

FIGURE 11-5 A transit truck mixer (courtesy of Construction Machinery Co.).

mixer may have one of three types of drum bodies. These types can be described as:

1. the horizontal-axis, revolving-drum type;
2. the inclined-axis, revolving-drum type; and
3. the open-top, revolving-blade or paddle type.

The different types have some obvious differences in their operation. Each has slightly different drum size, but these are found by similar standards.

The standard sizes of transit mixers are at half-yard increments from 3 1/2 to 9 cubic yards and then stepping up at whole-yard capacities to the largest 16-yard size. The volume of the different types of drums to have the cubic yard capacities (C) specified varies somewhat. However, it is basically set by TMMB as follows:

$$V \approx 47C - 19, \tag{11-2}$$

where V = the drum volume in cu ft, and
$\quad\;\; C$ = capacity of mixer in cu yds.

With a simple conversion of the units it is found that the drum volumes are between $1.6C$ and $1.75C$. This shows that the mixed concrete, C, occupies a maximum of nearly 63% of the total drum volume. Sometimes the mixer receives pre-mixed concrete and is used only to agitate the concrete. The mixer is allowed to receive 1.30 to 1.36 times as much pre-mixed concrete for agitating as it can take in dry-batched material to mix.

Now a formulation for maximum truck mixer production can be developed, similar to the previous one for the S-type mixers and E-type pavers. Using the controls mentioned above,

$$q_t = \frac{60\,C}{c + R/r + e}, \tag{11-3}$$

where C = mixer capacity, cu yds;
$\quad\;\; c$ = charging and discharging time, minutes;
$\quad\;\; R$ = number of revolutions required;
$\quad\;\; r$ = speed of drum revolution, rpm;
$\quad\;\; e$ = total time empty between loads, minutes; and
$\quad\;\; q_t$ = concrete production, cu yds/hour.

For example, an 8-yard transit truck mixing a batch for 100 revolutions at 10 rpm, and estimating charging-discharging time as 2 minutes and time between loads as 6 minutes, could produce

$$q_t = \frac{60 \times 8}{2 + 10 + 6} = 26.7 \text{ cu yds/hr.}$$

To get this high a production requires that the time between loads (e) covers return travel and any waiting in line at the batch plant or after the 100 revolutions before dumping at the construction site.

The mixing action in a mixer is produced by baffles shaped in a helical fashion within the drum, which moves the material in a direction away from the charging end. If the mix is discharged at the opposite end, the drum does not have to be reversed; but when the discharge is out the same opening as it was charged, then the drum must be reversed to get the counter-action back to the opening. One particular drum arrangement is shown in the cutaway sketch of Fig. 11-6.

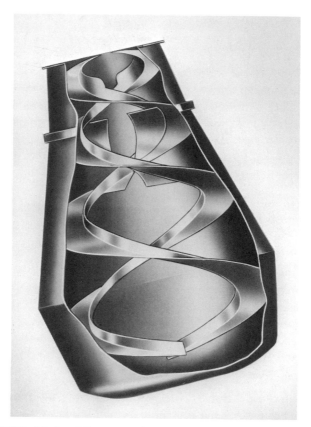

FIGURE 11-6 The inside of a concrete mixer drum (courtesy of Construction Machinery Co.).

Some specifications for mixer drums have limited the drum speed to a peripheral velocity of 225 fpm, which is approximately 10 rpm. This is based on the theory that at higher speeds some material would be held on the circumference of the drum by centrifugal force and not be mixed. Tests conducted in 1969 by the **National Ready Mixed Concrete Associa-**

tion proved that this was not a problem. That study[4] showed that mixing at 18 rpm is beneficial to the concrete. However, the high rate of drum speed produces more stress in the mixer, requires greater power supply, and may reduce the equipment life.

The major concern for uniformity of the concrete mix shows up in several key factors. ASTM specifies that a mix should have its coarse aggregate proportion within 6% of the designed weight, its air entrainment within 1% variation, the slump within a 1-inch range for workability, and its compressive strength within 7 or 8% of the designed value. It has been found that these measures of uniformity are helped by controlling the rate and number of revolutions of the drum during the mixing process. Also, the size of the batch for the drum and the lapsed time between batching and mixing of the concrete are important. Factors which have not been specified are the sequence of batching, the degree of blending of materials during the charging, and the shape of the drum and its baffle blades. It will be helpful to mention a result of study on the matter of blending the mix materials. The best concrete seems to result from ribbon loading of the mixer. This can be done by loading with a conveyor belt which helps to blend the sand, gravel, and cement even before it gets into the mixer drum.

Most mixers have some automatic controls to insure the desired uniformity and quality of the mixed concrete. A batchmeter controls the mixing time and also the action of the discharge mechanism. On a mixer equipped with a skip to charge it, the batchmeter is automatically started for each batch when the skip is raised to dump a batch in the mixer. A bell rings when the set mixing time has elapsed so that the operator knows when he can discharge the mixed concrete. Another part automatically controlled on many mixers is the water measuring and introduction to the mixer. In large, modern, permanent concrete plants the whole proportioning, charging, mixing, and discharging process is automated.

11.2.2   Batchers and Control Equipment

A batcher is a container for collecting and measuring loose materials before loading them into a mixer. As mentioned earlier in this chapter, the measurement of the separate batch materials for concrete might be by volume or by weight. In the early part of this century concrete materials were batched by volume. Now practically all batching is by weight.

The batcher, particularly for the loose aggregates, has a top opening to receive the material much larger than the discharge opening. It should have a capacity of at least a third more than the requirement of the mixer it is going to load. The weighed material is discharged through a restricted, gated opening at the bottom of the batcher. The gate may be operated manually, by electrical power, or pneumatically with compressed air. In any case there must be positive control to insure accurate amounts of material in each batch of concrete for the mixer. If the batcher discharges its load into the mixer's skip, its bottom opening must be narrower than the width of the skip. In other cases, where the load is delivered directly from the batcher into the mixer, a smaller, funneled discharge opening may be needed on the batcher.

Loose material is generally loaded into a batcher from one or more bins. These are hopper-bottomed storage units. Bins for aggregates can have an open top, but those for cement must be enclosed to keep moisture out. One or more batchers may be supported under a set of material bins.

Batchers are known as (1) single-material batchers and (2) multiple, or cumulative, batchers. For a single-material batcher there is need for only a simple, single weighing scale. To fill the batcher with the right amount of material, the operator has his scale set at the specified weight and opens the gate from the hopper bin bottom. If the batcher has all manually operated lever controls, the operator has to watch his scale carefully to avoid getting too much material in the batcher. In a batcher with some automatic gate controls, the operator has an easier time. He

FIGURE 11-7 Modern punch card batch plant controls (courtesy of the Koehring Road Division).

pushes a button to open the gate from the bin. When the specified weight of material is approaching, the bin gate is automatically closed. Then the operator must manually crack the gate to get the last amount to reach the specified weight. With that amount of material he is ready to push another button to discharge the batch material.

For proper mixing of concrete all the materials in a total batch must be charged into the mixer together. Generally, the separate materials include cement and at least two sizes of aggregate. If single-material batchers are used, there will have to be at least three of them, or a collecting hopper between the batcher and mixer. With three or more batchers to fill, the operator would be kept pretty busy. But this batching arrangement has the advantage that all the mix ingredients can be weighed at once to save time. Certainly, automatic gate controls would be necessary. When all the materials for a batch of concrete are weighed, the discharge buttons for all batchers can be pushed simultaneously to load the mixer.

A multiple batcher, also known as a cumulative batcher, weighs the different aggregate materials one at a time and loaded on top of those previously weighed. The cement and water are still measured out separately for the various reasons previously given. The weighing scales for a multiple batcher may be a dial-type or cheaper weighbeam scales. With a circular dial the first aggregate is loaded to its specified weight, say, 1400 lbs, and then perhaps 1000 lbs of the second aggregate is added on top to bring the cumulative dial reading up to 2400 lbs. This sort of cumulation is continued until all the different materials for a concrete batch have been added.

In the case of the weighbeam scales, the amount of each aggregate is weighed out on a separate weighbeam. That particular weighbeam is locked in and others are off, even though the materials are loaded one on top of the other in the multiple batcher. This has the advantage that the operator does not have to cumulate his weights as on the dial scale. He merely has to set each weighbeam initially to tally the specified weight for its material.

Multiple batchers are made with capacities from one to six cubic yards of concrete mix. Therefore, their volumes are greater than these sizes to hold the loose materials. They can be arranged to receive and weigh two to six different specified materials. Larger batchers are made by special order for the large, permanent concrete plants.

The scales for batchers are carefully checked and controlled for the purchaser's benefit. For governmental construction the regulation may be by federal specifications or the National Bureau of Standards. One of the more practical specifications is advocated by the American Road Builders Association in its ARBA Bulletin 15. It requires that the scale accuracy be maintained within a variation of 0.4%. Primary U.S. government construction agencies require that the batchers deliver an amount of cement within 1% of the specified weight, and aggregate no more than 3% from the specified amount.

Controls for batching equipment vary all the way from strictly manual operation to fully automatic control. The accuracy of manual operation is entirely dependent on the operator's skill in getting the specified weights. A fully automatic operation calls on the operator only to start and stop the equipment. The automatic system will tend to be slow when the final amount of each material is being weighed into its batcher. It is for this reason that a semi-automatic system, where the operator controls the final weighing as mentioned previously, may be preferred. Graphical or digital recording by date, time, and number of batch helps to protect the producer or the purchaser who may suspect he is being shorted.

Other controls are helpful in the batching process. The use of computerized batch cards with the weights given and automatically set for specified mixes is a time saver (see Fig. 11-7). A moisture-sensing device which probes the fine aggregate bin is very helpful. It will automatically adjust the batch water for varying moisture in the sand so that the final mix has the correct total water content. Still other automatic control devices are being developed to improve the consistency and uniformity of the concrete material.

11.2.3 Portable Batch and Mixing Plants

An early and simple design for a portable batch plant is known as the trolley batcher. This has an integrated set of two to four open-top bins in one line horizontally supported close to the ground. Having low bins means that they can be charged by a front-end loader, as well as a clamshell or conveyor. The batcher is in a trolley arrangement running horizontally in line with and just under the bottom discharge openings for the bins. It is stopped under a bin opening to weigh the required amount of the given material for the batch. Then the trolley batcher is moved horizontally under the next bin. This is a cumulative, or multiple, type of batcher.

A portable trolley batcher built low to the ground is designed for small capacities. The bins hold up to 40 tons of material. This equipment is used for small concrete pours too remote for delivery of ready-mixed concrete. In that sort of situation the planner expects to rely heavily on manual efforts and controls. With the trolley batcher moving close to the ground it is most easily discharged into the charging skip of a small mixer. The skip is loaded when it is flat on the ground or supporting surface for the mixer. This makes it easy for men to load with sacked cement. Measuring the cement that way means that the batcher and bins are of sizes that will hold aggregate quantities which will balance with even-sack mixes of concrete. The batcher generally holds 1000 to 4000 pounds.

For simplicity and economy the trolley batcher is moved along the track manually. Though the times for operation will depend on the operator's efficiency, the following figures[1] will give some idea for batching productivities:

Aggregates	Batching Time	Maximum Productivity
2 materials	1 min 20 sec	45 batches/hr
3 materials	1 min 40 sec	36 batches/hr
4 materials	2 minutes	30 batches/hr

With a sufficient number of men emptying the sacked cement, that step should not delay getting the batched material on the mixer's skip. As in all concreting operations, the batching cycle should be fast enough so that the mixer is not delayed in its operation.

Larger, more mechanized and automatically controlled portable batch plants are now available. They are used for highway and other construction where mobility between various uses is important. This equipment consists of bins for aggregates and cement, conveyors for moving the material, batchers for controlled weighing, and the supporting frame. Portability is designed into the equipment.

The aggregate bins may have three to five compartments to hold up to 60 tons total material. These are loaded commonly by a 24" belt conveyor that will load at a rate of around 250 tons/hour. The cement bin, which must be enclosed for moisture proofing, is designed to hold from 150 to 600 barrels. To weigh granular material for the batches, a more portable possibility than using regular batching weigh hoppers is to make use of a conveyor belt designed to weigh the material it handles in any controlled pass. This batching method will generally allow weighing at the rate of at least 125 TPH, or better than 60 cu yds/hr. A water tank of perhaps 650 gallons capacity and a 350 to 600 gpm meter are part of the complete portable batch plant. It should be able to produce 60 to

100 batches per hour for a 34E dual-drum paver. The batch discharge is designed to be 9 to 12 feet above the ground for truck loading.

Portable batch plants of the type described above can be set up with a mixer mounted to receive the batches. In that arrangement the plant is a portable concrete mixing plant such as the one shown in Fig. 11-2. Such a plant can be designed to handle a 4- to 10-yard mixer. Thus, it will produce from 60 to 300 cu yds/hr.

An even more portable plant is the truck-mounted mobile batcher-mixer. This has bins for cement, sand, stone, or gravel and a water tank contained in its enclosed truck bed. The storage capacity is for about 5500 lbs of material and 50–60 gals of water per yard of concrete. The truck has an empty weight of 6,000 to 10,000 lbs and can be designed to mix 4 to 10 yards of concrete. The 4-yd unit has a GVW around 28,000 lbs and, so, can have a single rear axle; but the bigger units will have dual rear axles for operation on the highways. This mixer takes about 30 minutes to mix and pour 10 cu yds.

11.2.4 Central Mixing Plants

For mass production of concrete with a high degree of control, the setup of a central mixing plant is most desirable. A complete concrete plant of this sort will have a full set of aggregate bins, one or more cement silos, and all the necessary built-in conveyors for moving the materials in and out of these storage units. There will be water tanks and metering equipment to control their filling and the water weighed for each batch of concrete mix. Batchers will be located between the material storage bins, silos, and tanks and the equipment for mixing the concrete ingredients. Such batchers will have their appropriate scales, controls, and recording devices.

The smallest plants for central mixing of concrete are designed so that they can be moved rather easily from one project to another. Consequently, this type might be thought of as a portable mixing plant. It is designed compactly and can be "knocked down" into two to four separate parts of about 9-foot widths mounted on rubber tires for towing on the open highway with special permit. Moving such a plant can be economical where it is done within one day's time. At a maximum moving speed of 45 mph, this move could be up to 300 miles. The older movable plants were erected by a crane without too much difficulty. But that involved arranging and paying for the erecting equipment. Now most of these units are self-erecting with built-in hydraulic jacks and can be done by two men in a half-day.

Larger central mixing plants may be equipped with two or more mixers, as compared to only one mixer for the movable type of plant. Where there are a number of mixers, the plant may have a concrete collecting hopper under the mixers to hold up to 25 cubic yards. This arrangement will help to increase output from the plant, especially if the hauling units are of different sizes. A large central mixing plant is shown in Fig. 11-8.

A central mixing plant is designed to be used for three distinct purposes —ready-mixed concrete, a concrete products plant, or for mass concreting. The ready-mixed concrete plant receives a variety of orders for different concrete mixes from many customers in its vicinity. This plant is located for ease in receiving the separate materials for its concrete and to be centrally located for its customers. A concrete products plant is designed for less variety of concrete mixes but need for area to store finished products. Since the producer will probably deliver the products, there is not so great a desire to be located for the customer's convenience. However, this type of plant has to be located for economical delivery of the separate concrete ingredients. A mass concreting plant is designed and located to produce the concrete for a dam or other project requiring large quantities of this construction material. It generally will not have to deal with much variety of concrete mix but will be able to reach a high production rate for turning out concrete. A major concern in its location will be for productivity and least cost of delivered concrete.

High-Production Concrete Plants. To get the high production of concrete required from ready-mixed or mass concrete plants, the design must account for various important factors. A key feature in modern plants is the need for automating many of the steps in the processing of concrete. Automation leads to more accurate and faster batching of multi-material mixes and better quality of the mixed concrete.[5] The significance of this and other important factors to consider in the plant design will be discussed in the following paragraphs.

A ready-mixed concrete plant may deliver dry-batched or wet-mixed concrete. The dry-batched concrete materials may be delivered into ordinary dump trucks. This provides the cheapest transportation for the customer who must then have a mixer, such as a paver, to receive the dry batches. A more common receiver for dry-batched material from a ready-mixed plant is a transit truck mixer. An example of that equipment is shown in Fig. 11-5. It is also common practice for the plant to discharge wet-mixed concrete into a transit truck mixer. Another type of hauling equipment to receive wet-mixed concrete is known as a Dumpcrete truck.

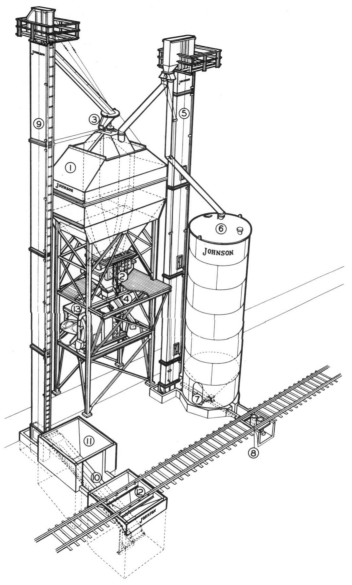

FIGURE 11-8　A central concrete mixing plant (courtesy of the Koehring Road Division).

These trucks for hauling concrete will be discussed in Sec. 11.3.2, and the transit truck was discussed as a mixer in Sec. 11.2.1.

For a simple concrete mixing plant on fairly level terrain the age-old method of handling the aggregate materials has been with a crane outfitted with a clamshell bucket. This equipment can unload open-top rail road cars and barges which transport the large quantities of material for the concrete. Clamshells can also handle the loose aggregate that might be dumped from hopper-bottom rail cars or dump trucks. Since a majority of concrete plants are not located on or near a navigable waterway, highway and railroad transportation are used most generally for the materials. A more modern way for handling the large quantities is by systems of conveyors. That type of equipment was described in Chap. 9. Conveyors can be set up with automatic controls, whereas operating a clamshell is almost entirely manual. Thus, a clamshell for unloading materials has its limitation, as shown in Table 11-2.

A central concrete mixing plant should have enough bin storage and ground area to store two days' to a week's supply of aggregate materials. It certainly must be able to store more than enough for the biggest concrete pour that could possibly be ordered for an around-the-clock operation. To take advantage of automatic operations with conveyors, it is usually necessary to unload the surface transporters into subsurface hoppers under the railroad track or roadway. Conveyors will pick the loose material up for transporting to storage. Where the bottoms of storage piles are at ground level, subsurface reclaiming tunnels are used. These have a conveyor running horizontally along the tunnel centerline to the end for transfer to another, perhaps inclined, conveyor line. The opening of hoppers to feed aggregate material by gravity to the reclaiming conveyors, and the starting and stopping of the conveyor belts, can all be done from a central control panel or housing.

As suggested by this discussion, the location of a permanent central mixing plant is often determined by the convenience for delivering the aggregates and cement. For instance, if the plant can be located on a navigable waterway, it may be worth doing so to have the cheaper barge transportation. In hilly terrain it is advisable to consider a location where trucks, moving out on a high-level ramp or bridge, can dump aggregate directly into the top of the bins. Or with advantageous differences of elevation, the conveyors can be set to move material with less vertical lift to conserve power requirements for the plant.

The material-storage requirements for direct batching of the aggregates and cement must be carefully planned. This results in the sizes of bins and silos needed. Generally, there should be enough of this form of storage to cover two to four hours of continuous material demand. The required volume capacity is determined by two key factors: (1) the maximum expected production of concrete in cubic yards per hour, from which tons of aggregate and barrels of cement can be calculated, and (2) the planned rate and dependability for delivery of the material into the bins and silos. In the Table 11-2 key data and comparisons will be found.

A ready-mix concrete plant not only has to deal with a great variety of orders but frequently has a particularly high demand for concrete at the start of the working day. This likelihood calls for extra aggregate storage capacity and the ability to load the bins at a high rate to keep the batching equipment fed with their demand for materials. Having several acres clear for aggregate storage in addition to the bins will serve this purpose. To have extra cement storage for high demand at the start of a day the plant operators might plan to have one or more full cement haulers ready to unload as the day's orders are filled. Then the ability to unload directly into the silo feeding the batcher can satisfy the early, high concrete demand.

Automation for Concrete Plants. The best way to reach the high capacities of modern batching and mixing equipment and still ensure quality concrete is with automation.[5] This means that the central mixing plant is designed with automatic controls for the weighing and measuring of the batch materials and for the cycling times of charging, mixing, and discharging. A single plant operator at a control panel performs all the batching and mixing operations with push-button controls (see Fig. 11-7). The automatic controls can include a moisture-sensing device in the fine aggregate. This determines the moisture content in the sand about to be weighed for batching and automatically adjusts the amount of water to be measured for the batch to have the specified water in the concrete mix. The amount of materials to be batched for a given concrete mix are controlled by a card punched for use in the computer-programmed controls. Other automatic controls are becoming available. The planner of a central mixing plant should seek the advice of experts and the developers of this equipment when designing a high-production concrete plant.

The advantages of automatic controls can be noted in comparing the production of an automated central mixing plant to that possible with more manually controlled equipment. For instance, 34E pavers must have most of the steps of their cycles controlled manually because they are receiving batch material from other mobile equipment and they are discharging concrete for other independent equipment to spread. A central mixing plant with an 8-yd mixer and 90-second mixing cycles can be com-

TABLE 11-2 Material Capacities for Concrete Plant Equipment[1]

Concrete, cy/hr	Aggregate Handling				Cement Handling—Main Silo/Aux.			
	Req'd. TPH	Clams. size	Bucket size[a]	Belt Conv.[b]	Req'd bbl/hr	Screw Diam.	Bucket size[c]	Auxiliary Silos, bbl
10	16	⅜-yd	10 × 6"	18"	13	8"	8 × 5"	350[d]
30	49	1-yd	10 × 8"	18"	39	8"	8 × 5"	400
50	81	1½-yd	10 × 8"	18"	65	8"	8 × 5"	650
80	130	2½-yd	16 × 8"	18"	104	10"	10 × 6"	1050
100	163	3-yd	20 × 8"	24"	130	10"	10 × 6"	1300
125	183	none	22 × 8"	24"	163	10"	12 × 7"	1625
150	224	none	none	24"	195	10"	12 × 7"	1950
200	325	none	none	30"	260	10"	14 × 8"	2600
250	405	none	none	30"	325	10"	14 × 8"	3250
300	488	none	none	36"	390	10"	14 × 8"	3900
400	650	none	none	36"	520	10"	2-14 × 8s	5200

a Belt-type bucket elevators with 6-inch-wide buckets are not recommended for aggregate larger than 1½ inches, nor 8-inch for larger than 3".

b For aggregate of 3"–6" a minimum of 24"-wide belt is recommended.

c For unloading cement 10" × 6" bucket elevators are recommended as the smallest size to reduce the time.

d Minimum storage recommended when the cement is delivered by railroad hopper-bottom car (if delivered by tank truck over highways, minimum reduced to 200 bbl).

pared to two 34E dual-drum pavers with one-minute (60-second) mixing cycle times. With automation the central plant can reach almost 100% efficiency and turn out 240 cu yds/hr. The 34E pavers should be able to equal that productivity, but instead they do well to turn out 180 cu yds/hr (see Table 11-1), at best. A mixing plant can best achieve high production by using automatic controls for the batching, mixing, and discharging steps of the process.

In addition to the higher productivity, it has been found that the maintenance record of an automated concrete mixing plant is better than one that is manually controlled. This is partly because the controls and action activating mechanisms are electrical or hydraulically-driven instead of having primarily mechanical drives. The driving mechanisms are sealed against dust and moisture and work with sealed, explosion-proof motors and control panels. These measures can result in 95% operating availability of the mixing plant.

An automatically controlled plant will need to be monitored to insure that it is turning out quality concrete. The means for this supervision has been modernized with several important communication schemes. One obvious need is satisfied by the use of an intercom telephone system between the operator at the batching and mixing control panel and a dispatcher at the point where the mixed concrete is discharged for transporting. In the case of a ready-mix plant, that would be at the truck loading point. In a mass concrete plant for a dam, locks, etc., the intercom system may terminate at the place for pouring the concrete. Another monitoring innovation for a high-production concrete plant is the use of a TV communication system. TV cameras can watch the concrete in the mixing drum or as it is discharged. In the case of a mass production plant for a concrete dam, TV cameras may watch all the batching dials, the location of transporting cars or containers, and the concrete bucket on a cableway. The extent of automatic controls and monitoring is generally governed by the Concrete Plant Manufacturers Bureau and the National Ready Mixed Concrete Association as well as the governmental agency for whom a large project is being constructed.

11.3 Concrete Hauling Equipment

Around the start of the 20th century concrete was just beginning to be used as a construction material. The quantity used in any one pour was generally small. To haul and pour the small amounts involved, constructors of those historical times used either wheelbarrows or horse-drawn dump carts. Since those early times millions of cubic yards of concrete have been hauled and poured, in a few cases tens of thousands of yards in a single pour. Consequently, many innovations and different pieces of equipment have been designed to do this important construction work.

The choice of equipment to use for hauling concrete from the mixer to the point for pouring, usually, depends on two basic considerations. These are (1) the location and volume of the concrete pour and (2) the methods chosen for mixing the concrete and pouring it into the concrete forms. Hauling will not usually govern the concreting methods chosen. The method for hauling the concrete will be one of several of the following:

1. hand-operated cart or wheelbarrow with one or two pneumatic wheels;
2. power buggy or cart with two axles and three to six wheels;
3. monorail dump cars to travel on an overhead rail beam;
4. hoist elevator bucket traveling in a vertical tower;
5. crane-handled buckets moved vertically and horizontally;
6. concrete belt conveyors;
7. concrete pump with rigid or flexible pipeline;
8. transit mixer or concrete dump trucks;
9. railroad cars, generally small-gauge, to carry buckets; and
10. a cableway to handle buckets on its hoisting lines.

The type of project being constructed will dictate the answers to the questions of location and method for pouring into the concrete forms. For instance:

1. concrete for buildings will generally be hauled by one or more of methods (1) through (8);
2. for paving and other ground-level work the concrete will likely be hauled by method (8);
3. for inaccessible pours, such as tunnel lining, concrete will probably be moved by method (6) or (7); and
4. for the mass concreting of a dam the concrete will likely be hauled by method (9) or (10).

All the modern methods for hauling, as a part of the handling and placing of concrete, are designed to minimize the segregation of the material ingredients between the final mixing and the pouring into the forms. The most frequently used hauling equipment will be discussed in the following sections.

11.3.1 Concrete Buggies or Carts

The buggy type of concrete hauling equipment is either propelled by manpower or powered with a small internal combustion engine. Small, hand-propelled wheelbarrows have but one wheel, which should have a pneumatic time to ease the man's work and reduce shocks that segregate the concrete when moving over rough ground. The wheelbarrows hold only a few cubic feet of concrete and should not be used where the move is more than 200 feet. Average hauling production is around one cu- yd/hr per wheelbarrow.

The larger, hand-pushed concrete cart with a single axle will have two tires for more lateral stability and a supporting foot under the handle for the at-rest position. It is unloaded by pivoting the material container or bed forward about the single axle. To ease the operator's work and extend the usefulness of this type of small concrete hauler, motor power has been added along with a second axle and other wheels. In this hauler the concrete bed is dumped separately by a powered mechanism similar to that for a dump truck. The smallest powered buggies are walking units where the operator follows along behind, as with the hand-operated haulers. Such small units have capacities up to six or eight cubic feet. They should not be used for hauls much over 200 feet, though the powered units have hauled 500 feet one way. The average productivity to expect from concrete buggies of this variety should be from 2 to 5 cu yds/hr for each unit.

Still larger powered concrete buggies are available on which there is a seat for the operator to ride (Fig. 11-9). This operator's position is over the rear axle, which may have one wheel on the smaller units or two for the larger ones. The rear axle serves for steering while the front axle under the concrete bed takes the majority of the load. The bed may hold from 9 to 14 cubic feet (1/3 to 1/2 cu yd) of concrete. To distribute the load from the larger units for lower pounds per square inch on a haul ramp or soft ground, four wheels or dual tires may be used on the front axle. A wooden runway for this type of concrete hauler should be at least five feet wide to take full advantage of their high speed and maneuverability as well as dumping over the side of the runway. The maximum distance of one-way haul for good concrete production with this type of powered buggy should be 1000 to 1500 feet. A maximum productivity should approach 15 cu yds/hr for each concrete buggy.

FIGURE 11-9 Powered concrete buggies in action (courtesy of Whiteman Manufacturing Co.).

11.3.2   Trucks to Haul Concrete

The trucks for hauling concrete materials may be one of several designs. Ordinary dump trucks can be used to haul dry-batched concrete material from a central batch plant. The specially designed trucks for concrete that has already been mixed are the Dumpcrete trucks and the transit mixer trucks. There is also the truck-mounted mobile batcher-mixer which was described in Sec. 11.2.3.

When an ordinary dump truck is used, it will be modified by inserting wooden sideboards to increase the height of the dump bed sides and to support swinging partitions. These partitions are merely flat panels extending the full width of the bed to divide the bed into three to six compartments for separate concrete batches. The panels hang vertically with each one supported at the top by a horizontal bar which serves as a hinge for the panel to swing on when the bed is raised for dumping. Each panel can be held at the bottom so that it does not swing and let the batch dump when the bed is in a raised position. By this sort of arrangement one batch can be dumped at a time when its rear panel is released. This hauling scheme is arranged particularly for delivering dry-batched material to a paver. The paving mixer has a big enough skip so that a dump truck can back into it and dump a batch of concrete material. The truck drives off the skip for it to be raised, as shown in Fig. 11-4, to load that material into the mixer of the paver. Then the skip is dropped back down to the ground for another batch. The dump truck repeats this procedure for its next batch, etc.

On some isolated projects, ordinary dump trucks have been used to haul wet-mixed concrete from a central plant. This has been limited to very short hauls where the concrete would not have much chance to become segregated by the vibration of the truck bed. There is also the problem of getting the concrete out of the lower corners when the bed is raised and a small gate opening is used to control the discharge of the material. The use of ordinary dump trucks is not recommended for hauling wet, ready-mixed concrete. In fact, their use is not allowed for most concrete jobs that are carefully controlled by specifications.

However, the similar but specially designed Dumpcrete truck is frequently allowed. This type of truck, introduced by Maxon Construction Company, differs from the ordinary dump truck in the shape and design of the dumping bed (Fig. 11-10). The Dumpcrete bed is like a tub with rounded corners and a hopper-type chute at the rear of the bed. It also has a pivoting open-trough chute extension that can swing through almost 180 degrees horizontally. This arrangement means that, with the bed raised for pouring, the concrete can be spread over an area of 200 to 300 square feet in back of the Dumpcrete truck. To facilitate the pouring from the newer versions of this type of truck, revolving baffles inside the bed agitate and help move the concrete out. A Dumpcrete truck should be able to have its bed unloaded in about 30 seconds. This can be done with its bed raised hydraulically to a 90° discharging angle.

A Dumpcrete truck may use single or dual rear axles, depending on the load distribution. Larger units with capacities to handle 8 cu yds of mixed concrete will have the dual axles to limit the load per axle to about 15,000 lbs for legal highway travel. The Dumpcrete truck is more expensive than an ordinary dumptruck because of its specialized design, but it is much less expensive than a transit mixer truck of the same concrete capacity.

The transit mixer discussed in Sec. 11.2.1 and pictured in Fig. 11-5 is also a truck for hauling concrete. It may receive dry-batched material or wet, ready-mixed concrete from a central plant. When it takes dry-batched material from the plant, the transit truck serves as the concrete mixer. The material it can take for mixing is limited, as given by Eq. (11-2). The truck must have a water tank and meter to add just the right amount of

FIGURE 11-10 The Dumpcrete truck used for hauling concrete (courtesy of Maxon Corporation).

water for the concrete as specified. In this use of the transit mixer truck, it will mix the concrete while enroute or wait until it reaches the site for pouring the concrete. If the latter timing is used, the travel distance for the truck is limited only to economic reasonableness. If the concrete is mixing while the truck is traveling, that time is saved from any delay when the truck reaches the pouring site. In this case, which is much more commonly followed, the concrete should be poured within the hour after the mixing was started. This is so that the concrete will not have started to set in the mixer drum. One feature of the design of transit mixer trucks to delay the setting of mixed concrete is its ability to keep the wet concrete agitated. The agitation is done by revolving the mixer drum slowly, as explained in Sec. 11.2.1.

The volume of concrete material relative to the drum's volume for the transit mixer truck was also explained in that previous section. If there is a need to increase the amount of concrete that a transit truck of given size can haul, that can be done by having the central mixing plant pre-mix the materials. By doing that, the transit mixer can take about 33%

more concrete than if it were dry-batch material. This means that the mixed concrete would fill about 80% of the drum's volume.

The completely self-contained transit truck mixer must be equipped with a water tank and delivery system. However, the Truck Mixer Manufacturers Bureau will permit in its standardized units any one of three possible water arrangements.[3] These designs for the water requirements are (1) mix-and-flush water tank(s) and water system, (2) flush-water tank only and water system, and (3) no water tank nor system on the truck, with water to be supplied from an external source. In case (1), the total tank capacity is not to exceed 50 gallons/cu yd. For a 6-yard transit mixer this amounts to 300 gallons, or about 2,500 lbs in addition to the water tank and system added on the truck. The system must include a water pump or other means for delivering not less than 45 gals/minute into the batch. Also, the system must be equipped with some accurate water-measuring device. The approved devices are the automatic cut-off siphon type, a water meter of the automatic shut-off type, and sight gauges of acceptable form. Any of these must be able to measure the water in the tank(s) accurately within 1% when the truck mixer is stationary and essentially level. The same requirements apply in case (2) for a flush system used to keep the mixer drum clean. Generally, sight gauges will be good enough for that case.

A completely equipped and loaded transit mixer with its drum, driving machinery, water tank(s) and system, as well as the truck's power plant, creates quite a load problem. To distribute the total load satisfactorily requires at least a dual rear axle under the drum. With the tandem-drive axle design some states will limit the transit truck to a 7 cu yd load. The truck's load limit can be increased by several yards and still be within the legal load per axle limit in most states by using a tri-axle design under the drum. One design for the third axle is an air-suspended tag-along axle behind the regular dual-axle drive wheels. This additional axle not only increases the load limit that can be legally run on the highway but also helps lower the pound-per-square-inch load application for off-the-road movement. Otherwise, a heavily loaded mixer truck could get mired in soft ground. To handle the largest transit mixers will require the use of a truck with both tandem rear drive axles and a dual-axle trailer to support the back, discharge end of the big drum.

11.4 Concrete Buckets and Other Pouring Equipment

The equipment used for pouring concrete must satisfy one important feature. It cannot allow the concrete to suffer harmful segregation while it is being poured into its final form. To insure this, the pouring equipment should be able to place the wet concrete as near its final resting place in the forms as possible. Also, the concrete should not be allowed to flow freely, without the confinement of a container, chute, or pipe with surfaces on almost all sides perpendicular to the direction of flow. This means that the concrete should not be moved any distance by the use of a vibrator in an open-top form. Certainly, the concrete cannot be freely or loosely dropped any significant distance. Some specifications limit the free drop to a distance such as two or five feet.

A considerable variety of equipment will satisfy these important restrictions. If the previously described trucks for hauling concrete can get close enough to the forms, they can also serve to pour the concrete. With the use of suitable temporary runways, concrete buggies can pour their loads directly into the forms. Where the concrete haulers cannot pour their material for final placement, other handling equipment, such as concrete buckets, conveyors, and pumps, must be used. The design features and considerations for this variety of concrete pouring equipment will be discussed in the following sections.

11.4.1 Buckets for Pouring Concrete

The buckets used for pouring concrete are supported by hoisting lines from cranes, crane-like hoisting towers, or cableways. Common versions of these pieces of lifting equipment were discussed in Chaps. 7 and 8. In those discussions it was obvious that the weight of the swinging load at the end of a hoisting line is an important consideration. The weight of a bucket and its load is a key feature in the design of this auxiliary type of equipment. This will be emphasized in the following paragraphs. Also, the means for discharging the concrete from the bucket is important. The discharge opening is made with a movable gate at the bottom of the bucket. This gate may be opened manually or by some other form of power. A manually operated gate has a lever to open the gate directly from its normally closed position. Since the bucket must be raised and lowered and allowed to swing freely, it is not feasible to have a permanent connecting line for power to control the discharge gate. The most effective way to operate the gate with extra power is by a compressed-air line. This can be attached quickly to an air hose coupling on the bucket and detached as simply with twisting action.

Concrete buckets are classified as (1) lightweight buckets, (2) standard-duty buckets, or (3) heavy-duty buckets. The standard concrete bucket is generally cylindrical and has been made of steel plate. This means that the empty bucket weighs about 20 to 30% as much as the concrete it can carry. The lightweight buckets are small ones with small gates and used for low-production concrete pours. They should not be used for concrete with larger than 3″-size aggregate. Capacities of lightweight buckets range from 1/3 to 2 cubic yards. To keep the weight of a loaded bucket as small as possible, they are frequently made of magnesium, which is only a quarter to a third as heavy as steel. A standard-duty bucket is used for stiffer mixes with low slump and has a larger gate opening than the lightweight. They are made in sizes ranging from 1/2 to 4 cubic yards. The heavy-duty buckets are extra large with specially designed discharge gates. These will take care of an almost-dry or low-slump concrete with aggregate up to 6″ maximum size. Heavy-duty buckets are built with sizes from 1 to 12 cubic yards.

Other special-design concrete buckets include laydown-and-rollover buckets that are not cylindrical. The purpose for such a design is to give a low height for loading the bucket from trucks and shorter hauling or mixing equipment. These special buckets are made in either lightweight or heavy-duty varieties and of sizes ranging from less than 1- up to 5-cu yd capacities. Another special feature for any concrete bucket is designed to limit the free fall and possible segregation of the concrete. This is an elephant trunk drop chute which is attached below the discharge gate of the bucket. It is especially useful for pouring concrete walls, columns, piers, and other structural parts with horizontal dimension very much less than that of the bucket width.

11.4.2 Conveyors for Pouring Concrete

Belt conveyors are made with certain special design features to allow their use for pouring concrete. They generally differ from the conveyors used for aggregates by having a more rounded trough and are contoured by having more rollers. These features help to reduce the possible segregation of the concrete by keeping the material compact and giving it a smoother ride. The belting has widths of 12 to 16 inches, and the lengths of the separate conveyor units may be from 24 to over 60 feet long. A regular concrete conveyor can be inclined to an angle of 30° from the horizontal and still move low slump concrete satisfactorily. With the use of special cleating the belt conveyor can be raised to a 40° angle to move concrete. For a 60-foot-long conveyor at this maximum angle, the concrete can be lifted about 40 feet above the bottom of the conveyor. At the discharge end is hung a 4- to 10-foot chute that can be swiveled through a full 360° circle under the end of the conveyor for more pouring range.

The concrete belt conveyors may be self-propelling with a seat for the operator. Their belts are powered to move at speeds up to 500 fpm. This makes it possible for the conveyor to deliver around 150 cu yds/hr without difficulty. To pour over a large area, such as a building floor or hangar slab, a series of concrete conveyors can be set up. This might be two to 20 flights of conveyor units end to end each to extend as much as 100 feet. The setting up of such a system would be done by a crew of perhaps five men at a rate of about 100 feet per hour. A few manufacturers of this type of equipment have ready-made systems of conveyors that move out on lines of track or ride piggyback with the head end of one conveyor unit on top of the foot end of the next. These special designs have their own individual power units for each set of conveyor flights. Or they may be designed with compressed-air lines and controls so that a simple coupling with the hose from a compressor will provide the needed power. A specially designed concrete conveyor system of this sort will not take nearly as long to set up as suggested above for a series of separate conveyors. The makers of concrete conveyor systems claim savings in the order of $1 to $3 per yard by pouring with such equipment, compared to using a crane-and-bucket method.

11.5 Concrete Pumping Equipment

The purpose of concrete pumping equipment is to move wet, ready-mixed concrete through an enclosed delivery line into its place for pouring. This is done by forcing the fluid concrete by some form of pump through lengths of pipe or hose. The pipe or hose line can be laid out with a combination of horizontal and vertical or inclined stretches. Consequently, it is a very flexible method for moving concrete into place on a jobsite. The pumping method for pouring concrete is said to get the wet concrete into its forms smoother and faster than the other methods described. In fact, concrete proportioned as described later is claimed to get better mixing, improved blending, and greater density by the pumping equipment. To observe these benefits, concrete samples and test cylinders should be taken at the discharge end of the delivery line.

The pumping of concrete was started for lining tunnels, where there did not seem to be any other suitable method available. It was originally introduced in Europe by a manufacturer of the equipment from Holland. The introduction in the United States was made in 1932 when the Rex Chainbelt, Inc., obtained a franchise for the Pumpcrete machine.[6] There have been a number of modifications to this type of equipment since that early version, so now they are commonly called concrete placers or pumps. The use of concrete pumps have been extended from tunnel lining to pouring bridge decks and building floors, long walls, etc. They are used generally wherever the pouring is into "hard-to-get-to" forms that would be inaccessible, congested, or crowded for a crane with buckets or buggies operating on a rampway. Figure 11-11 shows this type of concreting equipment.

FIGURE 11-11 A concrete pump in action (courtesy of Challenge-Cook Bros., Inc.).

11.5.1 Design Features of Concrete Pumps

The basic feature of concrete pumping equipment is its mechanism for pushing the wet concrete through the delivery line. Three pumping principles are in general use by the makers of this equipment. They can be simply described as (1) piston pump action, (2) pneumatic pump action, and (3) squeezing of the fluid concrete like toothpaste through a tube. In the case of a piston pump the concrete is moved along by the push from a reciprocating piston stroking back and forth at the beginning, or charging end, of a pipe or hose. A pneumatic pump uses the pushing power from air under pressure delivered by a compressor of 125 cfm capacity, or bigger. The squeezing action is a special patented arrangement of the Challenge-Cook Bros., Inc., using a more flexible hose for the tube. The concrete is moved along by a flywheel or rollers that press along the outside of a beginning section of the tube in a cylindrical vacuum chamber, squeezing on it and causing the concrete to move ahead of the rollers. By having such rollers work repetitively on the tubing in a semi-circular stretch of its length, they give a pumping type of action to the concrete.

To insure success in pumping concrete, the material mix should be given special consideration.[6] The concrete mix should meet the following limitations: (a) have five sacks or more of cement per cu yd, (b) use less coarse aggregate than 60% of the total, (c) have the fine aggregate meet the ASTM C33 specification, and (d) have a slump of 2″ to 7″ at the most. Lubricating of the lines is often done with a high cement and sand grout at the start of a day's use. Specifications will frequently not allow the pumping of lightweight concrete because the pressure on the fluid mix will tend to force the moisture into the pores of the lightweight aggregate, drying up the mix and causing it to stiffen and become unworkable. Any segregation of the concrete ingredients will cause difficulty, if not blockage, in the pump lines.

The wet concrete is fed to a pump through a 1/2- to 2-cu yd charging hopper, which is part of the pump design. It is pumped through rigid pipe of about ten-foot lengths, or flexible hose in somewhat greater-length sections, that are coupled together to make the total delivery line. The lines vary in diameter from the large 8-inch pipe with the original Pump-crete equipment down to the small 3-inch tubing or hose with some newer equipment models. These sizes will often be varied in any one equipment setup from the largest diameter at the pump end to take more concrete down to the smallest at the discharge end for the convenience of handling the line, which will be moved the most. Some pump pistons are as large as 9 to 12 inches in diameter, while the end hose may be only 3 or 4 inches in diameter. Changes in the size of delivery line cause pressure build-up in the moving concrete. The Squeez-Crete has the advantage of a constant-diameter tubing. The maximum size of aggregate in the concrete will govern the minimum size of hose that can be used. The largest aggregate should not be more than 40% of the inside diameter of the line. Therefore, the following size relations govern the selections:

a 3"-diameter line can take a 1" maximum-size aggregate,
a 4"-diameter line can take a 1 1/2" maximum-size aggregate; and
an 8"-diameter line can take a 3" maximum-size aggregate.

A concrete pump can move its wet, mixed material through delivery lines up to 2,000 feet horizontally and to 400 feet vertically, depending on the design of the mix and size of the line. Generally, one foot of vertical movement upward is thought to be equivalent to six feet horizontally, so it should be noted that a concrete pump cannot reach its maximum horizontal and vertical limits on the same job setup. Some special self-contained concrete pumps have the complete equipment, including its jointed pipe sections, all mounted on a carrying vehicle. These models have 3- to 6-inch lines 50 to 70 feet long supported on two or three boom sections jointed with great flexibility for pouring range (Fig. 11-12).

The delivery lines are made up of pipe sections and lengths of hose to be coupled together on the site. The pipe has less frictional resistance

FIGURE 11-12 Self-contained concrete pumping system (courtesy of J. I. Case Company).

to movement of the fluid concrete, but hose is much easier to handle and move at the discharge end for pouring the concrete directly into place. The simplest way to support the pipe line is on scissor frames made of wood. These can be erected easily and moved simply for changing the line location. For vertical reach a pipe line has the rigidity to stand with extra support. A vertical line of hose can be supported most easily by a crane boom or other hoisting frame. It should be a more permanent frame, if the line will not need to be moved much. On many jobs for a concrete pump it is an advantage to be able to move the lines frequently while extensive areas or many parts of concrete are being poured.

11.5.2   Productivities and Costs for Concrete Pumps

The production rate that can be expected from a concrete pump will vary between 10 and 100 cu yds/hr with a single pump. Actual productivity will depend on the type of pump used, the size of the delivery lines and the efficiency of the operation. In the case of a piston-type pump the variation in production is small for a given size because the concrete delivered is dependent on a fixed piston displacement. At the other extreme, a pneumatic pump can have its productivity varied considerably for a given size of line. No matter what pump is used, the concrete should not be kept in the lines for more than an hour.

With the original Pumpcrete having a 6" or 8" pipe line and a single piston, q_m = 15 to 35 cu yds/hr. Using a similar Pumpcrete except with double parallel pistons delivering to the 8" pipeline, productivity can be increased to maximums of q_m = 50 to 65 cu yds/hr. For the smaller sizes of newer mechanical or hydraulic piston pumps, generally with dual pistons and 3" to 4" lines, the maximum productivity, q_m = 15 to 50 cu yds/hr, can be expected. Larger ones with 5" to 7" lines get up to q_m = 100 cu yds/hr. A pneumatic pump outfitted with 4" to 8" lines can be expected to move maximums q_m = 15 to 75 cu yds/hr. Using a Challenge-Cook Squeez-Crete placer with 3" flexible tubing, a maximum production, q_m = 20 to 30 cu yds/hr, can be expected. If a Squeez-Crete placer with the larger 4 1/2" flexible tubing is used, the productivity can reach q_m = 80 cu yds/hr. This type of concrete placer also has the capability, with its constant diameter of tubing, of being used successfully to move light-weight concrete. The productivity variables are covered on the Concrete Pump Performance Estimator chart, shown in Fig. 11-13.

As mentioned previously, the advantages in using a concrete pump show up where the job would not be ideal for a crane with bucket nor for buggies operating on ramps. If the concrete for a job can be poured directly

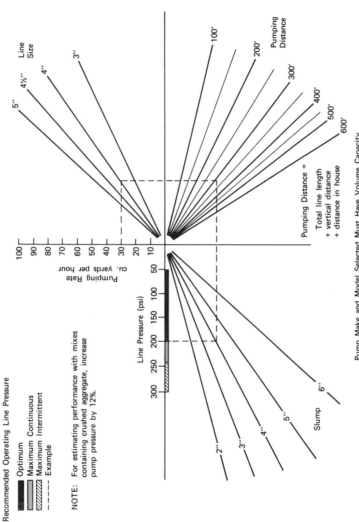

FIGURE 11-13 Concrete pump performance estimator (courtesy of Challenge-Cook Bros., Inc.).

into place from a transit mixer or other concrete truck, none of the above equipment alternatives would be economical. When the concrete cannot be poured directly, labor to handle the concrete is a major expense. A number of men are needed to spread concrete dumped by the bucketful. Each concrete buggy requires an operator, and other men are needed to spread the concrete and move sections of the ramp.

With a concrete pump the major labor requirement is for moving the delivery line and its supports as areas of concrete are completed. The discharge end of the line can be swung or have a pouring chute to place the concrete close to its final resting place. Using a concrete pump can reduce the manpower requirement up to 30%, compared to the main alternative methods.

Equipment cost for a concrete pump is also economical. The original investment in a concrete placer may be in the order of $30,000 to $50,000. This might be compared to the crane of sufficient size to handle a bucket for equivalent productivity which might cost $200,000 originally. Of course, the crane has many more uses than a concrete pump. As such, it might be used for many more operations, and its higher cost might be spread over more working days or hours. Cost per hour of these major pieces of equipment for a concrete pouring operation is another basis for comparison. A concrete pump may cost between $5 and $10 per hour, with the delivery line adding perhaps $2 or $3 per hour for 100 feet. This equipment cost compares to a crane with bucket, which probably costs around $40 per hour. So, it is evident that a concrete placer can be an economical piece of equipment for pouring concrete.

References

1. "Concrete Mixing and Placing," A Reprint from *Construction Methods and Equipment,* McGraw-Hill Book Company (New York, N.Y., 1955).
2. "Concrete Mixer Standards," Mixer Manufacturers Bureau of the Associated General Contractors of America (Washington, D.C., 1956).
3. "Truck Mixer and Agitator Standards," Truck Mixer Manufacturers Bureau (Silver Spring, Maryland, 1968).
4. Bloem, D. L. and R. D. Gaynor, "Factors Affecting the Homogeneity of Ready Mixed Concrete," National Ready Mixed Concrete Association (Silver Spring, Maryland, 1970).
5. "Modern Construction Operations," *Engineering News-Record,* McGraw-Hill Publications Company (New York, N.Y., December 9, 1965).
6. "Concrete Placers," A Reprint from *Construction Methods and Equipment,* McGraw-Hill Book Company (New York, N.Y., 1965).

Chapter 12

■■■■■■■■■■■■■■

■■■■■■■■■■■■■■■■■■■■■■■■■■■■■■■■■■■■

COMPACTORS
AND PAVING
EQUIPMENT

■■■■■■■■■■■■■■■■■■■■■■■■■■■■■■

Introduction to the Paving Process

The paving of an area of the ground surface provides an artificial cover for one or more purposes. Important recognized purposes are to support superimposed loads, to protect the ground from water damage or weathering, to provide a protective lining for containing fluid material, and to give a solid surface for moving on with traction and a minimum of slippage. Construction of a paved surface may be done in various ways, which depend on the materials used for the paving. The most common paving materials are asphalt and concrete.

The process of paving with asphalt and concrete basically involves preparing the subgrade for the pavement, laying down the paving material, and then shaping and consolidating the paving surface. Preparing the subgrade starts by working over and making the ground into a dense, stable subsurface. This will generally require special compacting of

507

the fill or existing ground. Earthwork equipment was discussed in Chap.
5. The next sections will discuss the means for compaction.[1] When the
subgrade is prepared to the planned level or shape, it is ready to have the
paving material laid on top of it. If the material is laid down in a semi-
fluid state, such as wet concrete, side forms will be needed to contain the
material until it solidifies. When paving material is placed in two layers
they are of about equal thickness. To get the desired uniformity of
material, an upper layer is placed before the previous one is completely
solidified or with the application of a tack coat in asphalt paving. The
final step of basic paving will be done with special finishing or rolling
equipment, which will be described later in this chapter.

12.1   Equipment for Compacting Material

Centuries ago the Romans saw the need for compacting the road materials
placed for the Appian Way.[2] They used big, cylindrical stone rollers
pulled by many slaves. The result was a closely bound base and surface of
stones that could take the pounding hoofs and hard, loaded chariot
wheels without destroying the roadway. As the loads of later years became
bigger and put more stress on the traveling surface, other compaction
methods were needed. Sometimes it was necessary to add new material to
the *in situ* soils for blending to get the right compaction.

Generally, the specifications for the final condition of a pavement subgrade call for a specific density of the material. Such standards are set by
contract awarding authorities such as the American Association of State
Highway Officials (AASHO). The specifications used frequently call for
making field tests that are an attempt to match controlled laboratory tests
on each part of work as it is done. Parts of an embankment are the layers
or lifts of earth placed successively to build up a solid mass to the designed
height of section for the roadway. Specified field tests to determine if the
material has the required compaction may be rather laborious. They may
require cutting one or more samples out of each layer placed and compacted as specified.

The thickness of layers may be varied with the compaction method
used. That in turn depends on the material being compacted and the
construction operation being done. Thicker layers will make it more difficult to cut out a sample for testing. And where an earth fill is on top of
natural ground, there is a question about the compactness of the ground
to support the fill. Specified density of a base layer must account for the
density of the subsoil. The pressures and vibrations of heavier and heavier

moving vehicles have settled the subgrade and base material, causing pockets and chuck holes in pavements built years earlier. Moisture-density testing instruments of a seismometrical type, and similar instruments that do not require cutting samples out of the fill or layer of compacted material, are now being used. These tend to promote variation in the equipment that can be used to achieve the specified compaction. Then it is up to the construction planner to select appropriate, economical compacting equipment for the material and operation as specified. The following sections are intended to help with such a determination.

12.1.1 Design of Compaction Equipment

The material of the natural subgrade, fill, subbase, or wearing surface will be of various consistencies. To carry load, it will be mainly granular material. It may be well-graded, clean sand and gravel that drains water freely. In that case the degree of compaction is often specified, based on its dry density since its load-carrying capacity will be best when it is compacted in that condition. More often the load-bearing material will have some fine or binder material in it. The amount of those fines will be as little as 5 to 10% of the total material.

The well-graded, free-draining granular material, which has no sticky cohesive quality, can be compacted entirely by the right amount of vibration. It responds like the cereal flakes in a box that, when first opened, appears not to have been filled. At the other extreme are mastic, cohesive, or sticky materials that respond best to compressive force. They compact like the moist snowball by the pressure applied between two hands. The earth or other material to be compacted for construction is generally granular with some fines and moisture in it. So, the compaction equipment to choose from may be designed to apply pressure with its weight, or a vibrating dynamic force, or both.

The effort applied by compaction equipment can be identified in four distinct ways.[2] These are the efforts due to (1) static weight, (2) kneading action, (3) vibration, and (4) impact force. A heavy, cylindrical roller is applying static weight primarily. The compactive effort of load-carrying tires gives a kneading action in having their load distributed somewhat outward as well as downward on the supporting material. A vibratory action is built into some compaction equipment by the use of eccentrically revolving weights. Others have vibration due to different mechanisms. An impact force is most obvious in an air-piston ram for tamping a loose fill.

Following World War II with demands for pavements to carry heavier loads, there has been a continuous development of more advanced compaction equipment. The development has generally led the way to newer specifications for the compaction process. Bigger, more effective steel rollers have been made. The sheepsfoot-type roller has been built in many shapes. Grid-type rollers, which are designed to gain the compactive advantages of both the smooth cylindrical and the sheepsfoot-type rollers, have been developed in various forms. Most recently, vibratory rollers and compactors of all sorts have been manufactured. Many of the newly designed rollers, other than the small towed models, are self-propelled and articulated for greater maneuverability.

With the rapid development of compaction equipment it is difficult for the construction planner to know what to choose. A little more detailed explanation of the main types should help with his understanding. The following sections are given for that purpose. They will discuss six basic types of equipment designed for compacting material: (1) smooth-steel rollers, (2) pneumatic-tired rollers, (3) sheepsfoot-type rollers, (4) vibratory rollers, (5) vibratory plate compactors, and (6) impact tamping rams.

Smooth-Steel Rollers. A smooth-steel roller is the modern version of the Romans' solid cylindrical stone roller. Those used for construction are heavy and have been self-propelled by their own power unit, starting with a steam engine in the 19th century. The first rollers of this type had two axles, with power delivered to the back one driving two smooth-steel roller wheels mounted like a car's rear wheels. Today, with the same basic design, each of the rear rollers almost six feet in diameter are no more than a couple of feet wide, and the total roller width can easily fit within the width of a roadway lane. A modern version of this roller type is shown in the background of Fig. 12-1. The front axle supports a single, wider— but with a less than 4-foot diameter—smooth steel roller centered in the equipment's width. This is the roller axle that turns for steering. A roller of this form is called a three-wheel roller. The three rollers are lined up so that they give full coverage of the equipment's maximum width. The broken line between the edges of the front and rear rollers tends to leave an unevenness on the compacted surface.

A three-wheel roller may have either spoked wheels or hollow cylinders that can be loaded with water or other "fluid" ballast. They are probably the best-known static weight roller currently used. These self-propelled units range from 7 to 12 tons of equipment weight. Ballasting can add from 15 to 35% to the roller's weight. This gives a maximum of 200 lbs/inch of front guide roller width and around 500 lbs/inch of the drive roller. Each one-way trip, or pass, at a speed up to 6 mph on a surface to be compacted amounts to a single pass, in spite of the two axles moving over the material.

FIGURE 12-1 Smooth-steel rollers and a pneumatic-tire roller (courtesy of Galion Iron Works & Mfg. Co.).

The more modern development of a smooth-steel roller is the so-called tandem roller. One is shown nearest the camera in Fig. 12-1. It is called a tandem roller to indicate that one steel roller follows in the same path of another. Originally, these rollers had only two axles, so "tandem" was appropriate. Now this type is available with three axles and known as a three-axle tandem roller. All tandem rollers are made with ballastable wheels 3 to 5 feet in diameter. To account for the weight, both when empty and the maximum when fully ballasted, each roller is designated by two numbers. One may be shown as a 10–14-ton roller. The ballasting of a tandem roller may increase its weight 25% to 60%. With 4 1/2-foot-wide rolls the maximum weight gives around 350 lbs/inch of roll. It is desirable to have the ballasting result in equal weight on each roller. Each one-way trip at a speed up to 6 mph over the surface amounts to two passes for a two-axle tandem and three passes for a three-axle roller.

The smooth-steel rollers are generally equipped with scraper bars and sprinkling devices. These keep the roller wheels from carrying surface material around the full revolution and causing an extra irregularity. Trickling water from the ballast tank on the roller will help the compacting effort on some materials, but possibly will not be uniform.

Small two-axle tandem rollers in the 3- to 5-ton range are equipped with running tires on both sides of the piece of equipment between the roller axles. These tire wheels are raised for the roller operation, but can be lowered to lift the roller off the surface and serve as hauling wheels for towing the roller behind a truck.

One special, smooth-steel roller seen often now is a modification of the three-wheel roller. It has just one of the rear side roller wheels and is used to compact narrow strips such as a few feet of pavement widening or a backfilled trench.

Pneumatic-Tire Rollers. A pneumatic, rubber tire produces a kneading action that radiates out from under the tire to help consolidate the material. The pneumatic-tire roller is a specially designed piece of com-

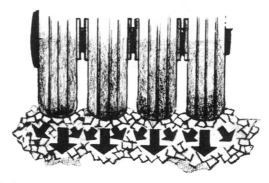

FIGURE 12-2 Kneading pressure under pneumatic tires.

pacting equipment that combines the kneading action with static weight. The combined effect is shown in Fig. 12-2. An example of one pneumatic-tire roller is shown in Fig. 12-1. The wheels are mounted on the two axles so that the rear tires track on lines centered on the spaces between the front tires. Thus, there will be an odd number of tires, commonly from 9 to 19, on a pneumatic roller. Individual wheels can move up and down to ride over hard lumps. With the sizeable body supported by so many tires, it is possible to heavily ballast this type of roller. Overballasting of any compaction equipment should be avoided to be assured that the material does not break down into smaller-than-specified sizes under excessive weight. Self-propelled units giving total weights from 5 to 50 tons are available, with the larger ones exceeding the highway pavement load limits. This type of roller can travel at speeds not over 20 mph.

The self-propelled pneumatic roller with, generally, 15″-rim tires are frequently used on hot bituminous surfaces. Fully ballasted with as strong as 14-ply tires inflated to 150 psig, this roller helps to give compactive pressures comparable to the maximums that might be applied on the pavement by heavily loaded trucks or buses. Some pneumatic rollers are designed to vary their tire pressure while they are rolling. A low pressure of 30 psi will give greater ground contact area for initial passes, then the tires can be built up to perhaps 130 psi for final compaction. These rollers may also be used on earth embankments or roadway subbase material.

For compacting granular embankment or subbase material a special variation of the pneumatic-tire roller used to be found. That is known as the wobble-wheel roller, in which the tire wheels are loosely mounted on the axle and can wobble, or pitch, from side to side. A wobble-wheel roller applies its load in more directions and also produces a vibrating effect.

Towed-type ballast boxes have pneumatic wheels with larger tires than those mounted on the self-propelled units. The towed unit has only one axle and probably no more than four tires to keep within a lane's width. Therefore, extra passes must be made to get full coverage of the material to be compacted. Fully ballasted, these towed rollers may weigh from 50 to 200 tons. They are used primarily on earth fills and embankments and can operate on the fresh, loose material dumped in 6″ to 24″ lifts. To move such a heavy roller through loose material will probably call for a crawler tractor. A design feature of these pneumatic rollers is the individual knee action. Each tire can move vertically over a mound of dumped material so that the whole roller does not bridge a part of the fill.

Sheepsfoot-Type Rollers.   The basis for using the original sheepsfoot roller is still applied in a variety of modern designs. Projecting feet or other shapes that can sink down about a half-foot into loose material is the key benefit of this roller type. It works best in sandy material with some clay binder. If the loose earth is dumped in lift thicknesses of 6 to 10 inches, that matches the sheepsfoot design just right. The feet sink down to knead and tamp the fresh material into the previously compacted layer, while the solid part of the roller applies pressure on top of the lift. As the lower level of the lift becomes compacted, the sheepsfoot-type roller rides up in the fill at higher and higher levels of the lift as the number of passes increase. This is what is meant by the roller "walking" out of the material. To produce a well-consolidated, cohesive mass, it is best not to compact each lift to the very top surface. If the top few inches are somewhat loose, the next lift can be bonded better to the lift below it.

The sheepsfoot or grid-type roller generally works by moving through loose material rather than rolling on top of it. That means that a major concern for this equipment's use is its rolling resistance to motion, which can run as high as 500 lbs per ton. Single 4- to 5-foot-long sheepsfoot drums, or pairs side by side, with a simple yoke frame for towing, have been available for many years. A single 4-foot-wide drum may weigh 1 1/2 to 5 tons, which can be increased by ballasting the hollow drum. This will amount to applying a vertical, compactive load between 100 and 300 lbs/inch of roller width.

Originally, crawler tractors were used to get enough tractive effort to pull sheepsfoot rollers at their specified low speed. To counter the low speed of travel for each roller pass, more powerful tractors were used to pull a train of the double-drum rollers. Each axle of drums amount to a pass over the fill for the full width of the drum assembly. Then, the balance to reach is between a train of rollers to match the required number of passes, or an even divisor of them, with a tractor that can just pull that train at the chosen speed. With a 10-ton double-drum sheepsfoot roller and rolling resistance, RR = 500 lbs/ton, the tractor needs to provide 5000 lbs of DBPP for each axle of drums.

More recently, the specifications and testing methods to assure enough compaction have emphasized the end results. That is, they have specified the required material density. This leaves the choice of equipment and its use to the construction planner. It is a more desirable approach which has fostered the development of many compactors.

The effect on the design of sheepsfoot-type rollers has been the introduction of various self-propelled compactors with different shapes of feet and grid patterns. An example of the sheepsfoot-type is shown in Fig. 12-3.

A self-propelled unit is articulated for a short turning radius and more maneuverability than the train of drum assemblies. It can be operated at speeds as high as 20 mph. These features help the compactor avoid interfering with the high-speed earthmovers. All of these self-propelled compactors have two axles with approximately even weight distribution between them. These modern sheepsfoot-type rollers apply static weights of 16 to 20 tons per axle. That amounts to between 300 and 500 lbs/inch of roller width, or several times larger than the older, separate sheepsfoot drum.

A special design feature of units that have two drums on one axle line, like wheels on a truck, is that each drum can move up and down independently to ride over mounds on the fill. They all have cleaning devices to scrape out material which is lodged between the feet. Both of these features help to reduce the rolling resistance that is a great power consumer in sheepsfoot-type rollers. Other improvements are bound to be made with this type of compaction equipment.

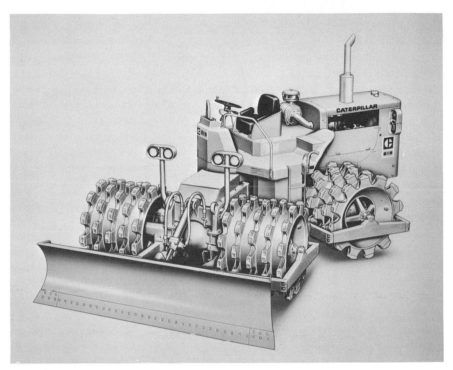

FIGURE 12-3 A sheepsfoot-type compactor (courtesy of Caterpillar Tractor Co.).

Vibratory Rollers. The benefit in vibration to help compact a granular soil was first noted with the use of crawler tracks on a fill. More recently, compaction equipment has been designed to apply vibratory action. The effect is to give deeper penetration of compactive effort on most granular materials than with just static weight and kneading action. This means that deeper lifts of loose material can be placed and compacted. However, it does not apply to soils with more than 15% clay or other cohesive material.

With the move to have specifications for compaction emphasize end results, the development of vibratory compactors has been the most obvious of all compaction equipment in recent years. A vibratory roller is made fairly simply by conversion from an ordinary smooth-steel roller or a sheepsfoot-type roller. The design is merely a matter of attaching one or more rotating eccentric weights applied to the roller's axle. The

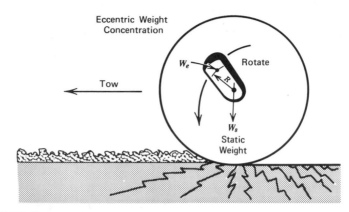

FIGURE 12-4 Schematic of vibratory mechanism.

principle of a vibratory mechanism is shown schematically in Fig. 12-4. The eccentric weight rotates on the roller's axle. It applies a centrifugal force due to eccentrically rotating weight. That force can be calculated by the following equation:

$$F = m\frac{v^2}{R} = \frac{W_e R n^2}{35,200},$$ (12-1)

where m = mass of eccentric weight (W_e/g);
W_e = eccentric weight, pounds;
v = velocity of rotation, inches/second;
R = eccentricity of W_e, i.e., the distance to its center of gravity or points of concentration from axle, inches; and
n = revolutions of W_e per minute, rpm.

This centrifugal force, F, always acts radially outward from its axis of rotation. It adds to the compactive effort of the roller's static weight, W_s, when the eccentric weight acts downward. In the sketch it is momentarily in an upward direction. At the moment when the F force is acting vertically upward, it has the greatest tendency to lift the roller off the surface.

One notes by Eq. (12-1) that for a given eccentric weight (W_e) and center-of-gravity location (R) in a vibratory compactor, the centrifugal force can be varied by changing the speed of notation (n). This term describes the frequency of the oscillations for vibration. Its value has a great influence on the compactive effect of the vibrations. Compaction of a granular soil generally increases with frequencies up to and beyond the natural frequency of the vibrator-soil system.

To better understand the vibrating system, it is necessary to know the terms "amplitude" and "velocity" of a sinusoidal oscillation. These are defined in relation to the diagram of Fig. 12-5. The frequency can be

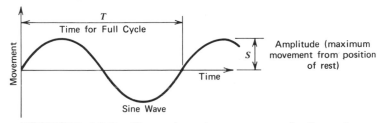

FIGURE 12-5   Time/motion wave of vibration.

shown as $f = 1/T$ in cycles per second, and the corresponding amplitude is s. Thus, the maximum velocity for the sine wave is $v = 2\pi f s$. The amplitude of the vibrating roller depends on the masses of the eccentric weight and the total drum cylinder. The so-called nominal amplitude is found as $s_n = mR/M$, where m and R were defined before and M is the drum's total mass.[3]

The actual amplitude for a vibratory roller differs from the nominal due to the centrifugal force, F, of the eccentric weight tending to lift the drum off the surface being compacted. Actual amplitude can be 50% to 100% larger than the nominal, s_n, if the vibrator-soil system is oscillating at its resonant frequency. That "natural" frequency is the one that will produce maximum vibrations. Since compaction will increase with oscillations up to the resonant, or natural, frequency, a vibratory roller should be operated at or above that frequency.

Each vibrator-soil system will have a different resonant frequency. It has to be found by field test. So, a determination with a vibration meter on the compactor is particularly helpful. In order to adjust to the desired frequency for a given soil, it is necessary to have a motor to rotate the eccentric weights that is separate from the compactor's driving motor. Then the vibratory motor's rpm can be varied to get the best frequency while the compactor's prime mover operates at its best power rpm.

A good way to tell if the vibratory compactor is in resonance with the natural frequency of the ground is to stand between 10 and 15 feet away from the machine and then vary the frequency of the compactor. When the vibration is at a maximum, as felt through the soles of the feet, the ground is in resonance and the compactive effort is the greatest. This is surprisingly simple, but it will tell more than all the sensors you can put in the ground.

The compaction by a vibratory roller is dependent on a good balance between its static weight and the dynamic effect of vibrations. The maximum depth to which a soil can be compacted depends on the total dynamic load, found as

$$W_d = W_s + F,$$ (12-2)

when the centrifugal force (F) adds to the static weight (W_s). The stress waves set up by the vibrations must be strong enough to overcome the shear strength of the soil. In partly saturated sand and gravel with only apparent cohesion it has been found that dynamic stress in the range of 7 to 14 lbs/sq inch (psi) is required to gain a compaction of 90% of the specified AASHO optimum density. For clayey soil the dynamic stress had to be about five times greater, approaching 70 psi, to reach the same relative level of compaction.[3]

A great number of different vibratory rollers are now available. Some are separate drums with the vibration-inducing attachments and motor built on them. These have to be towed over the fill by a suitable tractor. Other vibratory compactors are self-propelled pieces of equipment which may be easier to maneuver on the fill. The frequencies possible in the various units range from 1000 to 4800 vpm (vibration cycles/minute), with the lower values more common. They are designed for nominal amplitudes (s_n) of .012 to 0.1 inch or up to about 1/8 of an inch. When the amplitude is increased with resonant frequency, the average displacement amounts to about 3/16 of an inch. To produce the required compacted densities, a vibratory roller will weigh from 2 to 25 tons. Self-propelled vibratory rollers weigh at least 6 tons. Tractor-drawn compactors are generally designed to produce a centrifugal force which is 1/2 to 3 times the static weight of the roller. In all vibratory compactors an effort is made in the design to isolate the vibration from the part of the equipment combination where the operator sits.

It has been found by test that relative compaction is dependent on the total vibration time on the surface. To get the desired compaction at reasonable depth, the compactor must not pass over it too fast. Using a higher travel speed will mean more passes are needed. The depths at which effective compaction can be expected with vibratory rollers are between 20 and 40 inches, which will permit a relatively thick lift. Maximum dry density in coarse gravel or sand is found between 4 and 12 inches depth. The surface is likely to be "over-compacted," i.e., crushed or loosened by too heavy a compactor. The best surface compaction is obtained by a lightweight vibratory roller, a light- to medium-weight smooth-steel roller, or a vibratory-plate compactor.

Vibratory-Plate and Impact Compactors. In addition to the various roller compactors, there are two impact-type compactors. These are units which deliver impact blows in rapid succession on the surface of material to be compacted. Their operation is like a pogo stick with very short jumps. The blows may be delivered simply by an air- or ram-type piston onto a tamping plate. Or they may be generated by revolving eccentric weights inside a box-like container with a plate for the bottom to act on the material.

The best way to get an impact force in the vibratory-plate compactor is by having the eccentric weights operate in pairs. If a pair of eccentrics moves in opposite directions, as shown in Fig. 12-6, they combine their

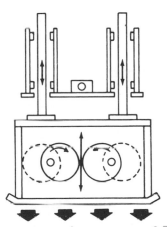

UP & DOWN
VIBRATION ONLY

VIBRATION COM-
PLETELY ISOLATED

NO SHAKING OF
BACKHOE BOOM. NO
NEED TO STOP COM-
PACTOR TO RE-
POSITION IT.

FIGURE 12-6   Force of eccentrics (courtesy of Racine
Federated Industries Corp.).

centrifugal forces [F calculated by Eq. (12-1)] in the downward and upward directions. This means that when acting vertically upward, if they are large enough, they will lift the plate compactor off the surface. When they both are directed vertically downward, their combined centrifugal forces add to the static and dynamic weights of the falling plate. During the moments of each rotation when the pair of F forces are directed more horizontally, they are in opposite directions and cancel each other.

FIGURE 12-7 Vibratory plate compactor working on a
trench backfill (courtesy of Racine Federated
Industries Corp.).

The application of a vibratory plate compactor for consolidating a trench backfill is shown in Fig. 12-7. In that case the handling equipment is able to move the compactor. Flatter plate compactors are sometimes used in tandem or line up with a lateral guiding bar to span across a roadway lane width. The guide bar ties them together and is moved by a tractor, motor grader, or other self-propelled equipment. The equipment they are attached to transmits the power to operate the eccentric weights for vibration. The vibrating mechanism may be driven by electric, hydraulic, or mechanical power. Compaction is done by vibration, with frequency ranging from 1200 to 6000 vpm, or impacts per minute. The lower values are for the heavy, self-propelled units, and the high frequencies for small plate compactors.

The vibratory-plate or impact-type compactor has many variations in its design and operation. In addition to the wide range of frequencies, the F force of one or more eccentrics can be directed to move the vibrator laterally. This way of moving is particularly useful for a single-plate compactor directed by a man on foot. By adjusting the rotation of the eccentrics, its direction and velocity of horizontal motion on the surface can be varied by the operator. It can be expected to move up to 80 feet per minute. With this power from the vibratory mechanism a man can compact up to 7200 square feet (1/6 acre) in an hour's time. The single plates, which are generally almost square on the contact surface, have dimensions that range from 15″ to 50″ on a side.

Field tests have shown some key points about the relative compaction with vibratory plates. For instance, it is possible to get 95% of the specified AASHO optimum density in moist sands 2 to 5 feet deep with vibratory plate compactors weighing 1 1/2 to 5 tons. Also, the effective compaction goes deeper using a vibrating plate compared to a vibratory roller of the same weight because of the impact forces. A plate compactor requires fewer passes than a vibratory roller to get the desired compaction. Generally, two passes with a vibrating-plate compactor will give the 90% or more relative compaction. To make this work, the same time effect as with a vibratory roller must be realized. A speed of 10 to 15 fpm for the plate must be used instead of the top speed mentioned earlier.

12.1.2 Operations and Costs for Compactors

Much has been said about the operations for compaction equipment because each type is designed for specific uses. A brief summary may be helpful at this point in the discussion. The compaction requirements can generally be divided into operations of embankment buildup, backfill compaction, base course preparation, and finish surface treatment.

An embankment may be several to many feet high, so for efficient, rapid buildup the material will be dumped in layers of 6″ to 24″ each. This calls for compaction from the bottom up, so penetrating rollers are needed. One may use sheepsfoot- or grid-type rollers for cohesive soil with layers not too thick, or a vibratory compactor for strictly granular material of any reasonable depth.

Backfill compaction is done in a manner similar to an embankment with layers of material dumped in the space to be filled. Generally, the material is of a granular consistency so an impact or vibratory type compactor will be used. If the backfilling is for a narrow strip such as a trench made for pipe, only a small compactor can be used. An air-powered hand-controlled tamper, as mentioned in Chapter 4, or a vibratory compactor as pictured in Fig. 12-7 would be suitable.

A base course is generally a few inches to not much more than a foot thick. Material for it to perform satisfactorily as a load-carrying layer must be well-graded and freely drain any water. Therefore, base course material that is granular, with coarse to medium aggregate particles and maybe 10% fine material as binder is best. To get good compaction for the base course, it will be necessary to use enough pressure from static weight and kneading action, with possibly some vibration. Most likely a pneumatic-tired or sheepsfoot-type roller will be the best compactor to use, though a smooth-steel roller with vibration may be desirable to get uniform compaction.

Compacting a surface coat, such as for an asphalt, soil, cement, or macadam pavement, is a matter of layers only a few inches thick. The materials are already well-mixed and have a positive binder in them. The object is to get all particles of material into the most compact condition possible so no void spaces for air, water, or other weakness can exist. Smooth-steel rollers or rows of vibratory plate compactors are most advantageous for this construction operation.

The costs for using compaction equipment are quite variable. Though compaction is a secondary operation, it is an important part of subgrade or pavement construction. Therefore, the variations in costs for effective compaction must be carefully planned. The following discussion should help to give a general overview of the costs of compaction equipment.

Small, impact-type or single-vibratory-plate compactors are quite economical. The piece of equipment will cost from $1 up to $3 per hour for the largest unit, so the operating labor expense is a more significant factor. At the other extreme of vibratory-plate compactors is the self-propelled string of plates attached to and powered by a wheel-mounted tractor. This equipment used for roadway base compaction will cost $8 to $15 per hour, not counting the operator.

The towed models of compacting rollers are also economical pieces of equipment. Pneumatic-tired and sheepsfoot-type rollers will cost from $1 to $3 per hour for a single towed unit. A grid-type towed roller will cost on the upper end of that range. The tractor to tow these units will cost $10 to $15 per hour, which is about the cost of a single-drum, self-propelled sheepsfoot tamping roller. Of course, double-drum, self-propelled units give two passes per trip over a fill for a somewhat higher hourly cost. If more passes are required for each layer of deposited fill, an economical equipment combination is the use of a large enough tractor to tow a train of sheepsfoot-type rollers. The length of roller train depends on the tractor's power and the number of passes specified.

The variety of self-propelled, pneumatic-tired rollers and the smooth-steel wheel rollers forces the planner to think in different terms. It seems desirable to reduce their costs to an amount per ton for each hour, i.e., cost/ton-hr. Self-propelled, pneumatic-tired rollers cost from 20 to 35 cents per ton-hr for the range of sizes from 35 tons down to 9-ton weight. This reflects that the larger the roller is, the more economical its cost per ton. A diesel-powered compactor is 10 to 20% more costly equipment, but for long use its operating cost will be less than gasoline powered.

Smooth-steel wheel rollers, which are all self-propelled, vary slightly between the three-wheel units, the two-axle tandems, and the three-axle tandems. The small three-wheel, two-axle steel rollers are the highest cost at about 50¢ per ton-hr for a 5-ton piece of equipment. Then the two-axle tandems cost 40 to 30¢/ton-hr for 1- to 14-ton rollers with the expected difference between gasoline- and diesel-powered units. The larger sizes of three-wheel rollers cost about the same as the equal-weight, two-axle tandem units. A three-axle tandem roller, which is made only in the larger sizes of 12–20-ton weights, will cost less than their smaller cousins, or about 30¢ per ton-hr.

The vibratory rollers cost more than the plain, static-weight rollers discussed above. It does not seem to matter much whether the compactor is to be towed or self-propelled. In either case there has to be an engine to drive the vibrating mechanism. These rollers will cost $1 to $3 per ton-hour, with the smaller units costing more per ton, as with the plain roller equipment. This cost range is for vibratory rollers from a half-ton unit up to 15 tons. Of course, to plan well for this type of equipment, as with other types, a detailed analysis with current costs must be made. Also, it is necessary to use the right compactor for the given material.

12.2 Subgrade Finishers and Automatic Controls

Equipment is often needed in pavement construction to finish preparing the compacted subbase before any surfacing material is put down. Such pieces are used to give the subbase an even, designed grade. This grade is essentially parallel or will allow a uniform, designed thickness of the pavement to the finished surface as planned. The equipment for this final preparation of the subbase may be called fine-grading machines or subgrade finishers.

The manufacturers of this equipment have seen a real opportunity to build into it the advantages of automatic controls. This is mainly because it works on a solid, nearly even surface. A constructor has other economic reasons for wanting automatic controls on subgrade finishers. Before such controls were available, the constructor had a lot of manual work in setting grade stakes, shooting levels, and taking readings. This is time-consuming and now, with the high price of labor, that way of finish grading is too costly. An automatically controlled finishing machine can follow a taut-string guide line running several hundred feet parallel to the pavement edge, and the manpower need is greatly reduced.

12.2.1 Subgrade Finishing Machines

A finegrader, or subgrade finishing machine, is designed with an earth-cutting part extending across its width at right angles to the direction of machine travel. In this way it is similar to a straight-blade dozer. But to have better control for fine grading by the cutting part, the subgrade finisher has leading wheels or tracks ahead of the cutter. This is like the supporting wheels located in front of the blade on a motor grader. A real uniqueness is found in the cutting part. It is generally a helical cutter rotating on a horizontal axle spanning most of the machine's width.

As the subgrade finisher moves in the direction of the pavement's length, the cutter trims the excess material from across the pavement subbase's width. The material thus trimmed may be thrown out by hand shovels. Such is the case with a small, single-lane finegrader working inside steel pavement forms.

On the modern, bigger subgrade finishers, as in Fig. 12-8, a conveyor is mechanically loaded with the trimmed material and conveys it to one or both sides of the machine. A large subgrade finisher, such as that pictured, covers a 28- to 30-foot lane width. It can trim up to 2 miles in a day's time. Adding the operation of spreading base material along with

FIGURE 12-8 A subgrade finisher in action
(courtesy of CMI Corporation).

trimming, it can still work over a mile in a day. These operations are
done at infinitely variable speeds of 0 to 75 fpm.

12.2.2 Automatic Controls for Paving Equipment

In spite of the immense size and considerable productivity of a dual-lane
subgrade finisher, it can still be operated theoretically to a tolerance of
1/16th of an inch. Actually, working to 1/8-inch variation vertically in
grade is satisfactory. The big machine can easily keep that tolerance with
automatic controls.

There were essentially no automatic controls on construction equipment before Honeywell introduced some in 1958.[4] One of the first pieces
of field equipment to be outfitted with automatic controls was the subgrade finisher. The solid base and economic reasons given in Sec. 12.2
encouraged this development.

An automatic control system on a piece of construction equipment
operating in the field has to react satisfactorily to movement of the total
equipment. It is based on sensing an unwanted variation in the relative
movement from a planned path or plane of motion. In the cases of subgrade finishers or pavers the path can be established by a taut-string line
along the edge of the pavement. For a motor grader the automatic control

may be based on the plane that the blade should maintain. With a string line the machine is guided by a sensitive, lightweight arm sticking out to the side following the line. When the arm wanders more than a tolerable amount from the string line, valves are actuated to adjust the machine's motion or travel automatically.

The most common controlling mechanisms are solenoid-operated hydraulic valves. The solenoids may be actuated by low-force, off-on switching through such a simple sensing mechanism as the lightweight guiding arm on the string line. Other control mechanisms, such as for slope control, may use a gravity-guided, pendulum-type switch to operate solenoid valves. On some equipment the solenoid-operated valves are replaced by servomotors. The sensors, in addition to the microswitch and pendulum-type, may include the rotary switch, mercury switch, or variable-resistance sensor.

An example of the automatic control arrangement for a subgrade trimmer and spreader is shown in Fig. 12-9 with the following explanation. "The heart of any automatic machinery is the control mechanism that

FIGURE 12-9 Automatic subgrade finisher (courtesy of the R. A. Hanson Co., Inc.).

senses from reference lines to guide (both course and grade control) the machinery in its work. Automatic machinery is never any better than the small control box that directs the action to be taken.

True control actually encompasses three different phases of work: 1. Automatic steering; 2. Automatic grade; 3. Automatic cross leveling.

Action of an automatic steering system involves an electric pickup from the sensing device . . . this is transmitted to solenoid valves that control the steering system itself. This electric hydraulic system is highly responsive to immediate changes in sensing.

Normally the same reference line used to control steering is used to control grade. In this case a second electric control mechanism is used to sense from the bottom side of a reference line or from the top side of a curb or road surface.

This sensing control, like the steering sensor transmits an electric impulse to solenoid valves, and they in turn activate control cylinders which in turn raise or lower the cutting or paving head.

Steering and grade can be controlled within $\pm 1/8$th inch and under present conditions this is closer control than can be normally measured with our present system of grade checking.

The maintenance of a heavy construction machine in a system of equilibrium, or at a predetermined slope is one of the most difficult of all control problems. If the job is one of highway construction where the use of string lines on both sides of the machine is practical and necessary, then cross level can be most easily maintained by the use of two grade control devices, one on each side of the machine, and each compensating and adjusting its side of the machine so that the operating head is always in the correct position with relationship to the road. Whenever dual lines can be used they will always provide the most accurate system of cross level control that can be established. With such systems machines can operate at speeds of 50–100 feet per minute."*

A majority of the construction equipment used in the field with automatic controls is for paving work. In addition to the controls for subgrade finishers, there are controls to guide the screeds of asphalt pavers and finishers, as well as those on concrete slipform pavers. These pavers will be discussed in following sections of this chapter. The use of automatic controls is promoted by two-thirds of the state highway departments in the United States specifying them for one or more parts of their work.

*From RAHCO booklet by R. A. Hanson Co., Inc., Spokane, Wash.

12.3   Bituminous Paving Materials

The construction of a paved surface using bituminous material is a highly specialized process. This makes use of carefully graded and controlled aggregates and a previously processed and contained binder material. Binders may be any of the various bituminous materials, including natural or petroleum asphalts, tars, or other bitumens. A bitumen may be defined as a mixture of hydrocarbons of natural or pyrogenous origin, which may be gaseous, liquid, semi-solid, or solid, and which are completely soluble in carbon disulfide. Generally, each place and use for this type of a paved surface calls for a preferred combination of aggregate and bituminous material. The success of such a paving depends on three key elements in the process.[5] These are:

1. the types and gradation of the aggregate;
2. the type and condition of the binder; and
3. the construction method used.

Aggregates in the bituminous paving material serve to give the surface mechanical stability. They support the loads on the surface and transmit their weight and other forces to the subgrade. The materials used for aggregate in bituminous paving include sand, gravel, crushed stone, slag, and mineral filler. There may be a variety of sources for any of these. Most of the aggregate material will have to be processed from its natural condition to make it suitable for paving material. The crushing and screening equipment to produce well-graded and controlled aggregate for a use such as this was discussed in Chap. 10.

12.3.1   Controls of the Materials

For use in a bituminous mix the aggregate is generally spoken of in terms of its gradation. There are three broad classifications[5] used:

1. dense-graded aggregate with close tolerance;
2. dense economically graded aggregate; and
3. selected aggregate giving open gradation.

The first two are designed to give a dense, tight paving mat with maximum mechanical strength. The second one sacrifices a little from the first in strength and stability to save on cost of the paving material. The third classification of aggregate covers a variety of material for special surfacing, such as seal-coat chips or open-graded macadam stone.

The control of the aggregate to insure close tolerance in the paving material is most obvious in the dense-grade aggregate. Measures taken to insure the specified gradation include careful selection of individual sources of aggregate, processing each aggregate at its source, blending the different aggregates at the mixing site, and rescreening the combined aggregates after they have been dried and before mixing with the binder. The tolerance variation allowed for this material is ±5% for the coarser sizes and closer on the fine sand and dust sizes.

To meet the requirements for a "dense-graded aggregate with close tolerance," it is often necessary to transport satisfying material quite a distance. This will run the cost of aggregate up considerably. In that case a "dense economically graded aggregate" may be preferred by those paying for the paving. This classification of aggregate is less expensive because it may use nearby crusher-run or pit-run aggregate, may eliminate the screening and reblending after drying, and make a few other economies. Obviously, this grade will not be as good as the first classification, but the decision between quality versus economy must be made, as with so many construction choices.

The binder serves not only to hold the aggregate particles together but for several other purposes. It protects the particles from exposure to moisture, which causes the ill-effects of weathering, and it acts as a cushioning medium. The only bituminous materials used as binder in a pavement are tar and the asphaltic materials. Paving asphalts are classified in one of the following designations: (1) road oils or slow-curing (SC) liquid asphalts; (2) cut-back asphalts—medium-curing (MC) liquid asphalts, or rapid-curing (RC) liquid asphalts; (3) asphalt cement (AC); (4) emulsified asphalts; or (5) powdered asphalts. A brief explanation of the major differences between these may help with the understanding of the equipment to handle the bituminous paving. The curing rate of liquid asphalts depends on the air temperature and the volatile portions of the bitumen. A higher portion of volatile material will mean it has a faster rate of curing; and as the air temperature rises, the rate of curing increases.

Asphalt cement is the heavy binder used most generally for hot bituminous mixes. Its suitability is judged by the binder's degree of hardness, which is measured by the penetration of a weighted probe into the material. The extent of penetration is governed by the amounting of fluxing oil kept in the asphalt cement. This bitumen is converted to a cut-back asphalt by the addition of a volatile substance. By adding kerosene, a moderately volatile product, the result is a medium-curing (MC) cut-back asphalt. The addition of a highly volatile naptha or gasoline-type distillate to asphalt cement makes a rapid-curing (RC) cut-back asphalt. Under a hot sun and in high-temperature air an RC may catch on fire.

Other mixes or forms provide different asphaltic materials for paving. It is important to realize that a good surface takes time to produce.

Usually, the specifications to construct a certain bituminous pavement will show one or more choices of binder material that can be used. The binder will amount from 3% to nearly 12% of the total mix. The larger percentage of binder is needed with finer grade aggregate (more surface to coat), such as in predominately sandy sheet asphalt. A high-type bituminous concrete will generally have binder amounting to from 4 to 9% specified for good mixing and paving results.

12.3.2 Paving Process with Bituminous Materials

The process of constructing a surface with one or more layers on top of a prepared subgrade or base is known as paving. When the material used to build up the surface is bituminous material, the finished product is most often called an asphalt pavement. The build-up may start with a prime coat to seal the subbase with a penetrating film of heated bituminous material. Then there generally will be a tack coat applied to the existing surface before another layer is added, to provide an adhesive between layers.

A full-wearing surface, or layer several inches thick, consists in the designed mix of aggregates, filler material, and bituminous binder. The aggregates give body to the built-up layer. They must be dry to make it possible for the binder to coat each particle and not be turned away by moisture. To be effective, there should be just enough bituminous material to fully coat all aggregate particles in the mix. This binder will coat during the mixing only if it is heated enough to flow evenly in the mixing step of the process.

The means for drying and grading the aggregates, introducing filler material, heating the bituminous material, and mixing all the materials is carried out by asphalt production equipment. That equipment will be discussed in the next section. Then the equipment to take asphaltic or bituminous material to a jobsite for surfacing needs to be given certain consideration. Finally, the jobsite handling of hot-mixed material to construct the bituminous pavement requires other special equipment. These pieces of equipment for the bituminous paving process will be discussed after the production equipment.

12.4 Asphalt Production Equipment

The process for making asphalt or other hot-mixed bituminous materials for paving requires a highly controlled plant. Such a plant has more than half a dozen key components to take care of the specific processing functions. Simply stated, these functions serve to handle graded aggregates, heat them and dry out moisture from them, regrade the hot aggregates for proportioning with heated bituminous material, and mix this combination to produce hot-mix paving material.

To grasp a quick view of an asphalt production plant, a schematic outline of the flow and relative proportion of quantities of material is shown in Fig. 12-10. The outline of an asphalt plant shows three general processes: (1) cold feeding and conveying; (2) drying and dust collecting; and (3) proportioning and mixing the aggregates and bituminous materials. A detailed discussion of the particular points for controls in the process and the equipment components needed will be covered in the following sections. The two common types of asphalt plants to be considered are the batch-type and the continuous-flow asphalt plants. A batch-type plant is easier to understand, so Sec. 12.4.1 will discuss that variety along with the general concerns for the production process. Section 12.4.2 will deal with the differences found in a continuous-flow plant compared to the batch-type.

12.4.1 Batch-Type Asphalt Plant

The basic components of a hot-mix asphalt production plant are the cold feed, aggregate dryer, dust collector, elevator and screens for hot aggregates, heater and pumps for asphalt or tar, proportioning devices, and pugmill mixer. Each unit of equipment in an asphalt plant will be discussed in the following paragraphs. A complete batch-type plant is pictured in Fig. 12-11.

Aggregate for the cold feed is generally taken from a set of relatively small bins that are loaded by a front-end loader or some other economical means. These bins are like those of a trolley batcher for concrete. The gates and conveying mechanism for the cold feed are set to draw the specified amounts of each size of aggregate to satisfy the required final mix. It is important to be feeding the right amount of each aggregate size for two reasons. It is costly for the plant to be delayed by starved hot bins above the proportioning equipment and pugmill mixer. And the overflowing of dried and heated aggregate is wasteful. The cold feed should particularly pass along the correct amount of aggregate passing a No. 8 screen because that size of fine aggregate is so important to the success of the final mix, especially the optimum asphalt content.[6]

Aggregate Dryer. An aggregate dryer is a long, hollow cylinder with its axis almost horizontal and open at both ends. The moist aggregate at atmospheric temperature; i.e., "cold," is fed into the dryer at the upper

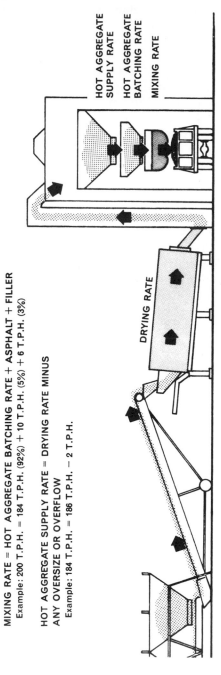

MIXING RATE = HOT AGGREGATE BATCHING RATE + ASPHALT + FILLER
Example: 200 T.P.H. = 184 T.P.H. (92%) + 10 T.P.H. (5%) + 6 T.P.H. (3%)

HOT AGGREGATE SUPPLY RATE = DRYING RATE MINUS
ANY OVERSIZE OR OVERFLOW
Example: 184 T.P.H. = 186 T.P.H. − 2 T.P.H.

HOT AGGREGATE
SUPPLY RATE

HOT AGGREGATE
BATCHING RATE

MIXING RATE

DRYING RATE

FIGURE 12-10 Outline for an asphalt plant
(courtesy of Barber-Greene Company).

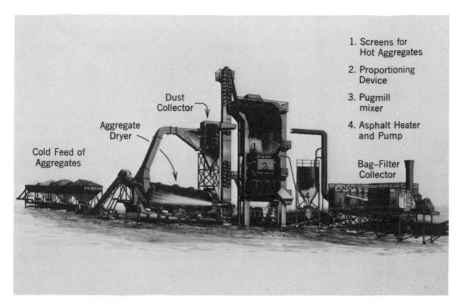

1. Screens for Hot Aggregates
2. Proportioning Device
3. Pugmill mixer
4. Asphalt Heater and Pump

Dust Collector

Aggregate Dryer

Cold Feed of Aggregates

Bag–Filter Collector

FIGURE 12-11 A batch-type asphalt plant (courtesy of Barber-Greene Company).

end. The drying flame with air and gas jets is introduced at the lower end of the cylinder. Some dust from the cold aggregate can be sucked out before it enters the dryer. This is done by a cyclone type of dust collector. Other dust may be blown out of the dryer by the air and gas jets. Instead of letting that dust blow out the exhaust stack and cause air pollution, it should be taken into the dust collector also.

Obvious and unhealthy air pollution conditions in large metropolitan areas have focused national attention on air contamination and dust nuisance, increasing the demand for more effective controls. This has resulted in laws and ordinances by many governments to restrict the amount of contaminants that may be released into the atmosphere.

The primary, cyclone-type dry-dust collector mentioned above and noted on Fig. 12-11 should operate with 70 to 90% efficiency. That means the dry collector will take out that percent of the dust particles suspended in the exhaust air and gases from the aggregate dryer. Particles of 20 microns (.02 millimeter) or larger diameter are generally removed by the dry-dust collector. The new air pollution laws generally require an efficiency above 90%.

To satisfy the stringent air pollution codes, either (1) a wet-type collector or (2) a bag-filter collector unit must be added to an asphalt plant.

The wet-type collector is an addition to the usual plant components, including a primary, dry-dust collector. It should bring the total plant efficiency up to 99% or more of the dust removed from the exhausted air. A bag collector, such as Barber-Greene builds for asphalt plants, eliminates the need for a primary dry collector by adding an elaborate purifying unit that can give essentially 100% dust-free exhaust air.

The wet-type collector works on the principle of wetting the dust particles so that they precipitate out of the exhaust air and are drained from the plant in the form of a sludge. One wet collector simply has a vertical spray bar with many nozzles to create a curtain of water mist in a vertical cylinder. The exhaust air from the dry-dust collector, moving at high velocity into this 10- to 20-foot-high and 3- to 7-foot-diameter cylinder, has the remaining dust coated with water. The air with wetted dust moves from the bottom of that "contactor" tank to an adjacent vertical cylinder almost twice as large as the first. This "separator" tank has a skimmer to precipitate the wet dust into the hopper bottom for removal as sludge while the cleaned, moving exhaust air swirls out the top, 20 to 40 feet above. An orifice plate with a central opening and otherwise filling the contactor tank's horizontal cross section may be used to improve the efficiency of a wet-dust collector.

The wet spray-type collector requires a water pumping rate, dependent on tank size, of 50 to 350 gpm at about 100 psi pressure head. Adding an orifice plate to get more concentrated turbulence of the dusty exhaust air necessitates a pumping rate of 150 to 900 gpm but at the lower pressure of 50 psi. In either case the motor to run the pump needs to be of a 15 to 50 hp size.

Now, returning to the aggregate dryer, the extra moisture in the cold aggregate is dried out as it passes through and around the insides of the cylinder, which is rotated slowly on its axis. The dryer should be able to get the aggregate's moisture content down to a single percent or two. A dryer's productivity with relation to its size and the initial moisture content of the aggregate is shown in the curves of Fig. 12-12. For an initial moisture content of 8 to 10% it may be necessary to pass the aggregate through the given dryer several times or to use tandem dryers in order to get the moisture down to an acceptable level. Obviously, the need for such an extra treatment will slow down plant production. In fact, the moisture content of the cold-feed aggregate is a major variable for the plant's ton per hour (TPH) capacity. The fuel flow for the heating flame in the dryer can be adjusted to different temperatures. However, there is a specified temperature limit for a given mix and bituminous material. The volume of air used is adjusted to the fuel flow to insure complete

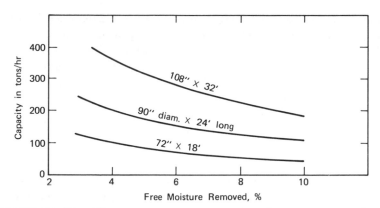

FIGURE 12-12 Variation in capacity of aggregate dryers.

combustion. One possibility for improving the rate for drying the aggregates is to increase the gas velocity moving through the dryer, resulting in moisture reduction and higher plant production. But this increase will be limited because gas velocities above 700 ft/min will result in an undesirable amount of dust removal.

The dried aggregate and fines collected for filled material are lifted in a completely enclosed bucket elevator. This unit of the batch plant is called the "hot elevator." It is designed to keep the heat in the aggregate, share some with the very fine material added to it and draw out through suction ducts to the dust collector material that will be a nuisance in the screens, bins, etc. Aggregate materials are elevated by it to a point for starting their vertical gravity feed through the screening, proportioning, and mixing part of the batch plant. This is the part to the far right in Fig. 12-10.

Screening and Storing Hot Aggregates. The hot aggregates are separated by a set of screens into two or more sizes which are stored in bins that may be insulated. This step in the process permits recombining the materials in a controlled manner for proportioning. Recombining this way insures more uniformity in the gradation of the aggregates from batch to batch than would be possible if the material were used directly from the dryer. The separation and temporary storage in these bins also helps to smooth out fluctuations in the cold feed of aggregates.

Screens in a batch plant are the flat, vibrating type in a multideck screen design. That type of equipment was discussed in Sec. 10.3 and 10.4 on aggregate equipment. The screens are decked on a slight incline, with the finer-mesh screens below coarser ones. In some cases one or more decks may be split to save space, similar to the 2 1/2-deck set discussed in Chap. 10. Screen sizes are selected to make separations that can be recombined for the batch-mix formula and for practicable production rates. The smallest, fine screen should be as coarse as can be tolerated and still meet the job-mix formula specified. Larger screens are designed to divide the rest of the aggregate material for good balance in the use of the hot storage bins and the proportioning equipment. If the batch plant operation can anticipate having a variety of batch sizes and demands, it may be desirable to have a twin-screen tower design. For large batches and high production both sets of screens would be used together. When there is less demand on the batch plant, only one set of screens is needed, and the other is a standby unit. This will result in savings of power to operate the unused screening as well as less maintenance and wear and tear of the screen cloth.

The hot bins are located below the deck of screens and above the batch proportioning mechanism. There are generally two to four separate compartments. Manufacturers will give a nominal rating based on the level, full gross volume for each bin. However, this may prove misleading, and it is better to speak in terms of their "live-storage" capacity. The live-storage rating considers the angle of repose of the aggregate as it falls from the screens. Only the net volume is considered after the space occupied by the overflow chutes in the bin is deducted.

Two other capacity factors should be figured for the hot bins. A bin's draw-down capacity is related to its live-storage tonnage. It is particularly important in the case of intermittent demand for aggregates, where the bin storage is used to minimize re-starts of the dryer and its cold feed. Hot-bin surge-batching capacity is very similar. It should be considered

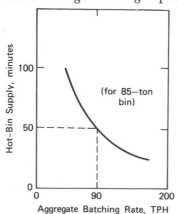

for the batch plant if short peak periods of demand are anticipated when production requirements are above normal. These periods will occur in the early morning or later at the height of production, and the hot aggregate supply rate from the dryer and screens is inadequate.

The advantage of good hot-bin draw-down and surge-batching capacity is shown in the following example from Barber-Greene.[7] An 85-ton Batch-pac hot bin will permit 90 TPH additional capacity for 50 minutes. This is shown in the adjacent curve for this equipment. The draw-down capacity is represented by the vertical time scale. The curve of draw-down time to batching rate gives the bins' surge-batching capacity. In this example, if the hot aggregate supply rate from the dryer and screens is 200 TPH, then an aggregate batching capacity of 290 TPH can be sustained for almost an hour with the given surge capacity.

Proportioning the Batch Materials. Proportioning of the aggregates is controlled by running gradation tests on the material drawn out of the hot bins. The desired gradation can be found by formulas, but trial-and-error methods are used more frequently. For an asphalt batch plant, the aggregates are normally proportioned by weight. A weigh hopper is mounted on scales and located under the discharge openings of the hot bins. Control of the weigh scales is very important and must be checked frequently. Many modern plants weigh the aggregates automatically, and the weights are checked electronically. If the weight is within tolerable variation limits of the designed job-mix amount, the automatic equipment proceeds with drawing the weight from the next bin. If the weight is not within the tolerance limits, the batching cycle is automatically stopped until the scales can be adjusted.

An operator's console for automatic control of a batch-type asphalt plant is shown in Fig. 12-13. The asphalt plant manufacturers provide

FIGURE 12-13 Automatic control console for batch plant (courtesy of Barber-Greene Company).

calibration charts for the proportioning, or feeder, mechanisms, but these should be checked with the specific aggregates being used. These aggregates flow from the bins by gravity, so their rate of flow is faster when the head of material is greater from a fuller bin. Satisfactory batching requires a fairly constant level of aggregate in the hot bins. Some plants are designed with interlocks that stop production when this level falls below a certain point.

The amount of bituminous material for a batch is even more critical. Generally, asphalt or tar is stored at the job site in tanks which hold only one or two day's supply. For batch plants it is proportioned by weight or by volume. If the bitumen is to be weighed, it is drawn into a bucket supported on weighing scales. Where it is proportioned by volume, the bitumen may be batched into a sized container or metered to the mixer by a calibrated pump. With either design, checks must be made throughout production to be sure that the quantity of bituminous material put into each batch is according to the job-mix and specified tolerance limits. A batch plant with automatic controls will have the necessary measuring devices tied into a console such as the one pictured in Fig. 12-13.

Bitumen is heated by steam coils or electricity, and no flame contacts the material. One must recognize that the volume of a bitumen will vary with temperature, so necessary corrections must be made with changing temperature for volumetric proportioning to get the specified weight in each batch. The specified quantities in a tank before heating are usually given in terms of the volume at 60°F temperature.

Mixing in a Pugmill. The mixing of accurately proportioned and heated batch materials is done in a tub-like unit known as a pugmill. This large, open-top-box piece of equipment makes use of built-in, twin-shaft paddle mixers. These shafts are parallel with their axes in the same horizontal plane. The paddles are on arms radiating out a foot or so from these shafts, which are rotated in opposite directions by a power unit. The materials are mixed by these parts with the paddles set at angles to move the pugmill's batch back and forth until it is discharged out a bottom opening in the vertical tower batch plant.

During the mixing cycle, the hot asphalt or tar is exposed to the air and so subject to oxidation, which leads to poorer pavement life. Therefore, mixing should be done in the shortest possible time. Oxidation of the bitumen also increases with increasing temperature. For this reason, mixing should be done at the lowest temperature that will allow sufficient coating and density of the pavement material. Generally, this calls for temperatures of 250°F to 325°F for asphalt and 175°F to 225°F for tar. When the bitumen's temperature has been set, it also sets that for the aggregates where the specifications may state they cannot be more than 25°F hotter.[6]

The ability of a pugmill to produce a homogeneous, completely coated mix of material in a short number of seconds depends on many design factors, including pugmill shape and proportions; number, dimensions, and arrangement of the paddles; size and action of the discharge gate; and the power input to the pugmill. The mixing capacity of a pugmill is obviously governed by the cubic content of the tub-like box, as well as the design features mentioned. Batch material to be mixed can fill the pugmill until just the highest tips of the paddles show. That amount of material gives rise to the pugmill's rated volume.

The manufacturer's association, CIMA, adopted the uniform-rating concept with their live-zone formula. "The live zone is defined as the net volume below a line extending across the top arc of the inside body shell radius. Shafts, paddle arms and tips are all deductible to determine the net cubic foot volume to the nearest foot." Then, to get the pugmill's rated capacity, the net live-zone volume is converted to a total weight by assuming that the material weighs 100 lbs/cu ft. Pugmills are manufactured in sizes from 3,000 lbs to 10,000 lbs. The batches mixed in pugmills generally take up 45 to 95% of the net live-zone volume.[8]

A long-standing tradition has governed the mixing time for a batch of bituminous materials. It has called for a 10- to 15-second dry-mixing cycle to mix the hot aggregates in the pugmill. When the hot bitumen has been added, another 30- to 60-second wet-mixing cycle has been required. This total mixing cycle of 40 to 75 seconds has been required ever since judgments about it were set in the 1920's and 1930's.[6]

The recent development of the Ross Count procedure to tell when a batch is mixed enough, by inspecting the coating of course aggregate particles, suggests new mixing times. It tends to show that in plants of modern design a total mixing time of 20 seconds with very little dry-mixing cycle time is enough. This would help the desire stated earlier for mixing the hot batch as quickly as possible.

As the time for mixing becomes shorter and control is more critical with higher production, an automatic control system is more and more necessary. A review of how this works for a batch plant should be helpful in understanding the system. It was stated very simply in an article on automation.[9] "Production of a batch is started by the operator pushing a button. The automatic equipment draws aggregate from each bin in sequence, dumps this into the pugmill for a dry mixing cycle, if called for, draws the asphalt, holds this until the dry mix cycle is complete, dumps the asphalt, continues the mixing until the specified time has been completed, and then dumps the batch from the mixer into the truck. The automatic equipment has a 'repeater' setting so that the cycle will be repeated for the number of batches required to load the truck. After the required number of batches, the automatic cycle is shut down and must be restarted for the next truck load." More will be said later about the trucks for hauling hot-mix bituminous pavement materials.

12.4.2 Continuous-Flow Plant

Several design and operational features distinguish a continuous-flow plant from a batch-type plant. Only the special features of a continuous-flow plant will be discussed in this section. Design features common to both plant types have already been covered in Sec. 12.4.1. A view of a continuous-flow plant is shown in Fig. 12-14.

FIGURE 12-14 A continuous bituminous plant (courtesy of Barber-Greene Company).

Basically, the continuous-flow plant operates as the name implies without the cyclic intervals between batches. This is possible mainly because the continuous-flow plant delivers all material in continuous streams. Bitumen is always measured by volume. The heated, liquid material is metered by a calibrated pump driven by a mechanism interlocking it with the aggregate feeders. Aggregates are fed into the pugmill mixer, proportioned by a calibrated feeder mechanism for each hot bin. These mechanisms are interlocked to the same drive shaft. All the interlocks must show that the various materials are being proportioned correctly, within allowed tolerances, for the job mix specified. If any one ingredient—a separate size of aggregate or the bitumen—is not right, the interlocks will shut down the plant until the deficiency is corrected.

Another distinguishing feature for a continuous-flow plant occurs with the pugmill. Material is fed into it at one end and discharged at the other end. Also, the hot elevator is inclined rather than vertical as on a batch-type plant tower. These features of the continuous-mix plant make it more portable.

12.4.3 Productivities for Asphalt Plants

The production of hot-mix batched material per hour by a plant may be expressed in tonnage (TPH) or in yardage (cu yds/hr). With the usual assumption that the hot mix weighs 100 lbs/cu ft, an easy conversion can be made from one productivity basis to the other. A 200 TPH rate is the same as 148 cu yds/hr. This is found by multiplying the TPH rate by 0.74 to find the cu yds/hr rate. To convert from yardage, multiply it by 1.35 (reciprocal of 0.74) to find the tonnage. Of course, these conversions are based on the assumed unit weight. Other weights will require different conversion factors.

The key moving parts of an asphalt plant each have their designed or specified rates in the processing. There are maximum rates for the cold feed, the dryer, the hot elevator, and the pugmill mixer. They should be fairly balanced for an ideal process. Only the hot elevator's rate is practically fixed. This is at a productivity about twice the average of the governing unit. Designing the elevator that way means it will meet the best possible rate from the other units and not hold them back. The governing unit in a well-designed plant is either the pugmill or the dryer.

The dryer's rate of discharging aggregate dried enough to meet the specification depends mainly on the moisture in the "cold" fed material. It can be 100% higher for input with 3% moisture compared to cold-feed material with 8% free moisture to be removed. This was shown in the curves of Fig. 12-12. The dryer's productivity for input with 3% moisture is a close match with the hot elevator's rate. At the extreme of high moisture content in the cold feed and a short mixing time specified, the dryer may govern the total plant's productivity. Of course, the cold-feed rate must be adjusted to meet the dryer's demand for fresh aggregates.

The pugmill's mixing rate and, so, productivity is the most variable one in the plant. It is the unit that should govern the total plant's output rate. Major variations are due to the specified mixing time and the batch size compared to the pugmill's net available volume. Previous discussion noted that the mixing might take, at most, 75 seconds and may be as little as 20 seconds. At a total batching-mixing-discharging cycle of 60 seconds, the pugmill can turn out 60 batches an hour, especially if it has automatic controls. Its production rate could be more than twice as high using the modern, quick-mixing concept.

Batch sizes mixed in a plant's pugmill may vary from perhaps 95% of the net, live-zone volume down to as little as 45% of that available volume. The ratio of batch to mixing volume varies for some of the same reasons as with concrete mixers. The loose aggregates take up much more space than when they are completely mixed and closely packed. Specifications will dictate the volume of batched material that can be mixed on a given job. This volumetric ratio can vary the pugmill's output rate by more than 100% from the smallest to the largest productivity with given equipment.

The total effect is to make a given pugmill's productivity vary by as much as 400%. This is from the lowest rate, with maximum cycle time and lowest ratio of batch to live-zone volume, to the highest rate that can be expected for the equipment. The pugmill could really press the capacity of its plant's hot elevator and bins. Such would be the case if the batch occupied 90% or more of the live-zone volume and its total mixing cycle were as low as 30 seconds. There must be a reasonable balance between all the components of a batch plant. The chance to select the parts individually, as in the Batchpac system, makes this possible.[7] Selection of the best combination of components for cold feed, drier, hot elevator, screens, hot bins, proportioning controls, and pugmill can be made for reasonably balanced productivities to meet the plant's expected demands.

12.4.4 Costs for Asphalt Production

The costs for producing asphalt or other bituminous paving mix will include the aggregates and bitumen materials, as well as the equipment, labor, and energy costs to run the plant. In comparing the two main elements of expense, the cost of materials will run about two to three times the cost for the plant. Of course, this will depend on the source of processed materials for the asphalt, but in a locality where this form of pavement can be justified the materials should be available at a cost of $2 to $3 per ton. At this same time the expense of the plant will cost about $1 to $1.50 per ton produced.[10]

Costs for the asphalt plant itself break down into the initial cost plus ownership and operating expenses, as they were discussed for all construction equipment in Chap. 2. An asphalt plant is obviously a major investment for the owner and operator. It is generally sized or rated by the pugmill capacity in pounds. Then the plant's productivity is figured, based on the pugmill mixer governing the output. To illustrate, a 3,000-lb plant should have a production rate of 90 to 150 TPH, depending on its mixing cycle and volumetric ratio of batch to live-zone capacity. The initial cost of an asphalt plant can then be estimated by a factor of about $1,000 per TPH for the plant's best productivity. Using this basis, the above-mentioned plant would require an investment of $150,000 to purchase and install for operation. The automatic control equipment for this or larger plants will add an initial expense of around $15,000.[9]

The ownership expenses of an asphalt plant, even if it can be amortized over as long a life as 6 to 8 years with high production in each of those, will be at least 50¢ per ton produced. For the balance of plant costs the larger part is consumed in the expense of energy to operate the plant components. Energy is needed in the form of dryer and other heating fuel, as well as electricity or other power to run the cold-feed mechanisms, blower, elevator, dust collector, screen vibrators, proportioning mechanisms, and pugmill. The cost for all this energy will be 25¢ or more per ton produced.

It is of interest to note that larger plants do not produce asphalt at a significantly lower cost than a smaller plant, even though they are turning out a greater TPH amount. One factor that bothers the efficiency of production for an asphalt plant with so many control points is the variability of productivities in each of its components. With a slight rain shower on the cold-feed storage, the dryer will receive wetter material and need to work on it longer. At the other end, there may be a temporary shortage of trucks to fill with hot-mix bituminous paving material. Such variations show the need for some surge storage for hot material. As mentioned in the discussion of the hot bins, they are designed to provide some surge capacity. However, it would be uneconomical for the basic plant to have hot bins that could hold much more than an hour's demand for aggregates. If a frequent need or planned use for storage is anticipated, a hot-mix surge storage setup should be designed.

A surge storage system built to serve an asphalt plant can take care of high demand for hot-mix at the start of a day and other peak times, while using a non-work time such as the lunch period to build back the material in storage. To be completely effective, this storage must be in insulated tanks sealed with an inert atmosphere that can keep the hot-mix in usable form for more than a day. With a satisfactory design of surge storage balanced with the asphalt plant there can be a saving of 20 to 50¢ per ton delivered on the pavement. This will account for the saving of at least three trucks when the surge capacity equals half a dozen to 40 truck loads.[10]

12.5 Bituminous Paving Equipment

The equipment for transporting and laying down bituminous materials for a pavement are basically trucks, pavers, and rollers. Trucks serving to transport the heated material are of three types—ordinary and bottom-dump trucks and tank trucks. To haul batches of material from a hot-mix plant, the most common hauler is an ordinary dump truck, but also a bottom-dump truck may be used. These basic haulers were discussed in Sec. 9.5. Any truck used for hauling the asphaltic or hot-mix material must protect it. The protection serves as insulation so that the temperature of material meets the specification when it is laid down and as protection against getting rained on before the mat is consolidated. A heavy tarpaulin used to cover the open-top of the dump truck bed is a minimum protection for these purposes.

A variation for hauling processed material looks like the long bottom-dump truck bed but can dump its load at the rear. It can haul 20 to 35 tons of mixed material. Controlled delivery of material at the rear is made possible by a horizontal conveyor-like drag chain running the length of the truck's hauling bed. Material can be dumped at any rate from zero to 20 tons per minute. This type of truck can be used like a bottom-dump for the windrow technique of laying down the bituminous paving mix. The mix is dumped in a windrow on the prepared grade for spreading by hand, by a motor grader, or by reloading into a paver.

An end-dump truck normally dumps the bituminous material directly into a paver for spreading on the prepared grade. The arrangement for dumptrucks working with a paver will be discussed later. Any truck bed should be oiled or coated with lime water to prevent the mix from sticking.

Rollers for consolidating the bituminous paving mat are of the pneumatic-tire type or the smooth-steel wheel variety. These were discussed along with other compaction equipment earlier in this chapter. Equipment for laying down bitumen or bituminous paving mats will be covered in the next sections.

12.5.1 Equipment for Spraying Bitumen

Bitumens are sprayed on a prepared grade as a prime coat or a tack coat. A prime coat is a layer of liquefied asphalt to protect the base from rain and construction traffic and to provide a bond between the base course and the pavement. Penetration into the base course is desired so light grades of medium-curing (MC) cut-backs, which do not lose their distillates too fast, are used. A tack coat is a lighter application of a heated bitumen on top of a paving layer to give a bond with the next layer. This may be one of several materials—RC cut-backs, emulsions, asphalt cement, or the heavier grades of tar.

The trucks used to haul and spray the liquified bitumens are generally two-axle tank trucks. They have built-in means for heating the bitumen to the range of 150°F to 175°F. Spraying is done through a spray bar supported at the back of the truck about a foot above the surface to be sprayed. The bar has many regular-spaced nozzles with varying angles for differing overlap along the spraying length. It extends the full width of the truck. To be sure of uniform and positive coverage, the bitumen is applied under pressure. This is done by a pump system on the tank truck with a tachometer to control the pumping rate in gallons per minute (gpm).[6] Manufacturers recommend setting the tachometer for a gpm about 10 times the length of the spray bar in feet. Then the truck's speed is selected to give the required bitumen quantity per square yard for the specified job application.

For a prime coat the quantity of bitumen to use will depend entirely on the condition of the base course. Generally, the amount of material will be adjusted so that the application will be absorbed in a period of 24 hours. The specifications will usually require between 0.1 and 0.5 gal/sq yd. If a quantity at the upper end of this range is absorbed within the 24-hour period, a heavier grade of bitumen should be used. For a tack coat the quantity of bitumen to use is 0.05 to 0.15 gal/sq yd. To get such a thin application, it is necessary to operate the tank truck at a fairly high speed, assuming the pump tachometer is not changed.

12.5.2 Bituminous Pavers

Specially designed equipment is used to receive hot-mix bituminous material from end-dump trucks and spread it for a pavement. These are called bituminous pavers or spreaders or finishers. One of these pieces of equipment is shown in Fig. 12-15. The hopper box to receive hot-mix material will vary from the smallest, with a capacity of 3 tons, to 12 tons

FIGURE 12-15 Bituminous paver in action (courtesy of Pioneer Division, PORTEC, Inc.).

for a large-size unit. Material is drawn from the hopper's bottom back to the screed unit by a flat conveyor. The screed is the most important part of a paver since it must spread the paving material evenly and accurately for a smooth, homogeneous surface. The screed extends the width of the pavement being laid. Hot-mix material moved back by the conveyor is then distributed laterally for the screed's width by screw-shaped augurs. More will be said about the screed's operation later.

The operation of a paver in laying a bituminous pavement starts with a dump truck of hot-mix material backed up to it. With the truck bed raised to gradually dump its load, the paver engages its back wheels ready to push. When material has filled the paver's hopper enough and a continuous flow to the screed is assured, the paver starts moving ahead and pushing the truck to be unloaded. This gives the paver a capacity of its hopper box plus the truck's bed volume for continuous operation. Obviously, there is a limit to the size of truck a paver can push. Some experienced bituminous pavement authorities believe that large semi-trailer trucks that carry 20 to 30 tons of mix cause trouble for an ordinary paver, especially to keep the correct pavement thickness.[6] The reason for this will be understood when the paver's way of controlling thickness is discussed.

To gain the necessary traction, special consideration has to be given for the paver. Some of this equipment is mounted on crawler tracks. That means the coefficient of traction (C_t from Table 3-4) is in the 0.8 to 0.9 range. The rolling resistance (RR) for the paver may be around 70 lbs/ton while that for the truck on tires may be slightly less. If the paver has an empty weight of 24,000 lbs and is carrying 8 tons of material in it, that could result in a total of 32,000 lbs tractive effort. Net tractive effort to push the truck should be ample. If the C_t is poorer and the RR higher, there still should be enough traction with the crawler tracks. With that type of mounting the paving speeds in the various gears run from 10 to 110 fpm. Travel speeds range up to between 2 and 4 mph for a maximum. That is not a very high speed for maneuvering between paving stretches on a job site.

Some pavers move on rubber-tired drive wheels. Their paving speeds will generally have infinite variation with a torque-converter-type transmission from zero to 150 fpm or perhaps 300 fpm as a top speed. Travel speeds for a wheel mounted paver will vary up to a maximum of 18 mph. The limitation for this form of paver is its traction pushing a heavily loaded truck. Its C_t on the base course may be around 0.5, and perhaps an upper limit of 80% of its weight is on the drivers. These estimates lead to a maximum tractive effort of 40% of the paver's weight. If traction is no problem, the maneuverability of a wheel-mounted paver is a real advantage. However, it must be noted that during the paving process the paver should move forward continuously and not back and forth. Reversing while paving is likely to damage the surface, so maneuverability is not always wanted in bituminous pavers.

Pavers are designed for a basic paving width of 6 to 10 feet. The width each covers can be changed with 2- to 10-foot extensions of the screen and augur spreader. Such variations lead to a paving width range from 6 to 19 feet for a single paver. By linking two identical pavers side by side, a width of more than 30 feet can be paved at one time. This has an advantage because it has been found that the pavement is smoother when the full width of roadway is paved all together than when the paving is laid down one lane of, say, 10- to 12-foot width at a time.

It has been said that a paver consists of two units—a tractor unit and the screed. That is an over-simplification, but it emphasizes the importance of the screed. It is pulled by two arms which are attached to the tractor at pivot points. The screed is then free to float on the paving mat and does not contact the surface that is being covered by the mat.

Adjustments for thickness from 1/4″ up to maximums of 6″ to 10″ can be made on the various bituminous pavers. The thickness of the mat is governed by the balance of the forces tending to rotate the screed about its pivot supports. Weight of the screed acts in one direction. In the opposite direction is the force resulting from the screed being pulled over the fresh bituminous paving mat. Blocks are used under the screed to start paving a mat. The force from the mat and, so, the pavement thickness can be adjusted by the tilt of the screed's supporting arms. Adjusting screws are used to control the tilt. Generally, the loose thickness should be between 1.15 and 1.25 times the required compacted thickness specified.

A paver's screed is equipped with several important design features. Heating units are needed to be sure the bituminous material is laid down at the specified temperature and that the screed plate is not too cold for that. The heaters are generally LP gas burners. A vibrating mechanism is used to dissipate air bubbles in the freshly laid mat and improve the pavement's density. The vibrator is electrically driven at 1500 to 4500 impulses per minute. It has been shown by a manufacturer that a tamping plate just ahead of the screed improves the density still more. Also, a study made to control roughness showed that a pan float ahead of the screed is very helpful.

Highway engineers have experimented to see if better bituminous pavement can be constructed by going over the mat with two or more screeds. A study concluded that the optimum for smoothness is three screed passes. Also, it studied the shape of a pavement, which for drainage requires that the mat be somewhat higher in the center than at the edge of the roadway. For smoothness it was found that there should be a constant cross slope rather than a curved, convex cross section. A constant cross slope means that the screed has to have no more than one break in its length instead of several segments to achieve a convex arch.

It is obvious that controlling the screed is very important to the success of the pavement. Manufacturers of modern bituminous pavers have included automatic controls for the screed.[9] These control the mat thickness or top surface height, if there is an old, irregular paved surface for the base. A taut stringline or some other temporary profile is used as the guide for the first layer of bituminous mat. For subsequent lanes or layers the edge of an adjacent mat layer can serve as the guide. The cross slope of the screed can be automatically controlled by a pendulum-type sensor. These controls are very much like the ones discussed for the subgrade finisher in Sec. 12.2.2.

The costs related to the use of bituminous paving equipment are not as high as those for a batch-type or continuous-flow plant. Using a paver and roller for, say, a 12-foot roadway width will cost between 30 and 40¢ per ton. For a road twice as wide several rollers as well as the wider paver equipment will be needed to achieve the higher production rate, so the cost will probably be in the same range. Labor to work along with the paver and rollers on the 12-foot width will cost about as much as the equipment for each ton. A wider pavement width will cost less per ton for labor. These are comparative figures that might be used with careful judgment and adjustments for the location and relative competition.

12.6 Concrete Paving Equipment

The variety of equipment needed to form, pour, and finish a concrete pavement is similar to that just discussed for bituminous paving. Differences are due mainly to the way the mixed material is consolidated and hardens, as well as its temperature when placed. Concrete is placed at any atmospheric temperature above freezing. It actually gains heat due to the hydration and chemical process of curing. After the needed days of curing it will cool down to the ambient temperature around the concrete pavement.

Specifically for concrete paving, there are the transit trucks, agitating trucks, or non-agitating dump trucks to haul the batched concrete materials. The dump trucks were introduced in Chap. 9 and those for hauling ready-mixed concrete were discussed in Chap. 11. Dry-batched concrete hauled by dump trucks to the paving site are discharged into the skip of a concrete paver. That type of concrete mixing equipment was also covered in Chap. 11. Wet, mixed concrete hauled in transit trucks, or other permissible ones, may be chuted into place for the pavement. The trucks should not drive on the prepared subgrade, so the pavement cannot be too wide in order to chute the concrete directly into place.

For acceptable quality of the concrete in a pavement there must not be segregation of the materials. To compensate for long chuting or other pouring methods which cause some segregation, a mechanical spreader is needed to help place the concrete. Following the concrete spreader on the side forms for a pavement there will be one or more finishing machines or screeds. There may be a transverse finishing machine and a longitudinal finisher. A slip-form paver may be used for pouring concrete on well-prepared subgrade. These key pieces of equipment for concrete paving will be discussed in the following sections.

12.6.1 Concrete Spreader for Paving

A concrete spreader is used to distribute the wet-mixed concrete across the width of the pavement. It bridges the roadway, riding on steel wheels that move on top of the metal paving forms where they are used. There are generally front and rear bridge members with the operating power mechanism spanning between these. The spreading is done by a one- to two-foot-diameter augur on a horizontal axis covering the width of the lane being paved. As shown in Fig. 12-16, the spreading augur is at the leading edge of the concrete spreader.

FIGURE 12-16 A slip-form concrete spreader in action (courtesy of Rex Chainbelt, Inc.).

The spreading action is designed to avoid the segregation of material that might occur if the concrete is moved around on the grade by some hand shoveling or vibrators. Spreaders are very helpful for pavements where reinforcing bar mats or wire mesh is to be placed at about mid-height of the slab's thickness. On a first pass over a new stretch for paving, concrete is spread for the bottom half of the slab. Then the spreader backs up, the reinforcing mat is laid on top of the fresh concrete, and the upper layer is poured and spread.

Following the spreader in the paving train is a concrete finishing machine. Where paving forms are used, this piece of equipment also rides on the metal side forms for the slab. It has two bridge members and two or more screeds. These are an oscillatory type of transverse screed. Vibrators are inserted into the concrete to consolidate it. The first screed for concrete paving is used as a strike-off screed to slice excess concrete from the top of the slab. A second screed will proceed with the leveling and finishing operation. Some finishing machines have three or four screeds. The second or third one may have a tamping bar to work the coarse aggregate down below the surface. Most transverse finishing machines have an arrangement in their screeds to put a slight arch or convex curvature on the slab's surface.

The final piece of equipment to complete the shaping of a concrete pavement is called a longitudinal finisher, or float. Two bridge members and wheels that ride on the side forms also describe the supports for this equipment. A strike-off bar spans between the bridge members, paralleling the direction of the roadway. It operates like a screed in the opposite direction cutting off high spots on the slab. The crown of the pavement is built into the supporting track for the strike-off bar.

12.6.2 Slip-Form Paver

The older, established equipment for concrete paving uses metal forms to hold wet concrete at the edges of the slab. As just described, the train of equipment also uses the side forms as rails. The time and manpower to place those forms firmly and accurately on line and grade has been a detriment to the efficiency and economy of a paving operation.

In recent years the development of a slip-form paver to eliminate the old hand-placed side forms has been a rapid innovation. A slip-form paver, as shown in Fig. 12-16, performs the functions of the previous spreader and finishing machines, as well as providing the slab's side forms. To insure that the wet concrete will hold its shape as the slip form moves away, the concrete poured with this equipment must have a really stiff mix. A typical specification limits the slump to no more than 1 1/2″ with a slip-form paver.[11]

The modern slip-form paver can pave a slab up to 12″ thick that has a width of 12 to 28 feet. In putting down a slab in that size range, it performs half a dozen or more paving steps. These steps, which follow one another in the paver from the front end to the rear, are:

1. spreading fresh concrete with an augur;
2. striking-off the excess with the primary concrete feed screed;
3. vibrating with internal-type spud vibrators inserted in the fresh concrete;
4. using an oscillating screed with frequencies that run from zero to 80 rpm;
5. shaping the final surface with an oscillating extrusion finisher; and
6. floating with a fine surface finisher.

To operate so many moving parts will take a prime mover with nearly 200 hp. In addition, some slip-form pavers will have small, individual auxiliary power units for each of four crawler-track drive assemblies. The need for these is explained next.

A problem that plagued the first uses of a slip-form paver was not having enough traction to move along the roadway. All the functions performed by this type of concrete paver comprise quite a load. Then the drag of stiff concrete along the inside faces of the slip forms is an extra load for the paver. The right balance of power for each crawler track is now possible with a load-leveling type of transmission, using hydraulic fluid flow. Another improvement that reduces the chance for the slip-form paver to get hung up on a curved section of roadway is an articulated paver. The Heltzel Company makes this equipment with the articulated joint running the width of the pavement between the two supporting bridges. At the limit of angle between the front and rear sections, this paver can produce a 150-foot-radius curve.

Slip-form pavers can generally work at variable speeds from zero to 30 fpm. The usual limit of actual paving is 20 fpm on a straight section of roadway. This can amount to 20 cu yds of concrete poured each minute. To insure that the slab does not slump at the edges when the slip forms leave it, there will probably be 40 feet or more of forms trailing the paver.

Maintaining alignment and grade is a key function in the use of a slip-form paver. This paver does not have the guiding advantage of carefully placed, fixed road forms as the older concrete paving equipment does. So a slip-form paver is an ideal application for automatic controls. These are designed to provide control for steering, maintaining required elevation for the pavement top, and cross leveling for the specified slope. The automatic controls are similar to those described in Sec. 12.2.2 They get their guidance from a taut string or piano wire set to parallel the roadway edge.

12.6.3 Costs for Paving Equipment

The thickness of concrete pavement varies from perhaps a minimum of 6″ to maximums over a foot thick. Therefore, the cost per unit area, such as a sq yd, is not meaningful except for a specific pavement of given thickness. It is more general to compare costs for concrete by the cu yd volume. Then one can also relate the cost of paving concrete to structural concrete. The total cost for the material, forms, equipment, and labor needed to build structural concrete may cost $25 to more than $100/cu yd. Paving concrete is very much more economical. It does not require much, if any, fixed formwork to build or install. Furthermore, paving concrete shows the economy of mass production.

This section will continue by stressing the equipment and labor costs for a cu yd of paving concrete.[11] For the older, regular paving methods with metal side forms the total equipment, excluding the mixer plant, may cost 66¢ to a little over $1 per cu yd. A wider pavement allowing greater mass production will result in the lower costs. The labor needed for setting the forms and pavement joint assemblies, as well as working with the fine grading and paving equipment, will add from $1 to $1.70 per cu yd. This combination of equipment and labor costs for paving runs between $1.50 and $3/cu yd. To batch and mix the concrete for delivery adds 35¢ to 55¢/cu yd for the equipment and 20¢ to 35¢/cu yd for the labor. The total costs, excluding that for the material, add up to between $2 and $4/cu yd. Proportioning a typical mix of cement, sand, and coarse aggregates amounts to about $11 or $12/cu yd. Adding the needed reinforcing steel and other materials increases this part to perhaps $15/cu yd.

It is obvious that the equipment and labor costs are not a major portion of the cost of paving with concrete. Yet this does not lessen the interest in reducing these costs by innovative methods or equipment. That has been the objective of the slip-form paver. It did not produce concrete paving at lower cost in its early years. But as labor costs have risen, the slip-form paver with automatic controls is showing some savings. It may eventually be the predominant equipment and method for concrete paving.

References

1. Morris, M. D., "Earth Compaction," *Construction Methods and Equipment,* McGraw-Hill Publications Company (New York, January, February 1961).

2. "Compaction Equipment—History—Use and Application," pamphlet, The Galion Iron Works & Mfg. Co. (Galion, Ohio).

3. Broms, B. B. and L. Forssblad, "Vibratory Compaction of Cohesionless Soils," Report of study made at the Swedish Geotechnical Institute (Stockholm, Sweden, circa 1970).

4. Bower, L. C. and B. B. Gerhardt, "Automatic Controls on Construction Equipment State of the Art," *Highway Research Record No. 316,* Highway Research Board (Washington, D.C., 1970), 1–14.

5. *Bituminous Construction Handbook,* 4th ed., Barber-Greene Company, (Aurora, Illinois, 4th printing 1970), 11, 20–23, 51–61, 83–115.

6. Aaron, Henry ed., "Construction Practice for Hot-Mix Bituminous Pavements," *Journal of the Aero-Space Transport Division,* ASCE, Vol. 94, AT1 (New York, N.Y., November, 1968), 31–55.

7. "Batchpac Manual," pamphlet, Barber-Greene Company (Aurora, Illinois, 1965), 38 pages.

8. "Asphalt Batch Plants," Pioneer Engineering (Minneapolis, Minnesota), Form 732, 15 pages.

9. "Modern Construction Operations: The Takeover of Automation," *Engineering News-Record,* McGraw-Hill Publications Company (New York, N.Y., December 9, 1965), 36–40.

10. "Dollars and Sense of Hot Mix Production," pamphlet, CMI Systems—CMI Corporation (Chattanooga, Tennessee, 1971).

11. *Estimating Guide for Public Works Construction,* Dodge/1971 annual edition no. 3, McGraw-Hill Book Company (New York, N.Y., 1971), 139–159.

INDEX
■■■■■■■■■■

555